The Green City

Jürgen Breuste

The Green City

Urban Nature as an Ideal, Provider of Services and Conceptual
Urban Design Approach

 Springer

Jürgen Breuste
Society for Urban Ecology (SURE)
SURE International Center Urban Ecology
University Hildesheim
Hildesheim, Germany

This book is a translation of the original German edition „Die Grüne Stadt" by Breuste, Jürgen, published by Springer-Verlag GmbH, DE in 2019. The translation was done with the help of artificial intelligence (machine translation by the service DeepL.com). A subsequent human revision was done primarily in terms of content, so that the book will read stylistically differently from a conventional translation. Springer Nature works continuously to further the development of tools for the production of books and on the related technologies to support the authors.

ISBN 978-3-662-63975-7   ISBN 978-3-662-63976-4   (eBook)
https://doi.org/10.1007/978-3-662-63976-4

This Springer imprint is published by the registered company Springer-Verlag GmbH, DE part of Springer Nature.
The registered company address is: Heidelberger Platz 3, 14197 Berlin, Germany

Foreword

Cities are the most important living space for the majority of people. Their demand for space at the expense of surrounding and urban nature in the interior continues to grow. Many urban dwellers understand their urban life as independent of nature. Although they often prefer urban life to other forms of living, they do not want to live without nature even in cities. However, the urban nature that surrounds them is often only present in a few remnants, pushed back in the competition for space, often neglected and dismissed as a less important component of a city. Yet urban nature is an essential part of cities. It is what makes urban life bearable and attractive. For this, however, nature is needed that is also accepted, embraced, and used by urban dwellers.

Urban nature is a conglomerate of different types of nature and includes remnants of rural nature as well as the specially designed nature of private and public gardens and parks. Remnants of "old wildernesses," such as woodlands and wetlands, where design according to human principles was not in the foreground, have been preserved alongside "new wildernesses" such as fallow land, which successively develop themselves after the abandonment of use. The private garden and the public park have established themselves almost everywhere and worldwide as the preferred urban nature. They meet the need to temporarily remove themselves from the stress of urban life and immerse themselves in a green oasis. This is what makes gardens and parks so attractive. They are temporary alternatives. Woodlands and gardens, individual trees and parks, wildernesses, and designed urban nature of all types, all belong in cities.

Various sciences are concerned with urban nature, each from a different perspective. The design of urban nature, which is first and foremost also the design of human living space, requires results from ecology, social sciences, psychology, urban studies, architecture, and urban planning to create "better," more livable cities with urban nature as a core structure. First and foremost, this means using, preserving, and further developing the existing urban nature potential in a sensible and targeted manner. Nature conservation in cities always means the use of nature. At the beginning of this process, however, there must be recognition and acceptance of all urban nature types and the will to do justice to their specific nature of use and to benefit from the services they offer. Urban dwellers not only enjoy urban nature by looking at it but also walk on it, do sports there, grow vegetables, or take time out from the hectic and noisy city. All this also contributes to their health and a fulfilled good life in cities. That is why urban nature is not only an ornamental addition to urban architecture, which could be dispensed with if necessary, but an essential part of cities, without which cities cannot be what they should be – the most attractive living space for people, the green city.

Jürgen Breuste
Salzburg, Austria
Spring 2021

Contents

1	**What is Urban Nature?**	1
1.1	What Does "Green City" Mean?	2
1.2	What Does Urban Nature Mean?	7
1.3	Components of Urban Nature	10
	References	16
2	**How Urban Nature Developed?**	19
2.1	Urban Nature as a Cultural Product	20
2.2	Urban Nature Is Establishing	29
2.3	Public Urban Nature for Beautification, Recreation, Public Health and Public Education	33
2.4	The Park Cemetery – A Place of Recreation for the Living	44
2.5	The Urban Forests and Forest Parks – How the Forest Stayed in the City	47
2.6	The Private Garden for Everyone Complements the Public Urban Nature	51
2.7	How the Waters Became Urban?	58
	References	64
3	**How Urban Nature Exists in the Context of Nature and Culture?**	69
3.1	The Relationship Between City and Natural Environment	71
3.2	Example Forest City: Integration into the Natural Environment	77
3.3	Example Desert City: Turning Away from the Natural Environment	78
3.4	Example Mountain City: Dealing with Extreme Natural Environment	82
3.5	Forest and Wilderness in the City: Places for Religion and Rituals	89
3.6	Gardens and Parks: Designed Nature for the Urban Recreational Landscape	93
	References	99
4	**What Services Urban Nature Provides?**	101
4.1	What Ecosystem Services Do We Expect From Urban Nature?	102
4.2	Which Urban Natures Provide What Kind of Ecosystem Services?	117
4.3	How Can Urban Ecosystem Services Be Valued?	124
	References	127
5	**What Urban Nature Provides Which Services?**	131
5.1	The Urban Park	132
5.1.1	Natural Element Urban Park	132
5.1.2	Services of Urban Parks	137
5.2	The Urban Tree Population (Urban Forest)	147
5.2.1	Nature Elements, Urban Trees and Urban Woodland	147
5.2.2	Services Provided by Urban Trees	152
5.3	The Urban Gardens	174
5.3.1	Natural Element Urban Garden	174
5.3.2	Urban Gardens as Services Providers	188

5.4	**Urban Waters**	200
5.4.1	Natural Element Urban Waters	200
5.4.2	Services of Urban Waters	204
5.4.3	Restoration of Ecosystem Services of Urban Waters	208
	References	217
6	**What Constitutes Urban Biodiversity?**	229
6.1	**Urban Biodiversity: A Paradigm Shift?**	230
6.2	**How Can Urban Biodiversity Be Measured?**	234
6.2.1	Integration Levels of Biological Diversity	234
6.2.2	Structure of Urban Habitats as Indicator and Basis of Spatial Assessment	235
6.2.3	Species Diversity as Indicator	238
6.3	**How Is Urban Biodiversity Perceived?**	244
6.4	**Urban Biodiversity and Ecosystem Services**	249
	References	251
7	**What Constitutes Urban Nature in the Green City Concept?**	255
7.1	**The Green City, a Conceptual Mosaic of Action Objectives**	256
7.2	**Green Infrastructure: The Local Basic Concept**	263
7.3	**The Concept of Urban Biodiversity**	266
7.4	**Wild Goes Urban: Wilderness as Part of Urban Nature**	283
7.5	**The Green City Concept in European and Global Perspective**	302
	References	309
8	**What Ways Are There to a Green City?**	315
8.1	**The First Steps**	317
8.2	**Maintain, Gain and Connect Space for Urban Nature**	321
8.3	**Making Urban Nature Accessible to All**	328
8.4	**Enhancing the Benefits of All Types of Urban Nature**	333
8.5	**Making Urban Nature a Space for Experiencing Nature and a Learning Place**	337
8.6	**Protecting and Using Urban Nature**	346
8.7	**Using Urban Nature for Climate Moderation**	359
8.8	**Solving Problems and Reducing Risks with Urban Nature – Nature-based Solutions (NbS)**	362
8.9	**Providing Guidance Through Good Examples**	367
	References	379
	Supplementary information	
	Index	387

What is Urban Nature?

Contents

1.1   What Does "Green City"
        Mean? – 2

1.2   What Does Urban Nature
        Mean? – 7

1.3   Components of Urban
        Nature – 10

        References – 16

© Springer-Verlag GmbH Germany, part of Springer Nature 2022
J. Breuste, *The Green City*,
https://doi.org/10.1007/978-3-662-63976-4_1

1

The term "green" is currently in danger of becoming inconsequential in everyday language, or at the very least, inconcrete and variable in meaning. This also applies to "Green City". Here, very diverse contents can be included, from transport to energy to nature. In this chapter, a limitation and concretization is made to urban nature as the main content of "green" in the concept of the Green City. Urban nature is explained and presented in its components as a concept for the consideration of very differently developed natural structures in cities. This introduces the multifaceted topic of characterisation, analysis, evaluation and handling of urban nature.

1.1 What Does "Green City" Mean?

The term "Green City" is now widely, frequently and with different meanings used in politics, planning, science and public debate, so that it has almost become commonplace. It expresses a positive goal that is either already achieved or at least to be strived for. As a goal, the Green City should also be achievable and concrete. Citizens, their representatives, media and politicians demand or postulate the goal of the Green City on a national, regional or local level. Cities are always local and concrete. The Green City must thus have a concrete content of "green" on a local scale. The Green City is a concept that should not remain just a vision, but can have realistic features of a programme.

What the contents of the concept are, however, is determined very differently. The keyword is "green". Its use in public debate is normatively mostly positive and multi-layered. Political groups and parties use it to describe their policy offer, energy producers their product, transport companies their service, food producers their products, universities their campus understood as innovative-positive. "Green" has become a catch-all term of positively apostrophized behavior in the public debate. This can range from saving resources to renouncing individual automobility, the consumption of certain foods (e.g., renouncing meat, consuming fair trade products, etc.), health, the achievement of climate and CO_2 targets and social goals (e.g., social balance, gender conformity). "Green" then often stands for general sustainability and environmentally friendly action.

"Green" as a substitute term for sustainability as a guiding principle of urban development has a wide range of approaches and serves as an orientation for achieving an ideal goal, often without a concrete, regulative model. The approaches are not necessarily set but are determined by actors, planners or scientists, partly in communicative processes. What can constitute a sustainable, future-oriented, or "Green City" is determined by the current and future challenges. This is based on well-founded assumptions and forecasts, but also visionary ideas of innovation.

Fields of action for sustainable urban development are:
- New ecological, technical and organisational solutions to problems,
- Use of renewable energy sources,
- Precautionary environmental protection,
- Urban mobility management,
- Socially responsible housing,
- Business development,
- Economical land management, and
- Resource efficiency, among others (Breuste et al. 2016).

Originally, the term "green" always had a connection to what is green in our living environment, i.e., to the green plants, nature. With this meaning, green as a synonym for nature, it shall be used exclusively here. This also includes abiotic nature as a prerequisite for biotic nature. Today, the term is also often used in this reduction to an essence. The term "Green City" is not intended to replace the terms "sustainability", "biodiversity" or "environmentally sound action". It is understood as a programmatic umbrella term with concrete content. This book is about urban nature, its types and structuring, its capacity for a healthy life in cities and how we currently and prospectively deal with urban nature or could deal with it better.

The Green City

The Green City is a city in which all forms of nature (living creatures, biotic communities and their habitats) have a high status as green infrastructure and are preserved, maintained and expanded for the benefit of the city's inhabitants. Urban nature is an ideal, a performer and a concept for urban development.

In this book, the Green City is understood as a metaphor for preserving existing nature and making it even more usable for people in the city, for increasing nature of all kinds in the city and bringing nature back into the city, for creating a new partnership between the built city and nature. Urban nature is also increasingly understood in Green Cities concepts as a valuable contribution to urban quality of life and as an indispensable element of our urban life. Urban nature belongs to the city. To preserve it and use it properly (*nature-based solutions*), we need understanding and knowledge about urban nature, good examples, realistic assessments and committed citizens and municipalities. There are already many examples of all this around the world. Nature in the city can no longer be seen as mere beautification and "aesthetic embellishment", but must be understood as an important functional contribution to a viable city, documented in green (and blue) infrastructure concepts.

1

Organizations Want to Be "Green" – Green Campus 2020 Strategy of the University of Copenhagen

With a strategy for resource efficiency and sustainability, the University of Copenhagen (UCPH) aims to become one of the greenest universities in the world. Knowledge, responsibility and sustainability are the three Green Campus goals:

» "PHYSICAL SETTINGS UCPH must have a sustainable physical environment: buildings, facilities, technology and infrastructure.

» LIVING LAB Development and demonstration of the sustainability solutions that UCPH itself researches and teaches. UCPH should be a living laboratory for the development of tomorrow's sustainability solutions.

» CULTURE We must create a university with a sustainability culture in which all staff and students encounter and practice sustainable behaviour in everyday life. Sustainability and resource efficiency should be integrated effectively and meaningfully in the organisation and management of the university" (University of Copenhagen 2017).

The sustainability goals for 2020 are:
CO2/Climate
- Reduction of CO_2 emissions for energy use and transport per full-time employee unit

Energy
- Reduction of energy consumption by 50% per full-employee unit

Resources
- Reduction of waste volume by 20% per full-time employee unit
- Recycling of 50% of the waste
- Reduction of water consumption by 30% per full-employee unit

Chemicals
- Procurement and construction activities free of health hazards and environmental contamination
- Reduction of total pollution and chemical waste generated by the university.

Organization/Behavior
- Sustainability and resource efficiency in all major decisions and activities
- Awareness as a sustainable university and sustainability as a daily practice

The campus as a field of experimentation
- Development and demonstration of sustainable solutions for the future on campus (University of Copenhagen 2017)

Companies Offer "Green City" as a Support Concept in Energy, Mobility and Climate Protection Management on a Customer-Specific Basis – Green City Project Munich

With the target groups of municipalities, organisations and companies, Green City Project supports climate-friendly mobility and energy and participatory urban and regional planning. The portfolio ranges from the development of mobility and energy concepts to consulting on climate protection targets and climate adaptation strategies. The development and implementation of innovative communication strategies in the form of campaigns, participation processes and event formats are offered (Green City Projekt GmbH 2017).

Green City e. V. – Citizens Organise Themselves in an Environmental Organisation to Make Their City "Greener"

Green City e. V. is a non-profit environmental organisation that has been working in Munich since 1990 and has over 1500 members and activists. Green City organizes events in Munich, including the Street Life Festival (■ Fig. 1.1). In 2014, Green City launched a greening project for Munich facades and house roofs. After more than 25 years since its foundation, Green City (registered association) is now considered one of Munich's most respected environmental organizations. With its subsidiary Green City Energy AG (founded in 2005), Green City aims to finance the expansion of green electricity supply in Munich through a capital investment company.

■ **Fig. 1.1**   Street Festival May 2017 in Munich. (© Photo Gleb Plovnykov, Green City e. V. 2017)

1

Cities Refer to Their Energy-Based Sustainability Concept as Green City – How Much Green is There in Freiburg's Green City Concept?

Freiburg describes its environmental policy under the name "Green City Freiburg" as an ambitious commitment to renewable energy sources with a variety of ecological, technical and organisational solutions. The city claims to have made a name for itself worldwide as a Green City (Stadt Freiburg 2017).

Freiburg is one of the greenest cities in Germany with forests, vineyards and a variety of biotope types and natural spaces. In the Green City Concept, Freiburg presents its nature concept as a sub-area and recommends itself as a "green feel-good city" with 660 ha of urban green spaces, landscape and nature conservation areas, parks, allotments, children's playgrounds and cemeteries, which have been maintained close to nature for more than 20 years without the use of pesticides. Biodiversity is improved by mowing the meadows only twice a year, local microclimate by 50,000 street and park trees. Green space design is done with citizen participation. Traditional forms of gardening in allotment gardens (4000 allotment gardens) are supplemented by new forms of gardening and promote the individual design of a privately usable open space, social interaction between different population groups and contact with nature. Almost 50% of Freiburg's area (6966 ha, 46%) is urban woodland and landscape conservation area, 683 ha are under nature protection, 200 ha are specially protected biotopes, 3623 ha are part of the European protected area system "Natura 2000". The heart of Freiburg's green space is the urban forest and with about four million visitors per year, it is the most

important recreational area close to the city. With the "Freiburg Forest Convention" (2001, updated in 2010), the city is committed to the ecological, economic and social responsibility of sustainable forest management (■ Fig. 1.2). Timber harvested from the forest benefits construction in the region as a resource. Nature and environmental education are actively promoted.

Freiburg's urban nature offer includes different nature areas and biotope types: Mountain meadows and forests to dry-warm biotopes with a variety of rare species, which are protected by a municipal species protection concept and a biotope network concept. Freiburg is part of the Southern Black Forest Nature Park and the Black Forest Biosphere Reserve. Through its precautionary protected area policy, the city has created new nature recreation and experience areas for the people and is involved in the alliance "Municipalities for Biological Diversity". Air pollution control, soil protection and near-nature watercourse design, since 2014 designation of flood hazard areas, precipitation water retention and precipitation water use, are components of the city nature concept.

With the "Rieselfeld" (70 ha), Freiburg has realized the largest new urban district project in the state of Baden-Württemberg with 3700 apartments for 10,500 people with over 120 private developers associations and investors. Comprehensive and needs-based public infrastructure, intact district life, civic engagement, open space and green amenities, energy concept and its location directly adjacent to a 250 ha

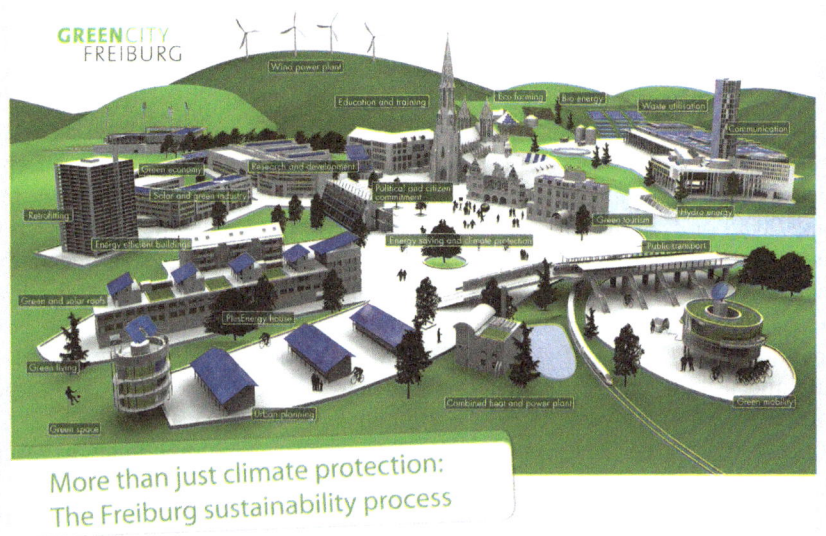

Fig. 1.2 The Freiburg sustainability process. (FWTM GmbH & Co. K.G., City of Freiburg 2017)

nature reserve, which serves as a local recreation area, make the district attractive.

However, the urban nature concept is only presented on two pages (p. 16–18) in the 25-page strategy paper for the Green City and thus takes a back seat in terms of its importance relatively behind energy, mobility and climate protection as sub-areas (Stadt Freiburg 2017).

1.2   What Does Urban Nature Mean?

What is nature? Understanding nature as the "totality of things that make up the world" ("all-nature") leads into a philosophical discussion that is of little use at this point since the concept of nature has now dissolved into various individual terms and made room for different "natures" (Leser 2008). To understand "natural" as "not influenced by humans" and to define this as "pure" nature would make nature hardly detectable anymore (Breuste et al. 2016). Nature that is normatively perceived as "good" is manifold, decentralized, uncontrolled, is perceived spontaneously, and thus has the sympathetic traits of a social role model (Trepl 1983).

Isolated nature ("partial nature"), as the nature of natural science, is a "mental isolate" of an unknowable wholeness of reality (Trepl 1983, 1988, 1992). The symbolic nature ("culture-nature") of cultural history determines our image of nature (Breuste et al. 2016).

1

> "We experience nature as something given – and yet it is a projection of cultural ideas and social ideals. Thus it is not only an ecological system, but also an ambiguous symbol: 'locus amoenus' and 'locus terribilis', on the one hand wilderness and on the other grandiose, homely, heroic, idyllic landscape" (Kirchhoff and Trepl 2009b, front text).

What nature is in the city can be answered very differently depending on the general understanding of nature (Brämer 2006, 2010; Breuste 1994a, 2016; Reichholf 2007). Usually, nature is not seen in cities, but the ("untouched") landscape (forests, coasts, moors, mountains). Also in the agricultural landscape, one dedicates oneself to its "rediscovery" (e.g., Müller 2005), finds "wildernesses" in our cultural landscape (e.g., Rosing 2009) or even discovers the landscape (Küster 2012, also 1995, 1998) in the middle of Europe as a "new" phenomenon of nature. Alongside the scientific-analytical effort to "understand nature" (e.g., Brämer 2006, 2010; Trepl 1992), there is always also the "feeling for nature", which can be found in all forms of art, especially in the Romantic period (Kirchhoff and Trepl 2009a, b). Understanding nature, feeling for nature and using nature should be well understood in cities to "offer" the "right" nature in the "right" places to enrich the human living space "city".

Understanding and feeling nature are the most important approaches to accepting, turning to and using urban nature in the city today. They are not mutually exclusive. However, many studies (e.g., Schemel 2001; Schemel et al. 2005) show that the aesthetic-emotional affection for nature in the city dominates for most urban dwellers and users of urban nature. This also corresponds to a predominantly urban horticultural, landscape-architectural design of nature in the city and a far-reaching aversion to unshaped, spontaneous nature in the city, since human order and "cleanliness" are not realized here. Unordered, spontaneous nature is also considered to be riskier to use (e.g., health hazards, social risks, etc.), which often does not correspond to real conditions.

The concept of nature in the city must of course also include "spontaneous to anthropogenic nature" (Leser 2008, p. 214). After all, "designed" or "anthropogenic" nature dominates everywhere in the human habitat and especially in cities. It would make little sense to exclude it in particular in the understanding and endeavour to understand nature, and instead to look primarily for little or no design nature that can also be found in cities.

Definition of Urban Nature

Urban nature (◼ Fig. 1.3) comprises the totality of natural elements present in urban spaces, including their ecosystem functional relationships and concerning their use.

Urban nature can thus be understood as all living beings, biotic communities and their habitats in cities. Urban nature can thus be found spontaneously ("wild") or as a result of human decisions (e.g., trees, plantings) accompanying almost all urban uses. It exists primarily in open space, but also on, around and in buildings. Unused or designated areas are determined by vegetation as dominant urban nature. They are also used (e.g., meadows, pastures, parks, gardens, urban forests, etc.) or have been eliminated from previous use (e.g., fallow land, certain wetlands and forest areas) (see also Naturkapital Deutschland – TEEB DE 2016, p. 15).

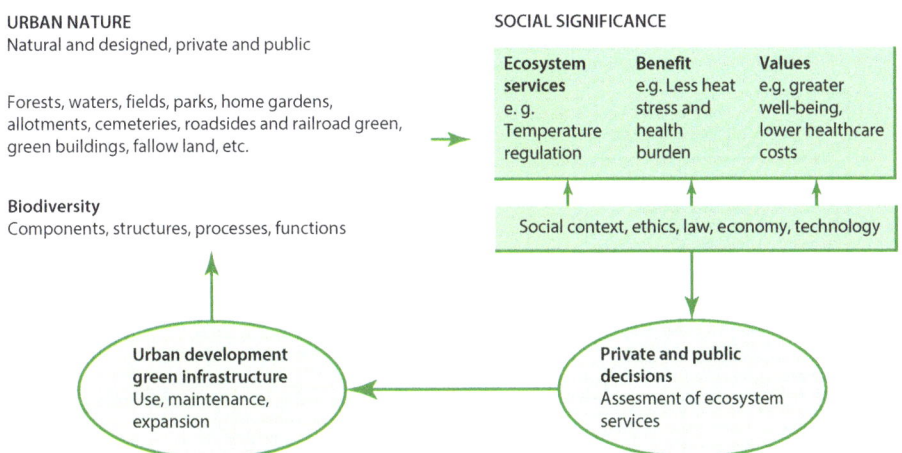

URBAN NATURE
Natural and designed, private and public

Forests, waters, fields, parks, home gardens,
allotments, cemeteries, roadsides and railroad green,
green buildings, fallow land, etc.

Biodiversity
Components, structures, processes, functions

SOCIAL SIGNIFICANCE

Ecosystem services e.g. Temperature regulation	Benefit e.g. Less heat stress and health burden	Values e.g. greater well-being, lower healthcare costs

Social context, ethics, law, economy, technology

Urban development green infrastructure
Use, maintenance, expansion

Private and public decisions
Assesment of ecosystem services

☐ **Fig. 1.3** Concept of urban nature. (Breuste and Endlicher 2017)

Urban nature is ecologically explained by spontaneous distribution and establishment of nature according to the diverse habitat conditions in cities based on cultural, -historical and utilization conditions and their history. It has a symbolic character and embodies positive values (affection) or partially stands for negative values (aversion, wasteland, dirt, risk, etc.) (Breuste 1994b, 1999, 2016; Hard 1988).

In a broader sense, urban nature naturally also includes the abiotic conditions that shape habitats, such as climatic parameters, hydrological characteristics and material components of the soil and the earth's surface. In the sphere model, they are grouped as atmosphere, hydrosphere and pedosphere, although of course they are not inanimate but are permeated by the biosphere. Together with their processes, feedbacks and interactions, they form the "natural system" of the city, which is studied selectively or holistically by various sciences (☐ Fig. 1.4; Breuste 2016).

When the concept of nature is linked to the urban concept, there is a lack of clarity concerning the inclusion or exclusion of certain concrete areas or components of nature. Usually, either the political urban area or the urban region is understood as a "city" in this sense. It is recommended to always link urban nature with its integration into urban structures and uses. It is thus also found outside the administrative cities, often spreading far beyond their periphery. In this sense, the spatial expansion of cities alone turns more and more "nature" into "urban nature".

It should be understood that urban nature is not an "ideal", even if there are repeated human efforts to present precisely this nature in the city, which is seen as "ideal" in terms of cultural history, and to bring it close to human users for "edification and instruction". Such nature offsets can be found as reminiscences of the idealization of nature from the Baroque park to the English landscape garden, the allotment garden, the eco-park to the "urban wilderness" as a space for experiencing nature (Reidl et al. 2005; Rink and Arndt 2011; Schemel 2001; Schemel et al. 2005) or the community garden. All these expressions of nature have still or again found new supporters and users.

1

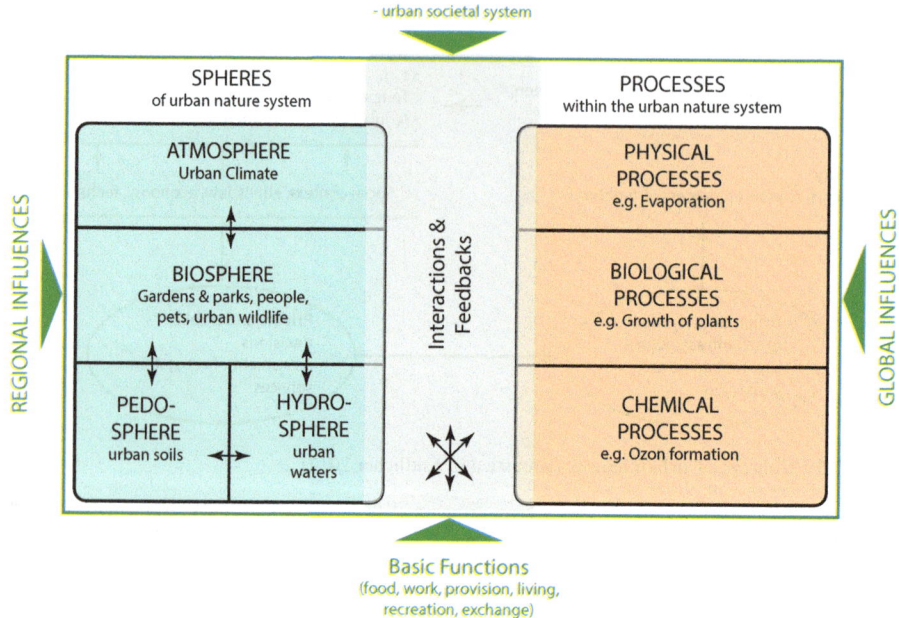

□ **Fig. 1.4**   The urban nature system. (Breuste and Endlicher 2017)

1.3 **Components of Urban Nature**

Shaped agricultural landscape and the natural landscape "woodland or forest" as "original nature" form the original opposites of cultural appropriation of nature in most regions of the world. The designed nature showed that one developed independence from the "wilderness", for which symptomatically the forest stood and mostly still stands, "culture". As symbols, both can be found everywhere in cities, shear grass from the cattle-rich cultivated floodplain landscapes, the urban kitchen garden from the rural living environment, trees and shrub plantings from the natural forest (Breuste 2016). Urban nature thus has cultural-historical justification and symbolic character (Breuste 1994a, 1999; Hard 1988).

Compared to rural landscapes and large forests, urban nature is diverse and rich in species.

» "'Urban nature' is more diverse, less threatened and less exposed to pressures. Above all, it is closer to us! Why is urban nature the way it (has) become? What are the reasons for the surprising diversity of species we find in cities? What opportunities does 'nature in the city' offer us?" (Reichholf 2007, p. 7)

The reasons for this difference are, on the one hand, in the progressive reduction in the species richness of rural landscapes due to intensive cultivation and, on the other hand, in the special ecological conditions of cities themselves as habitats. Through differentiated use and design, people who shape cities create a variety of

■ **Table 1.1** Original locations of apophytes in the settlements of Central Europe (Wittig 2002, p. 37)

Original location	Examples
Quarry forests, floodplain forests, high herbaceous vegetation on river banks	*Aegopodium podagraria, Calystegia sepium, Elymus repens, Equisetum arvense, Galium aparine, Glechoma hederacea, Humulus lupulus, Lamium maculatum, Poa trivialis, Urtica dioica*
Mud, sand and gravel areas on inland waters (pioneer meadows)	*Bidens tripartita, Chenopodium glaucum, Ch. rubrum, Corrigiola litoralis, Plantago major, Polygonum lapathifolium, Polygonum persicaria, Potentilla anserine, P. reptans, Rumex obtusifolius*
Dunes and rocks on sea coasts	*Atriplex postrata, Elymus repens, Sonchus arvensis, Tripleurospermum perforatum*
Glades by windthrow in forests	*Cirsium arvense, C. vulgare, Verbascum* species
Loose rock (scree slopes, sand dunes), rocks	*Chaenorhinum minus, Galeopsis segetum, Sedum telephium agg. S. acre, Tussilago farfara Asplenium ruta-muraria, A. trichomanes, Sedum album, Bryum capillare, Orthotrichum anomalum, Tortula muralis, Caloplaca saxicola, Lecanora muralis*

habitats that can be colonised, settle plants themselves and create habitat offerings for migrating animals and plants from other habitats (■ Table 1.1).

The general causes of urban natural diversity and species richness are explained in (■ Table 1.2):
— Structural richness of the urban habitat (structural diversity, different building structures, uses and intensities of use)
— Supply of nutrient-poor, dry and warm habitats
— Favouring pollutant and disturbance tolerant species
— Food and habitat availability and reduced competition for many animal species
— Introduction and spread of neobiota (Breuste 2016; see also Reichholf 2007)

Kowarik (1992) developed a simple way of making urban nature accessible in a manageable way by dividing it into four "nature categories" (■ Fig. 1.5). These represent the special features of urban nature (flora and vegetation, fauna) well and allow the diversity of anthropogenic, urban nature forms to be summarised and made manageable by only four "nature types". Detailed studies can be better classified based on this (Kowarik 1992, 2011, 2018; Breuste 2016; see ▶ Chap. 2).

■ **Remnants of Pristine Nature**

Nature of the first type (Kowarik 1992) covers the relics of old working or "pristine" landscapes such as natural forests and wetlands, which can be idealised as "pristine natural landscapes". They are the "old wildernesses" (■ Fig. 1.6), which still have something original, unshaped about them and which are characterised

1

◘ **Table 1.2** Effects of human interference from the "point of view" of plants. (Wittig 1996, modified after Wittig 2002, p. 17)

Human influence			Effects from plant perspective compared to the surrounding area
Art	Object	Effects[a]	
Indirect	Climate	Warmer (especially milder winters), drier, air more polluted	Favouring heat-tolerant loving and drought-resistant species; increasing the chance of survival of frost-sensitive species; hardly any possibilities of existence for strongly (air-)water-dependent species (hygrophytes); extension of the growing season. Favouring toxic-tolerant species; disadvantaging sensitive species
	Soil	Nutrient-rich, alkaline, contaminated less wet	Favouring basophilic species depending on high nutrient status Competitive advantage for pollutant-resistant species
	Water	Groundwater lowered, surface water drains off faster	Advantage for drought-tolerant and/or extreme deep-rooted plants; hardly any possibilities for hygrophytes to exist.
	Water	Enclosed, canalised or piped, polluted	Little chance for marsh and aquatic plants (helophytes and hydrophytes)
	Entire site	Disruption, destruction, creation	Favouring annual species (therophytes) with short generation cycle (several generations per year), high seed production, effective dispersal mechanisms (e.g., wind dispersal), long-lived seed bank; reducing competition; better opportunities for newcomers (neophytes).
Direct	Plant	Combat	
		Physical damage	Advantages for regenerative species; disadvantages for delicate and fragile species

[a]Compared to the surrounding area

especially by spontaneous development. The term "wilderness" is not a scientific concept. In its cultural meaning, however, it is easily understood and supports communication.

》 "'Wild' is not a characteristic that can be described in scientific terms, and 'wilderness' is not a scientific object. Rather, these terms denote meanings of nature that have arisen in society" (Kirchhoff and Trepl 2009a, p. 22, translated).

▪ **Agricultural Landscape**

The nature of the second type covers agricultural landscapes that continue to fulfil their agricultural functions as remnants embedded in urban structures or suburban areas. These are meadows (◘ Fig. 1.7), pastures, arable land and

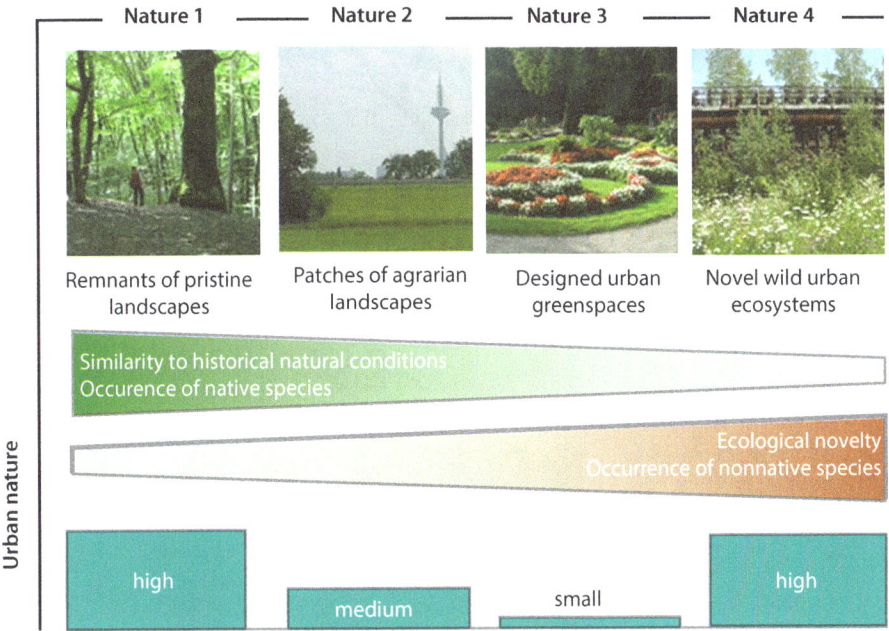

Fig. 1.5 Urban nature categories. (Kowarik 2018)

Fig. 1.6 Rabeninsel floodplain forest (NSG) in Halle/Saale. (© Breuste 2006)

associated accompanying landscape elements such as hedges, heaths, drifts and dry grasslands. Often, this type of nature is already urbanised and characterised by intensive management.

■ **Urban Gardens and Parks**

Kowarik (1992) refers to the nature of public and private green spaces, parks and gardens designed for urban use, the "symbolic nature of horticulture" (Breuste et al. 2016) as the nature of the third type.

1

■ **Fig. 1.7** Orchard meadow on the outskirts of Halle/Saale. (© Breuste 2006)

■ **Fig. 1.8** Hans Donnenberg Park, Salzburg, Austria. (© Breuste 2017)

As kitchen gardens, this nature type was originally economically justified. As an ornamental and landscape garden ("urban garden" or park; ■ Fig. 1.8) it is originated from the countryside estates of the nobility and was further developed as a "public park" in the expanding and beautifying city. Home gardens, green places, allotments (■ Fig. 1.9), traffic greens, urban parks, large landscape parks for recreational, individual urban trees, and tree-lined avenues make up this ornamental green of the city. Management, use and design are subject to fashions and economic justifications. Spontaneous development is usually pushed back or not allowed at all. Aesthetic design is in the foreground.

■ **Novel Urban Ecosystems**

Novel urban wildernesses in the form of urban-industrially based habitats that (re)enable the spontaneous development of nature constitute the fourth type of nature (■ Fig. 1.10). This type of nature developed under the more or less anthropogenic influence, but always in close relation to the strongly anthropogenically altered site conditions (soil, water balance, microclimate, etc.) after the abandonment of uses. Here, pioneer communities, spontaneous shrubs, and even new urban woodland as succession stages and adaptations to sites and distur-

▣ **Fig. 1.9** Allotment garden in Hildesheim, Germany. (© Breuste 2018)

▣ **Fig. 1.10** Spontaneous vegetation on derelict land, Greifswald, Germany. (© Breuste 2006)

bances. They are often the subject of urban ecology research (e.g., Rebele and Dettmar 1996; Wittig 2002 and many others). The understanding of this "new" urban nature species is still evolving (e.g., Kowarik 1993, 2018; Wittig 2002; Breuste 2016, and many others).

The Green City is a designed landscape with interconnected nature inside. It can also be understood in a larger scale as an urban landscape with neighbouring cities and communities, such as the Ruhr area (Vogt and Dunkmann 2015). In the Green City, all types of nature can contribute to biodiversity and nature experience (e.g., Reidl et al. 2005; Rink and Arndt 2011).

1

Urban planners and landscape architects often refer to urban nature as "green spaces" and distinguish them according to their location, accessibility and importance for urban dwellers. These classifications usually concern urban nature only in public or at least publicly accessible open space. The private open space of gardens and parks on private properties is seldom considered, as it is mostly not the subject of urban planning. Common utilization categories such as park, sports fields, allotment gardens, cemeteries, forests, etc. are used in planning. They are intended to stand as types for specific nature and handling of urban nature (Gälzer 2001; Gälzer and Hansely 1980, p. 43):

1. **Green spaces related to individual houses:** The garden at the dwelling (house garden, ground floor and tenant garden), play areas for small children, recreation and exercise areas for parents with small children and older residents with limited mobility. Maximum distance 500 m or 5 min walking distance (stroller distance).
2. **Green spaces in residential areas:** Playgrounds for older children and exercise places for young people, recreational and exercise areas (parks) for families and groups of adults, allotment garden areas (tenant gardens) for certain groups of people (families with small children, older people), facilities for frequently practised sports (school or youth sports grounds). Distance not more than 1000 m or 15 min on foot.
3. **District-related green spaces:** Allotment gardens, public outdoor swimming pools and sports facilities, cemeteries, larger parks with a wide range of usability. Easily accessible by public transport.
4. **Urban or regional green spaces (local recreation areas):** Recreation areas, including weekend recreation, allotments, larger sports facilities (e.g., stadiums, facilities for special sports), campsites, botanical and zoological gardens.

References

Brämer R (2006) Natur obskur: Wie Jugendliche heute Natur erfahren. Oekum, München

Brämer R (2010) Natur: Vergessen? Erste Befunde des Jugendreports Natur 2010, Bonn

Breuste J (1994a) "Urbanisierung" des Naturschutzgedankens: Diskussion von gegenwärtigen Problemen des Stadtnaturschutzes. Naturschutz und Landschaftsplanung 26(6):214–220

Breuste J (1994b) Flächennutzung als stadtökologische Steuergröße und Indikator. Geobot Kolloq 11:67–81

Breuste J (1999) Stadtnatur – warum und für wen? In: Breuste J (ed) 3. Leipziger Symposium Stadtökologie: "Stadtnatur – quo vadis" – Natur zwischen Kosten und Nutzen (= UFZ-Bericht 10/99, Stadtökologische Forschungen 20), Leipzig, pp III–V

Breuste J (2016) Was sind die Besonderheiten des Lebensraumes Stadt und wie gehen wir mit Stadtnatur um? In: Breuste J, Pauleit S, Haase D, Sauerwein M (eds) Stadtökosysteme. Springer, Berlin, pp 85–128

Breuste J, Pauleit S, Haase D, Sauerwein M (eds) (2016) Stadtökosysteme. Springer, Berlin

Breuste J, Endlicher W (2017) Stadtökologie. In: Gebhardt H, Glaser R, Radtke U, Reuber P (eds) Geographie – Physische Geographie und Humangeographie. Elsevier, München, pp 628–638

Gälzer R, Hansely H-J (1980) Grünraum, Freizeit und Erholung. Probleme, Entwicklungstendenzen, Ziele. Magistrat der Stadt Wien, Wien

Gälzer R (2001) Grünplanung für Städte. Planung, Entwurf, Bau und Erhaltung. Ulmer, Stuttgart

Green City e. V (2017) Wikimedia. www.wikipedia.org/wiki/Green_City. Accessed 15 Dec 2017

Green City Projekt GmbH (2017) Green City Projekt. München. www.greencity-projekt.de/de/. Accessed 15 Dec 2017

Hard G (1988) Die Vegetation städtischer Freiräume – Überlegungen zur Freiraum-, Grün- und Naturschutzplanung in der Stadt. In: Osnabrück S (ed) Perspektiven der Stadtentwicklung: Ökonomie – Ökologie. Osnabrück, Selbst, pp 227–243

Kirchhoff T, Trepl L (2009a) Landschaft, Wildnis, Ökosystem: Zur kulturbedingten Vieldeutigkeit ästhetischer, moralischer und Naturauffassungen. Einleitender Überblick. In: Kirchhoff T, Trepl L (eds) Vieldeutige Natur. Landschaft, Wildnis und Ökosystem als kulturgeschichtliche Phänomene. transcript, Bielefeld, pp 13–29

Kirchhoff T, Trepl L (eds) (2009b) Vieldeutige Natur. Landschaft, Wildnis und Ökosystem als kulturgeschichtliche Phänomene. transcript, Bielefeld

Kowarik I (1992) Das Besondere der städtischen Flora und Vegetation. Natur in der Stadt – der Beitrag der Landespflege zur Stadtentwicklung. Schriftenreihe des Deutschen Rates für Landespflege 61:33–47

Kowarik I (1993) Stadtbrachen als Niemandsländer, Naturschutzgebiete oder Gartenkunstwerke der Zukunft? Geobot Kolloq 9:3–24

Kowarik I (2011) Novel urban ecosystems, biodiversity and conservation. Environ Pollut 159(8–9):1974–1983. https://doi.org/10.1016/j.envpol.2011.02.022

Kowarik I (2018) Urban wilderness: supply, demand, and access. Urban Forestry Urban Greening 29:337–347

Küster H (1995) Geschichte der Landschaft in Mitteleuropa. Beck'sche Verlagsbuchhandlung, München

Küster H (1998) Geschichte des Waldes. Beck, München

Küster H (2012) Die Entdeckung der Landschaft. Einführung in eine neue Wissenschaft. Beck, München

Leser H (2008) Stadtökologie in Stichworten, 2., völlig neu bearbeitete Aufl. Borntraeger, Berlin

Müller J (2005) Landschaftselemente aus Menschenhand: Biotope und Strukturen als Ergebnis extensiver Nutzung. Elsevier, München

Naturkapital Deutschland – TEEB DE (2016) Ökosystemleistungen in der Stadt – Gesundheit schützen und Lebensqualität erhöhen. In: Kowarik I, Bartz R, Brenck M (eds) Technische Universität Berlin, Helmholtz-Zentrum für Umweltforschung – UFZ, Leipzig

Rebele F, Dettmar J (1996) Industriebrachen. Ökologie und Management, Ulmer, Stuttgart

Reichholf JH (2007) Stadtnatur. Eine neue Heimat für Tiere und Pflanzen. Oekom, München

Reidl K, Schemel HJ, Blinkert B (2005) Naturerfahrungsräume im besiedelten Bereich. Ergebnisse eines interdisziplinären Forschungsprojektes. Nürtinger Hochschulschriften 24, Nürtingen

Rink D, Arndt T (2011) Urbane Wälder: Ökologische Stadterneuerung durch Anlage urbaner Waldflächen auf innerstädtischen Flächen im Nutzungswandel: Ein Beitrag zur Stadtentwicklung in Leipzig (= UFZ-Bericht 03/2011). Helmholtz-Zentrum für Umweltforschung – UFZ. Department Stadt- und Umweltsoziologie, Leipzig

Rosing N (2009) Wildes Deutschland: Bilder einzigartiger Naturschätze. National Geographic, 5thed edn. G + J NG Media, Hamburg

Schemel H-J (2001) Erleben von Natur in der Stadt – die neue Flächenkategorie "Naturerfahrungsräume". Z Erlebnispädagogik 21(12):3–13

Schemel H-J, Reidl K, Blinkert B (2005) Naturerfahrungsräume in Städten – Ergebnisse eines interdisziplinären Forschungsprojekts. Naturschutz und Landschaftsplanung 37(1):5–14. www.naturerfahrungsraum.de/pdfs/ner_ziegenspeck_02.pdf. Accessed 14 Dec 2017

Stadt Freiburg (2017) Green city Freiburg. Freiburg. www.freiburg.de/greencity. Accessed15 Dec 2017

Trepl L (1983) Ökologie – eine grüne Leitwissenschaft? Über Grenzen und Perspektiven einer modischen Disziplin. Kursbuch 74:6–27

Trepl L (1988) Stadt – Natur, Stadtnatur – Natur in der Stadt – Stadt und Natur. Stadterfahrung – Stadtgestaltung. Bausteine zur Humanökologie. Deutsches Institut f. Fernstudien an der Univ. Tübingen, Tübingen, pp 58–70

Trepl L (1992) Natur in der Stadt. Natur in der Stadt – der Beitrag der Landespflege zur Stadtentwicklung. Schriftenreihe d. Deutschen Rates f. Landespflege 61:30–32

1

University of Copenhagen (2017) Knowledge, responsibility and sustainability GREEN CAMPUS
    2020. Strategy for resource efficiency and sustainability at the University of Copenhagen.
    greencampus.ku.dk/strategy2020/english_version_pixi_GC2020_webversion.pdf. Accessed 15
    Dec 2017
Vogt C, Dunkmann N (2015) Green City. Geformte Landschaft – Vernetzte Natur: Das Ruhrgebiet
    in der Kunst. Kerber, Bielefeld
Wittig R (1996) Die mitteleuropäische Großstadtflora. Geogr Rundsch 48:640–646
Wittig R (2002) Siedlungsvegetation. Ulmer, Stuttgart

How Urban Nature Developed?

Contents

2.1   Urban Nature as a Cultural
        Product – 20

2.2   Urban Nature
        Is Establishing – 29

2.3   Public Urban Nature
        for Beautification, Recreation,
        Public Health and Public
        Education – 33

2.4   The Park Cemetery – A Place
        of Recreation for the Living – 44

2.5   The Urban Forests and Forest
        Parks – How the Forest Stayed
        in the City – 47

2.6   The Private Garden
        for Everyone Complements
        the Public Urban Nature – 51

2.7   How the Waters Became
        Urban? – 58

        References – 64

© Springer-Verlag GmbH Germany, part of Springer Nature 2022
J. Breuste, *The Green City*,
https://doi.org/10.1007/978-3-662-63976-4_2

2

For a long time, urban nature was exclusively designed nature in an urban context. It emerged with the development of cities and towns and thus also took on different forms of design and acceptance. It began as predominantly private nature in personal surroundings, as gardens for the nobility and rich elites, and became their status symbol as decorative nature subject to fashion and trends. As the useful nature of orchards and agricultural areas, it remained in the cities for a long time, until the value of its areas forced it to the outskirts of the city in competition with other uses, or caused it to disappear altogether.

Trees were among the first purposefully established natural elements in cities. As avenues or individual trees they were often already placed in the urban public at an early stage. Even today, they are still the most intensively perceived individual elements of urban nature with high appreciation by the urban dwellers.

Public urban nature, usable and accessible to everyone, is a European idea born with the Enlightenment and the political upheavals in the wake of the French Revolution. This public urban nature aimed to contribute to beautification, recreation, public health and public education. It spread worldwide as public parks in the nineteenth century, encouraged by European colonization in other continents. In addition, new forms of urban nature emerged as new uses (for example, golf courses and horse racing courses) or established, previously non-urban uses (for example, park cemeteries, forests) were incorporated into the growing cities of the nineteenth century.

For a large number of cities, waters, mostly flowing waters were the original localisation reason, and still they are part of their natural features. Over long periods, they were important natural components, supplied water and energy for a variety of uses and discharged waste and sewage. When these tasks were solved differently and more effectively (for example, water pipes, sewage systems, new energy sources), urban water bodies changed in their functions. They became part of recreational areas and provided habitats for independently developing nature with flora and fauna that could only be found there, which attracted the interest of the emerging nature conservation. Where this change of function has not yet taken place, for example, in many cities of developing countries, the multifunctional use and overuse of water bodies persists.

2.1 Urban Nature as a Cultural Product

Urban nature was always either useful nature, decorative nature or (residual) wild nature. Thus, it served either the satisfaction of material needs (animal or vegetable food, firewood, medicine, etc.) or ideal needs (the beauty of the living environment, spiritual experience, recreation, etc.) of the urban dwellers. Wild nature places without the urban dweller's need, as never used and currently unused nature in the urban environment, has not existed for a long time (see ▶ Sect. 7.4). Wild nature as urban nature refers to ecosystems that have merely been "co-used" (game, timber, forest fruits, etc.) or that have developed independently for some time without any targeted use, but often certainly through ideal use (visiting, recreation). In many cities, this type of nature is completely absent or has only just emerged in the last few decades through abandonment of use. Wild nature is thus a special type of nature that is used without targeted claims to use, but also (as yet) without an ornamental character. A new ethical relationship is only just emerging out of respect for nature, which is developing independently of man. Here, a new "ethical concept" for urban nature may emerge, which is to be assigned to the idealistic needs.

Useful and decorative nature are part of a concept of quality of life. Where urban dwellers are released from the production of food for their consumption, interest in urban nature focuses primarily on ideational needs. In developed industrial and post-industrial societies, this is largely the case. In other societies, other cultures, and among urban populations with often extremely low incomes, food production continues to play a major role as the main use of nature, including in the city. In many Asian, African and Latin American cities, commercial food producers also produce close to their urban customers, in the city or on its periphery. This useful nature is usually not urban planned but urban related. Planning of urban nature, as an urban task, often refers only to park and ornamental nature in publicly accessible spaces at all. Many cities worldwide either have no urban nature planning (green planning) or it is largely ineffective (see ▶ Sect. 3.1).

Urban nature can therefore be assigned to either a utility concept or the decorative concept. The transitions are seamless if, for example, both productive and non-material needs are satisfied (for example, allotment gardens). The degree of use (intensive or only accompanying) and the ownership of the areas with urban nature are also important.

Urban nature was initially private nature in private gardens (see ▶ Sect. 5.3) and parks (see ▶ Sects. 5.1 and ▶ 3.6) or agricultural land. Public space was largely "nature-free". This was as true of the early cities of the Near East, the Indus culture or China as it was of Egypt, Greece or Rome. This was still the case in the European Middle Ages. A kitchen garden ensured the supply of vegetables and sometimes fruit at the place of residence in the city (see ◨ Fig. 2.1). Until the nineteenth century, urban dwellers in Germany, for example, sometimes had vegetable gardens, orchards, vineyards or small fields. Those who did not depend on them

◨ **Fig. 2.1** Former orchard areas within the historic city walls of Kotor, Montenegro. (© Breuste 2017)

2

afforded a garden only for decoration and pleasure. Urban nature was almost 100% private nature. It was located both inside and mostly outside the city walls. Large agricultural or garden areas were only enclosed within the walls if this was necessary for supplies in times of war or seemed possible due to the topography.

The ancient and medieval European city was largely treeless, except fruit trees in orchards. Individual trees, however, found their way into towns and cities as decorative elements, shade-givers and religious natural objects (see ▶ Sect. 5.2). Old, planted individual trees stood, for example, in churchyards, in monastery gardens, at places of jurisdiction, culture and assembly, and places for traditional competitions, festivals and games. Some city trees preserved in Europe and America are up to 400 years old (e.g. Salzburg/Austria, Kotor/Montenegro, Surce/Bolivia), in other cultures even much older (for example, in Iran over 4000 years, in China over 1000 years) (see ▪ Figs. 2.2, 2.3, 2.4, and 2.5). The town square as a public space was originally free of trees.

Even today, urban dwellers as property owners decide on the largest share of urban nature in terms of area. Their private utilization decisions are only limited to his by-laws and public regulations (only in highly regulated societies) (for example, legal provisions, tree protection statutes, distance regulations, planting

▪ **Fig. 2.2** About 1000-year-old Gingko trees (*Gingko biloba*) in the Confucius Temple (孔庙 "Kǒng Miào") in Qufu, China. The trees are believed to have been planted in the Song Dynasty (960–1279). (© Breuste 2009)

◘ **Fig. 2.3** 350-year-old (plantation 1667) black poplar (*Popolus nigra*) from the garden of the former Franciscan monastery on Pjaca od Kina in Kotor, Montenegro. (© Breuste 2017)

regulations, etc.). In private open space, the urban dweller's personal interest in urban nature, shaped by fashions and preferences, is evident. The state of nature that best meets the needs of the owner is maintained or adapted to changing fashions. The private kitchen garden exclusively as a source of food or as a food supplement dominates in Europe today among urban dwellers with low incomes or special lifestyles.

The private ornamental garden largely dominates as an expression of aesthetic sensibility. This is strongly assimilated worldwide and leads to "similar" natural features with often the same, non-native species. Urban ornamental nature becomes standardised due to aesthetic standardisation. This also applies to the urban nature of public spaces.

In the early modern period, urban nature in public space was limited to a few trees, cemeteries, fairgrounds, or parade grounds, some of which were already outside the city walls. For a long time, it was not the subject of planning, which took it up only late, for example, in the case of planned new town planning (e.g. St. Petersburg/Russia 1703; Karlsruhe/Germany 1715) and systematically only in the nineteenth century. Green planning of public space thus has a short history. It was not until the planned development of the city in modern times that public nature was accorded great importance in public education and urban hygiene.

2

◘ **Fig. 2.4** *"El árbol milenario",* 400-year-old West Indian cedar (or cedar) (*Cedrela ordorata*) in the Convento de La Recoleta (founded 1601), Sucre, Bolivia. (© Breuste 2015)

» "Green planning in urban development has had a higher status in history than it has today. Its disregard leads to the alienation of man and nature. A lack of quality of life is the consequence" (Krauskopf 2003, p. 78, translated).

Nature design is subject to green planning as part of urban planning. Wherever planning takes place, functions and corresponding natural features are assigned to public spaces. Urban nature is "established" (planted) and "looked after" (maintained). The experience, availability and financial scope of the urban gardeners employed by city administrations determine which natural features are suitable for which areas and which uses. In some cases, this is also left to private gardening companies, which take on this task commercially on behalf of the municipality in a more or less controlled and commissioned manner and act not only aesthetically, but also economically and according to the availability of planting material.

The goal is to design public space to best meet the needs of urban residents. These needs naturally change through cultural transformations and fashions, resulting in a constant adaptation through changing equipment, use and maintenance of the public urban nature.

What we find today as urban nature in cities is thus the result of a process of adaptation by city administrations to given circumstances and needs, and not least to available budgets. The needs of the city's inhabitants are by no means met in the

Fig. 2.5 Abarkooh cypress (*Cupressus sempervirens*), more than 4000 years old, national natural monument since 2003 in Abarkooh, Iran, height 25 m, trunk diameter 3.14 m, crown diameter 14.07 m. (© Breuste 2018)

same way; some parts of the city are given more public space and more, often also different, urban nature, others less, often with a lower maintenance budget. On the one hand, this is due to the history of the city, but on the other hand, it can also be traced back to unequal treatment of the different social classes or the political intentions of the city administrations. Green planning and maintenance is thus part of urban policy, and the Green City is not least a statement of municipal policy (see ▶ Sect. 8.3).

In this field of tension, urban nature came into the city and asserts itself there or was displaced by other interests as less important in the course of urban development. Urban nature was and is thus eagerly fought over economically, politically, socially and aesthetically everywhere. The result of this is visible and can be analysed accordingly. The process of creation is often opaque, sometimes difficult to reconstruct and often veiled. The arguments put forward in debates about urban nature are usually intended to make other interests less clearly visible. This is particularly true at present of "ecological" arguments, which are put forward by a wide variety of parties and are intended to conceal other, less socially accepted, often individual interests.

When in the Western philosophy of the eighteenth-century nature was normatively developed as original and as an ideal model ("back to nature"), this was

2

transferred to the design of the English landscape gardens of the landed gentry. As a positive model, it also determined the era of the dawning modernity in urban development and green planning. The landscape garden was integrated into urban planning and design and green became part of urban planning. This was an essential step towards a new understanding of the city, in which nature was seen as an immanent part of the city. Existing cities were increasingly understood as "unnature", although the absence of nature or its neglect had not been noticed or complained about until then. Rapid and ruthless industrialization, especially of the nineteenth century, created unplanned cities whose growing numbers of urban dwellers had little influence on urban design or urban nature. Reduced to a minimum of physical existence for the majority of its inhabitants, the industrial city became the antithesis of a good and healthy life.

> » "The great city appears as a symbol, as the strongest expression of culture turned away from the natural, the simple, and the naïve; in it, to the disgust of all well-meaning people, there piles up know-nothing hedonism, nervous haste, and disgusting degeneration into a ghastly chaos... One chides the unspeakable ugliness of the cities with their desolate noise, their filth, their dark courtyards, and their thick, murky air" (Endell 1908, p. 20, translated).

In the eyes of optimists like Endell (1908, p. 30, translated), the great city can reclaim the "hidden beauty" of the city as nature, "precisely because this beauty is almost always overlooked, because one is not at all accustomed to looking at a city as one looks at nature, as one looks at forests, mountains, and the sea." What the modern metropolis lacked was planned design, even against land speculation and profit maximization, in the interest of society's overall well-being. This was expressed in the demand for "space for light, air and sun" (Krauskopf 2003, p. 78).

The health of city dwellers seemed to be endangered even earlier, when the industrial city was just emerging. As early as the eighteenth century, idealized nature appeared to be a suitable means of promoting the health of city dwellers. With the general acceptance of the health-promoting effects of nature, a new, immense field of activity opened up for architects, urban planners and garden artists to improve the city through nature in it and to justify their designs to the decision-makers. The metaphor "green lungs" had already stood for this since the second half of the eighteenth century and is still in use today (for example, Thorén 2008).

The modern city as a liberal city, first London and Manchester in England, was to have a government concerned with the welfare of the governed. That included the welfare of health.

> » "The sanitary economy of the town was like that of the body. Both were characterized by a dynamic equilibrium between living organisms and their physical environments" (Joyce 2003, p. 65).

The shaping of this "physical environment", the existing urban nature, was already a matter of concern in the eighteenth century. Nature was hardly present and public in cities of the eighteenth century. Already since the seventeenth century, access to the Crown's property had been established as a common law almost everywhere in England (Hyde Park, for example, in 1637) (Kostof 1993, p. 167). The Royal Parks in London were used for exercises, horse racing, betting, rides, walks, etc. The restriction of this right of use repeatedly led to fierce, also politically used disputes in the House of Commons.

The opposing pair of city and nature was seemingly irresolvable, at least in the European tradition. Mystical concepts and effective buzzwords served urban planners just as well as scientifically based arguments to overcome the obvious contradiction. In the nineteenth and twentieth centuries, urban planners in Europe and the New World continued to develop new plans that were touted as the most "nature-friendly" solution, either mitigating the obvious physical and health burdens of the city through "nature additions" or completely redefining the city in relation to nature (Krauskopf 2003). Nature, by all means designed nature, was assigned a function to promote the health of city dwellers.

Social reformers such as the architect Edwin Chadwik (1800–1890) sought to improve the intolerable health burdens of working people living in growing poverty in large industrial sites in England in the 1840s through a system of sanitation (Chadwick 1843). Public spending on healthy living conditions, it was argued, would pay off in the long run by reducing disease and poverty. This included sewage and waste disposal, clean drinking water, sanitary measures, and also sanitary effects of public green spaces. The demand for public parks, raised as early as 1803 by the Scottish landscape architect John Claudius Loudon (1783–1843) (Kostof 1993), was integrated into the Public Health Act (pleasure grounds) passed in 1848 (Hansard 1848).

What the Metaphor "Green Lungs" Means – "Lungs" Cleaning the Air for Better Health in Cities

The health theory of the eighteenth century was based on pathogenic miasms, based on putrefactive processes in water and air and the resulting air pollution. This was not yet primarily concerned with, for example, industrial air pollution in the modern sense. Air without miasmas was considered beneficial to health. In 1771, the theologian and chemist Joseph Priestley (1733–1804) had discovered the gas oxygen, which he called "dephlogisticated air", and in 1779, he had demonstrated its regeneration by photosynthesis (Willeford 1979).

From the beginning of the nineteenth century, large green spaces that produced oxygen and reduced air pollution, especially parks, were recognized as antidotes to disease-causing miasmas in public debate. The Scottish botanist and noted landscape architect John Claudius Loudon (1783–1843) described the Squares of London as *"of greatest consequence to the health of its inhabitants"* because they promoted the *"free circulation of air"* (Loudon 1803, p. 739).

Priestley found that "...*no vegetable grows in vain...but cleanses and purifies the atmosphere"* (Raven et al. 2005, p. 116). So it was not about more oxygen, which is not lacking even in polluted air, but about less sheep matter, the "purification of the air" from miasmas. That this function was attributed to vegetation was correct, to link it to the lungs, probably more of a misinterpretation, but one that

2

is upheld and accepted by everyone to this day: Lungs purify the air.

A system of "lungs" distributed throughout the city was seen as a means of preventing the spread of disease by giving urban dwellers access to the vegetation-cleansed air of these green spaces. In the industrial city, there could only be larger green spaces rich in vegetation. The idea of establishing parks to "produce clean air" as a "counterbalance" to air-polluting industry was born. Reducing air pollution from industry was not up for debate, as this was seen as an impermissible restriction on private action. In the Public Health Act passed in 1848, the establishment of parks was enshrined for the first time in the United Kingdom with the green lung argument (Crompton 2016).

» "A metaphor can be a valuable tool for raising awareness. The lungs metaphor was effective in the nineteenth century because it aligned parks with the prevailing political and social concerns of that era" (Crompton 2016,12).

The "Green Lungs" – How a Metaphor Came Into Being and is First Documented in 1808: It Is About the Right to Public Greenery – A Topical Issue

The term "green lungs" has stood for the positive health effects of urban greenery and urban nature in general for more than 240 years. It is worth questioning, since lungs generally do not produce vital oxygen, but CO_2. But that's not the point of the term either. The now-greened Squares and Royal Parks, Hyde Park and adjoining Green Park and St. James's Park, were considered the "green lungs" of London as early as the eighteenth century. The metaphor can be traced back to William Pitt the Elder (1708–1778), first Earl of Chatham, and was held up as a powerful argument for the preservation of Hyde Park to the Tory Chancellor of the Exchequer, Spencer Perceval (1762–1812), by the Irish Wigh politician Lord Windham (1750–1810), Secretary of State for War and the Colonies, in a parliamentary debate on Hyde Park on June 30, 1808. The debate seems as if it could have taken place 200 years later.

The metropolisation of London led to increasing building expansion and land speculation. Negotiations by a consortium with the Royal Treasury to build eight houses on part of Hyde Park had become public in 1808 and, if royal ownership of the park was recognised, the rights of the general public to use the park were seen to be threatened as a result.

Lord Windham:

» "Now, if in addition to these a number of houses should be erected, the power of vegetation would be completely destroyed. The park would no longer be that scene of health and recreation it formerly was. **It was a saying of Lord Chatham, that the parks were the lungs of London.** He could devise no mean more effectual for the destruction of these lungs than the proposed plan. … He had heard of parks decorated with grottos and temples, but here was a plan to decorate a park with houses; as if a citizen, who should leave Whitechapel on a Sunday evening to get a little fresh air, would feel much gratified when he arrived at Hyde-park to see nothing but houses" (Hansard 1812, 1124–1125).

The plan to partially build on Hyde Park was defeated by 36 votes against 23 in favour (Crompton 2016; George 2017; Hansard 1812). The metaphor of "green lungs" has been reapplied countless times since then in the struggle for urban nature, just as successfully up to the present (e.g. Corbin 2005; Murray 1839).

2.2 Urban Nature Is Establishing

By the eighteenth century at the latest, existing private urban nature (rulers' parks) in large European cities became accessible and (semi-)public natural elements were inserted into the urban residential areas of the upper classes. The latter serve more as decoration and to increase the value of the properties than as a real use through the residence. The London Squares are an example of this. Greened town squares were laid out in London as early as the seventeenth century (squares and, from the second half of the nineteenth century, semi-circular crescents). Their sizes varied from 0.6 to 28 ha (Grosvenor Square 1725–1735). By 1780, the tree-lined square became standard. Today, most squares are still private green spaces that create attractive residential locations; some are open to the public (Mader 2006).

Amsterdam was the first city in Europe to plant trees on a large scale. Already before 1600, lime trees and later elms were planted on both sides of every newly dug canal. The trees were planted not only for aesthetic reasons, but also to strengthen the embankments and later for their air-purifying effect (Amsterdam 1999–2018).

The emerging bourgeoisie, especially in the eighteenth century, took over the rural-aristocratic leisure pleasure of "pleasure walking" in stately gardens and parks, transferred it to its urban living space and made it a "Spazier-Gang oder lustigen Zeit-Vertreib im Grünen" ("walk or a fun pastime in the countryside") (Zedler 1743). The walk became and still is a basic leisure activity. It became an element of bourgeois lifestyle in the eighteenth century. It takes place in the countryside outside the still small towns and in the city in the specially created space of promenades or parks. The emergence of parks or promenades in towns, the place where the bourgeoisie lived, is directly related to walking and the bourgeois way of life. Contemplative as well as social components come together in this (enjoying impressions of nature, spending time in a healthy environment, thinking, making contacts, having undisturbed conversations). In addition, there were horseback rides or carriage rides.

The avenues of the Italian gardens of the fifteenth and sixteenth centuries, like other landscape gardening elements, were adopted in the urban design of the European Renaissance and Baroque. These included representative avenues, especially as connecting elements between princely palaces and hunting grounds or pleasure palaces outside the cities (e.g. Untern den Linden, Berlin; Champs-Élysées, Paris; Hellbrunner Alle, Salzburg; Paseo de la Reforma, Mexico City).

2

Boulevards and Promenades

Wide tree-lined public promenades were laid out in towns to meet the need for strolling, and especially carriage rides in shady surroundings. They often linked the city and the residential areas of their users, the propertied middle and upper classes. With the defortification of the cities, the space needed for this was created on former city fortifications and filled-in city ditches. The French word *boulevard* denoted a city wall and derives from the Dutch *bollwerc*.

Between 1668 and 1705, Nouveau Cours, tree-lined boulevards that took their name from the Grand Boulevard bastion north of the Bastille, were built on the open spaces of the removed northern Parisian city fortifications. The Grands Boulevards are avenues at least thirty metres wide with mostly tree-lined pedestrian walkways (*trottoires*) and multi-lane carriageways. They are the oldest boulevards in Paris and form a semi-circular, almost three-kilometre-long traffic axis between Place de la Madeleine in the west and Place de la Bastille in the east.

In 1667, the garden architect André Le Nôtre (1613–1700) designed a Grand-Cours, 70 m wide and 1910 m long, in the extension of the central avenue of the Tuileries Gardens, from which the Champs-Élysées emerged, a west-facing tree-planted show axis.

Between 1852 and 1870, during the Second Empire, the Parisian prefect Georges-Eugène Haussmann (1809–1891) used the Grand Boulevards as a model for the redesign of Paris, which at that time still had a medieval structure. From 1853 onwards, Haussmann had further boulevards laid out as broad connections across the city. Large parts of the densely built-up inner city were demolished for this purpose. The engineer, urban planner and garden designer Jean-Charles Alphand (1817–1891) took over the design and layout of the boulevards for the transformation of the two suburban forests Bois de Bologne and Bois de Vincennes into park forests (Alphand 1867–1873).

The Vienna Ringstrasse was built from 1864 to 1865 (officially opened on May 1, 1865) on the former fortifications, ditches and glacis around Vienna's old town as a promenade 5.2 km long. The 6.5 ha Stasdtpark, completed in 1862, and extensive tree plantings along the street are part of the Ringstrasse project (Wagner-Rieger 1972–1981).

The approximately 15 km long and 60 m wide Paseo de la Reforma, realized under the direction of the Austrian officer Ferdinand von Rosenzweig (1812–1892) from 1864 to 1867, is the main artery of Mexico City. It connected the residence of Emperor Maximilian, Chapultepec Castle, and the city center. Today it is one of the city's main shopping streets (see ◘ Fig. 2.6).

◘ Fig. 2.6 Paseo de la Reforma, Mexico City. (© Breuste 2005)

After acquiring the inner ring of fortifications in 1881, the city of Cologne used the 104 hand-sized open space to build a magnificent boulevard in the style of Parisian urban planning and the Vienna Ringstrasse. When it opened in 1886, Cologne's prestigious Ringstrasse consisted of a chain of representative street spaces, between 32 m and 114 m wide. Unfortunately, its originally planted central strips and part of the tree population have not been preserved.

The 2.5 km long Hellbrunner Allee in Salzburg, which was laid out in 1613–1618, is the oldest preserved stately avenue in Central Europe and probably worldwide. It leads from Hellbrunn Palace is to the Freisaal palace.

Several manorial estates lie along the road, which was intended as a carriage route. The tree population consists of black poplars, English oaks and copper beeches. Numerous oaks from the time of origin (approx. 400 years old) have survived to this day. They represent the largest and most valuable old tree population in the city of Salzburg. The trees of Hellbrunn Alley are also of outstanding importance as a habitat for bats, tree-nesting birds and wood-dwelling beetles, many of which are protected. It is therefore also of interest in terms of nature conservation. The boulevard has been a natural monument since 1933 and a protected landscape area since 1986. (Medicus 2006; see ◘ Figs. 2.7 and 2.8).

2

■ **Fig. 2.7**   Hellbrunner Allee, Salzburg, Austria. (© Breuste 2018)

□ **Fig. 2.8** Park Alameda Central, Mexico City. (© Breuste 2010)

2.3 Public Urban Nature for Beautification, Recreation, Public Health and Public Education

The idea of public parks in cities (see ▶ Sect. 5.1) is a European idea that was exported from here to cities around the world. It was first implemented in North and South America, and later in European colonies worldwide as a lifestyle element and part of modern urban development in other parts of the world (for example, Kolkata, India: Maida Park, 1847; Tokyo, Japan: Ueno Park 1876; Shanghai, China: Public Garden/Huangpu Park 1886; Fuxing Park 1909, Bangkok, Thailand: Lumphini Park 1947).

2

Public parks arose in European cities with different, sometimes interconnected, intentions, on the one hand from the idea of public education and on the other from the idea of public health. The older idea is that of public education. It was intended primarily for the poorer city dwellers to experience "edification" and "recreation" by visiting a landscaped park and to meet the "upper" classes on the neutral grounds of the park in the city, thus raising the "spiritual character of the lower classes of society" (Kostof 1993, p. 169). Such parks were referred to as "Volksgarten" or "Volkspark" in Germany in the late eighteenth century.

In Europe, public parks first developed from **existing manorial parks,** which were initially opened for a limited period of time and later generally for public use, but remained in private ownership. Often there was still a spatial connection to a castle or country estate. There are many examples of this with the Royal Parks in London (Hyde Park 1637, Green Park and St. James' Park open in the seventeenth and eighteenth centuries, Regent's Park open in the nineteenth century) and many manorial parks in the rest of Europe (for example, St. Petersburg, Berlin, Vienna, Prater 1766) and Asia (e.g. in Tokyo) (see ◘ Figs. 2.9, 2.10 and 2.11).

◘ **Fig. 2.9** Avenue in the park of Schönbrunn Palace, Vienna, Austria. (© Breuste 2018)

Fig. 2.10 Valdštejnská zahrada, Prague, Czech Republic. (© Breuste 2017)

Fig. 2.11 Englischer Garten, Munich. (© Breuste 2011)

None of these parks, however, had been designed and laid out for public use. The first verifiable park that was intended for public use from the outset was the **English Garden in Munich.** It was commissioned by the Bavarian Elector Carl Theodor with the name Theodorspark on August 13, 1789, as a **people's park** and officially opened on April 1, 1792. It was followed as a people's park by the five-hectare *Vienna People's Garden,* built in 1819–1823 on the site of the former castle bastion of Vienna's former fortifications by order of the "Gardener-Emperor" Franz II. (I.). It was the first park planned for public use in the Habsburg Empire to be laid out in a regular design for "disciplined use" and opened on March 1, 1823 (Hajós 2005). In 1830, **Magdeburg's Herrenkrugpark** was completed as a public city park, designed by the Prussian garden architect Peter Joseph Lenné (1789–1866) (Hesse 1907; Hoke 1991).

2

The Volkspark

The Volksgarten or Volkspark is a form of public park. It originated at the end of the eighteenth century. Access to nature and recreation for all urban dwellers were the tasks of public parks. They were laid out on public land that was available and could be acquired as cheaply as possible, with a connection to the urban residential areas, on behalf of the city authorities. The features include woodland and meadow areas arranged in the style of English landscape gardens, tree plantings, ponds, water features, resting places, monuments and pavilions. The focus is on aesthetic design. With the integration of new functions such as play and sport, the Volksgarten developed further into a Volkspark at the end of the nineteenth century. Typical of Volksparks are central, large and contiguous, accessible play and sports areas and lawns (e.g. Volkspark Jungfernheide, Berlin) (Kostof 1993).

In 1842, in east London, Victoria Park was expressly dedicated to the working class by Act of Parliament as a counterbalance to the Royal Parks of the West End (Kostof 1993). Location and neighbourhood at least represent an idea of social balance in the provision of urban nature for urban dwellers.

Near the emerging industrial metropolis of Liverpool, the English gardener Joseph Paxton (1801–1865) was also commissioned by the city to design Birkenhead Park, which opened in 1847 and was surrounded by more than 100 luxurious terraced and detached houses for the upper-middle classes, the sale of which generated the costs of the park. This was thus also an early garden city design. Birkenhead Park in Liverpool is often inaccurately cited as the first-ever public park (Kostof 1993). In the true sense, however, it is neither the first nor a typical public park for the wider urban population. The 91 ha Birkenhead Park was located far from Liverpool proper on the other side of the Mercey, in a new urban development area for upper-middle and upper classes (see ◘ Figs. 2.12, 2.13, and 2.14). Birkenhead Park is an example of urban nature development in the wake of deliberate social segregation. In its intent and design, however, it became famous, not least because Frederick Law Olmsted (1822–1903), the most famous landscape architect in the USA, took it as the model for many other parks, in the USA, for example, Central Park in New York (Brocklebank 2003; Greaney 2013; Kostof 1993; Metropolitan

◘ **Fig. 2.12** Birkenhead Park, Liverpool. (Kostof 1993)

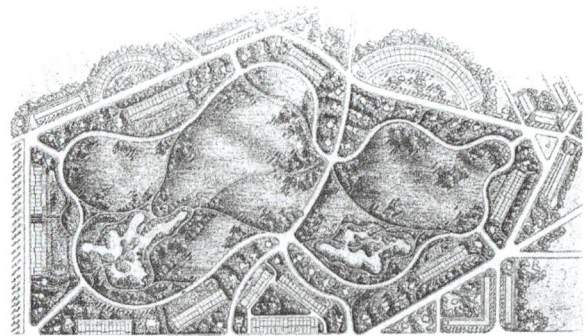

Fig. 2.13 Central Park, New York, USA. (© Grimm 2013)

Fig. 2.14 Plaza Mayor in Antigua Guatemala. (© Breuste 2012)

Borough of Wirral 2008). By 1865, most towns in England and most boroughs in London had public parks (Mader 2006; Wiede 2015).

Around the middle of the nineteenth century, parks became central elements of new urban development concepts for the first time, especially for dynamically growing cities. Nature was idealized and brought into the city in a "walkable landscape painting". "Happy landscapes" were redesigned through "aesthetic adornment".

2

» "Even the happiest landscape…can be aesthetically embellished and economically improved by the proper application of the art of gardening…" Peter Joseph Lenné (Wieland 2012, p. 59, translated).

In the course of the bourgeois takeover of European and American cities, urban planning and beautification were initiated. In many European cities, city-beautification associations were founded, which drew up proposals and plans, provided funds, acquired plots of land and induced the city administrations to undertake city-beautification measures or carried these out themselves, since green space offices did not always already exist as administrators. The aim of the associations was to "beautify" their city through planted and designed, but also preserved (e.g. forests) urban nature and to make these natural oases available to the public for recreational use.

Many areas and plantings secured by beautification associations for the Green City are still essential and valued components today, indeed they form the core spaces of the today existing urban green infrastructure.

At the end of the nineteenth century, new needs for the use of public open space, such as sports and games, demanded their realization in public parks as well, independent of areas specifically designated for this purpose. The Volksgarten developed into the Volkspark or city park. In addition to the contemplative walk, this also offers itself as a space for sport, play and exercise (Breuste 2010) (see ■ Fig. 2.15).

■ **Fig. 2.15** Pestalozzi Park Halle/Saale, Germany, **a** Classical park design of the 1920s and **b** integrating sport grounds 1960s. (Greiner and Gelbrich 1975, p. 132)

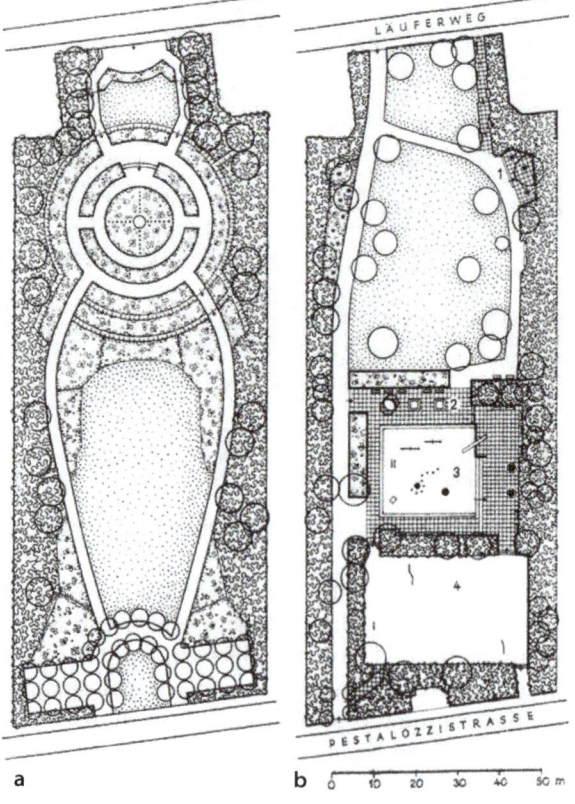

a   b

The First Park in the New World Adorns Nueva España's Capital Mexico

Various sources report that in the New World, the viceroy of Nueva España Luis de Velasco y Castilla (1539–1617), in connection with the construction of his capital Mexico City, had the public park Alameda, an area of 200×100 m planted with poplars, laid out on the outskirts of the city in 1592 (see ◘ Fig. 2.8). In 1791, its area was doubled to 8.8 ha, and fencing made it accessible only to the use of the upper classes. This is the first documented foundation of a public park in Latin America (Cocking 2016).

The English Garden in Munich

The "Englischer Garten" (English Garden) in Munich was laid out by order of the Bavarian Prince-elector Carl Theodor in 1789 by the Minister of War Benjamin Thompson (1753–1814) as a public-public garden between 1789 and 1792 for the then approximately 40,000 citizens of Munich. That corresponds to approximately 94 m^2 of park per inhabitant, an order of magnitude which is hardly reached anywhere today. At 375 ha in size, it is still one of the largest parks in the world (see ◘ Fig. 2.11). It was one of the first large parks in continental Europe to be open to the public. As one of the most extensive inner-city parks in the world, the English Garden enjoys enduring popularity among both Munich residents and tourists. In its design it has always adapted to the needs of its users without losing its landscape character. In 1972 the park administration opened the meadows of the park for public use, and in 1982 two areas were designated for naturism. Since 1990, public performances have been offered in the amphitheatre. From 2000 to 2011, an eight-kilometer stretch of the Isar river landscape was restored to its natural state.

Landscape gardener and designer of the park according to the English model was the court gardener Friedrich Ludwig von Sckell (1750–1823). Initially he worked under the direction of Thompson, but from 1804 until his death he was solely responsible. Under his direction, a landscape ensemble of wooded areas, meadows, hills and watercourses was created along the Isar floodplain, decorated with amphitheatres, statues and ornamental buildings and made accessible by a network of paths for pedestrians, but also carriages and riding horses.

In 1807, Sckell formulated the following as the mandate for use: "Here the people want to be seen, to please, and to be admired; all classes must therefore assemble there and move in long colourful rows, and the happy youth must hop among them" (Sckell 1807; quoted in von Freyberg 1989, p. 97; von Freyberg 1989, 2001; Landeshauptstadt München 2000, translated).

2

Hyde Park in London

Hyde Park is one of the largest and most famous city parks in the world, covering 140 ha. The park was used as a royal property (Royal Park) from the sixteenth century until 1768 as a royal hunting ground. Already in 1637 it was opened to the public. It is thus also one of the first royal parks to be open to the public. In its design and function, it is a model for European and North American parks in particular.

Visitors can row and, with permission, fish in the 11-acre Serpentine Lake, which was created in 1830. There is a 1384 m horse riding track (Rotten Row), a bowling alley and extensive lawns for picnicking and relaxation.

The Crown Lands Act 1851 regulates public access to the royal gardens. This set of regulations is regularly adapted to local developments concerning the use of the park (Hennebo and Wagner 1977).

A Hyde Park for New York

In 1873, the "Hyde Park for New York" (Rosenzweig and Blackmar 1992; Schwarz 2005a; Taylor 2009), the Central Park on an area of 341 ha, was opened (= 2.3 m^2/ew.). It was preceded by decades of public debate and media warfare by opponents and proponents, both on the side of influential landowners, merchants and politicians. The New York Tribune, with its publisher Horace Greeley (1811–1872), and poet and editor of the journal Post, William Cullen Bryant (1794–1878) had published idyllic pictures of rural Jones's Wood (Louvre Farm), a farm on the east side of Manhattan. Adjacent, they suggested, could be a large park. Park proponent Bryant stimulated public discussion, citing that because of a lack of other alternatives, thousands of "outdoor-hungry" families would picnic among graves at Green-Wood Cemetery in Brooklyn every Sunday (Burns et al. 2002). Andrew Jackson Downing (1815–1852), the first American landscape architect, also argued for the need for a public park for New York in publications beginning in 1844. Arthur

Tappan (1786–1865), editor of the Journal of Commerce, argued on behalf of landowners who wanted to turn their properties into valuable building land: "There is no need to turn half the island (meaning Manhattan – the author) into a recreational forest for loafers forever' (cited in Burns et al. 2002, p. 108).

As early as 1811, the Planning Commission for New York envisioned a grid of lots with single-family row houses with backyard gardens for the entire island of Manhattan (Commissioners' Plan). The long shoreline was to be used for dispersal and recreation. Between 1821 and 1855, however, New York had already quadrupled its population. By 1870, nearly 1.5 million people lived in houses rarely more than five stories high with obstructed backyards in extremely cramped quarters with only a few tiny green squares south of 23rd Street. The idea came up to save part of Manhattan Island for health and recreational purposes and now to act.

In 1853, urban land acquisition began between 59th and 106th streets on the northern city limits of 750 acres (304 ha). Five million dollars was acquired for 34,000 lots, and 1600 black, Irish, and German immigrants received eviction notices (Burns et al. 2002). In 1857, Board of Commissioners of the Central Park competition to design a park was won by Frederick Law Olmsted (journalist and farmer, 1822–1903), a student of Downing, and Calvert Vaux (architect, 1825–1895), who was oriented toward European, especially English, models. Neither Olmsted nor Vaux had ever planned a park or landscape before. Olmsted described their successful "Greensward Plan" in 1858 as *"of great importance as the first real park made in this country – a democratic development of the highest significance..."* and the key *"to the progress of arte & of esthetic culture"* (quoted in Scobey 2002, p. 20). Their design was a social-reform concept of a public space in which designed nature was to balance out the harsh social contrasts of the metropolis, "*places and opportunities for reunion where the gazes of rich and poor, educated and uneducated meet and where those may find each other again*" (Burns et al. 2002, p. 109) were envisaged.

Between 1858 and 1873, over 4000 workers had created a "new landscape" as modeled terrain, with ponds, plantings (25,000 trees), with paths passable by carriages, bridges, pavilions (6 million bricks were used), etc., making Olmsted the "father of American landscape architecture" and the park the "greatest American work of art of the nineteenth century" (Schwarz 2005b, p. 135). Everything, according to Olmstedt, was placed where it served its purpose, an artfully laid out kaleidoscope of nature (see ▫ Fig. 2.13).

Great expectations were associated with Central Park. It was expected to have a therapeutic and healing effect on the social and health problems of large sections of the urban population. It was supposed to be a place of encounter with nature for all strata of urban society, where the different social classes, the "rowdies" and "ruffians," could learn from the behavior of the park's fellow users, the middle and upper classes. This social illusion was translated into a multitude of visitor rules from behavior to dress codes and their control. Ultimately, however, Central Park in the nineteenth century was a park for the middle and upper classes, who integrated it into their elite lifestyle with their carriage rides and horseback rides and promoted it in many ways (Breuste et al. 2016; Wiede 2015).

Even today, looking at the luxury apartments in the residential towers around the park, "*There is no better expression of power and wealth than a view of Central Park*" (Wilner 2017).

"*In the distant future,*" Olmsted's vision says, "*New York will finally stand, all leveled and built over. Then the impressive rocky landscape of this island will have turned into rows of monotonous, straight streets, lined with masses of tall buildings. Nothing will remind us of its present diversity, were it not for the park. Then one will appreciate the inestimable value of the area's now unusual architecture and better understand how well it serves its purpose*" (Burns et al. 2002, p. 113).

2

Committed Citizens Change the City – The Example: Beautification Association (Verschönerungsverein), Stuttgart

Through citizens' commitment, areas are acquired on which urban green spaces were created for the public. Stuttgart, Germany, is an example. The Stuttgarter Verschönerungsverein was founded in 1861. Today it has about 580 members and is the owner of almost 40 green spaces, lookout points, monuments, fountains, lookout towers, shelters and old houses (most of them on its own property) in Stuttgart. From 1861 to 1902, the Verschönerungsverein (Beautification Association) was active in green planning and design instead of a municipal horticultural department. It was not until 1902 that an Urban Garden Inspectorate was founded in Stuttgart to take over the public task of green design and management. This meant that all the green spaces created in the city during this period had been created by the Verschönerungsverein. However, only a small proportion of these parks were built on land owned by the association. The green spaces owned by the association have remained the property of the association to this day and are therefore private, although they are open to the public according to the association's statutes. The other accosication grounds on municipal or state land continue to be maintained by the association, but have unfortunately also been reduced by municipal construction measures. The most recent green space of the Verschönerungsverein is the Chinese garden "Qingyin – Garden of Beautiful Melody" (Bruckmann 2018).

The Parque Central in Latin American Cities – The Example: Plaza Mayor in Antigua Guatemala, a City Square Turns Green

The city of La Antigua Guatemala, which today has about 35,000 inhabitants, was the capital of the Spanish colonies throughout Central America from 1543 until it was completely destroyed by an earthquake in 1773. The planned city layout had an approximately 1 ha large city square (100 × 100 m) with the representative buildings framing it as its centre. As in all Latin American cities, the plaza is the focal point and most important social meeting place of the city. Originally, the plaza was unpaved for flexible public use and without any facilities. Bullfights, horse races, public punishment, cattle markets and farmers' markets took place here regularly. In 1704 the plaza was paved, and in 1738 it was decorated with a fountain in the center (see ◘ Fig. 2.14). In 1912, as part of city beautification measures initiated by the local middle classes, the market was relocated and an attractive urban garden design with trees and plantings, paths and benches were begun. All other plazas in the city were now also beautified with planting, especially shade trees. The plaza, originally completely devoid of vegetation, became a public park, following European models. It has been so for about 100 years and enjoys great popularity and intensive use by the city's residents and visitors. In a similar way, plazas in many other cities in Latin America were transformed into parks (Bell 1993).

Racecourses and Golf Courses – Special Sports in Special Urban Green Spaces

Special leisure interests led to public spaces established for them, and often to special green spaces. One example is horse racing as a particularly widespread amusement in Great Britain and its colonies, which led to horse racing tracks in many cities of the Commonwealth. Horse racing had been common in England since the seventeenth century, and professionally from 1700 on. From 1750 racing was controlled by jockey clubs. Municipal horse racing tracks were established everywhere in British colonies (Colombo, Nuwara Elya/Sri Lanka, Hong Kong, Shanghai etc.). Shanghai's horse racing track was transformed into the city's now-famous central park, People's Square, after the founding of the People's Republic. Horse racing is held in the highest esteem in Canada, UK, Ireland, the USA, Australia, New Zealand and in South Africa. Germany also has 50 trotting and gallop racecourses in cities.

The game of golf, which originated in Scotland, spread via Great Britain to Europe in the sixteenth century as entertainment for the aristocratic upper classes. In 1764, the first 18-hole golf course was already established in St. Andrews. The first London golf course (Blackheath) was opened in 1603. With the British Empire, the game spread to cities around the world in the nineteenth century (Bengaluru/Bangalore, India 1820; Ireland 1856, Adelaide 1870; Montreal 1873; Cape Town 1885; New York 1888; Hong Kong 1889). In 1900 there were already 1000 golf clubs in the USA. Today, large areas of land are in use for golf around the world, many of them in and around cities. Great Britain is for this sport the country in Europe with by far the most golf courses (6935) (followed by Germany with 732 golf courses) and golfers. London is the golf capital of Europe with 67 golf courses. The golf courses are private greens accessible only to members. Many are located in cities or urban environments with special, often intensive, nature management. China's elites are currently implementing the sport and its courses rapidly and extensively in the country's major cities (e.g. 20 golf clubs in Beijing) (Golftoday 2018; Johnson 2018).

Company's Garden Cape Town, South Africa – From Vegetable Garden to Recreational Park

In 1652, Dutchman Jan van Riebeeck (1619–1677, first Governor of the Cape Colony 1652–1662) established a supply station in Table Bay, Cape Town, South Africa, for the merchant ships of the Dutch East India Company (Vereenigde Oost-Indische Compagnie; VOC) on their route to India. The food supply for the ships required a larger garden area, the Company's Garden. The present park (8.5 ha) is probably the oldest African city park in the sub-Saharan region. It was mentioned in 1652, the first year of the colony, and still occupied about a third of the total town area at that time 100 years later. The garden was watered by springs in the lower slopes of Table Mountain, dammed by a dam. While at first, the garden was primarily a fruit and vegetable garden, it changed with time and more varied and better supplies for the ship's crews. It was used in English colonial times for the acclimatisation of tropical plants from Asia before their onward transport to England. Today the park is the main public green space for Cape Townians to relax and walk in the middle of the city. In 1929 the rose garden was established. The park still contains the oldest cultivated pear tree, which is now more than 350 years old (Worde et al. 1998, see ◘ Fig. 2.16).

◘ **Fig. 2.16**   Company's Garden, Cape Town, South Africa. (© Breuste 2006)

2.4   The Park Cemetery – A Place of Recreation for the Living

With the takeover of burial rights from the church by the municipalities at the beginning of the nineteenth century, first in France, burial grounds were henceforth no longer allowed to be established as churchyards within the city limits, but were placed under the supervision of the political municipalities. In 1803, an existing extensive garden site belonging to Father François d'Aix de Lachaise in Paris was acquired for this purpose. The first burial took place in 1804 at the Cimetière du Père-Lachaise, now the largest cemetery in Paris and also the first burial ground in the world to be laid out as a park cemetery. With around 3.5 million visitors a year, it is one of the most visited sites in Paris.

Park cemeteries emerged in North America between 1830 and 1860, with the Père Lachaise Cemetery in Paris and the English landscape gardens serving as their models. Mount Auburn Cemetery, built in 1831, was the first such cemetery in the United States. The approximately 70-acre site is located about 6.4 km west of Boston (Reps 1992). It was the largest rural and wooded cemetery in the world at the time and the first of its kind in the United States. However, park cemeteries were also rapidly adopted as urban park features. Green-Wood Cemetery was established in 1838 by engineer David Bates Douglass (1790–1849) on 1.9 km² as a park cemetery in the then independent city of Brooklyn, since 1898 a borough of

New York. It is the oldest park in New York and became the model for many similar parks in America and Europe. After 1860, Green-Wood rivaled Niagara Falls as the largest tourist attraction in the United States. Green-Wood's popularity spurred the creation of other public parks, including New York Central Park.

A **park cemetery** is the name given to an urban burial ground whose design is based on the concept of the English landscape garden and whose complementary task is the respectful use for recreational purposes, especially walks. Characteristic landscape design elements have been integrated and supplemented by adapted pathways, landscaped mounds and ponds, and woodland-like areas.

A **forest cemetery** is an urban burial ground with a relatively dense and well-maintained stock of trees, giving it a forest-like character. In some cases, forest cemeteries have been integrated directly into existing forests, in others the trees were planted later. The term is not clearly distinguished from that of the park cemetery.

The City Graveyard in Halle/Saale, Germany – A Cemetery Became an Urban Nature Space

When, in the sixteenth century, burial grounds in many German cities were moved to the outskirts outside the city walls for hygienic reasons, a Renaissance cemetery masterpiece north of the Alps was built in Halle/Saale from 1557 onwards over a period of more than thirty years, modelled on the Italian Camposanto complexes, especially the Camposanto in Pisa. The complex has the shape of an irregular rectangle of five to six-metre-high walls measuring 113 × 123 × 129 × 150 metres with 94 flying arches with open crypts inwards. After war damage in 1945 and subsequent decades of neglect, the complex was restored by the private initiative of a "Bauhütte Stadtgottesacker" and with further private and public funds. The Stadtgottesacker is a listed building and is one of the most beautiful cemeteries in Germany. It was not until 1822 that burials also took place in the inner open space of the grounds, which consists of a forest park with an old stock of trees. It was not until the nineteenth century that urban nature moved into the cemetery, with trees now over a hundred years old. A design plan of 1818 mainly provided for "beautification measures" such as park-like design, grave planting and tree planting under the direction of an association.

The partly very distinctive closed plant cover (ivy, species-rich herb layer), mosses, lichens and ferns on walls and the old deciduous tree population, which is maintained horticulturally, account today for the value of the site as a natural oasis in the middle of the city (von Schweinitz 1993; Tietz 2004; Därr 2018; see ◘ Fig. 2.17).

2

◘ Fig. 2.17   City graveyard, Halle/Saale, Germany. (© Breuste 2018)

Forest and Park Cemeteries

Skogskyrkogården in Stockholm

The 108-ha Skogskyrkogården ("Forest Cemetery") in the Stockholm's southern district of Enskede was laid out in a pine forest by the Swedish architects Gunnar Asplund (1885–1940) and Sigurd Lewerentz (1885–1975) between 1917 and 1940. It has been a UNESCO World Heritage Site since 1994. It is a significant example of the fusion of architecture, culture and nature to form a special part of urban nature, the forest cemetery. It has influenced the design of burial sites in a designed natural environment throughout the world (Jones 2006; see ◘ Fig. 2.18).

The Ohlsdorf Cemetery in Hamburg

With 389 ha, Ohlsdorf Cemetery, opened in 1877, is **the largest park cemetery in the world and the largest cemetery in Europe** (Vienna Central Cemetery 250 ha). It is also at least comparable in size to large parks such as Hyde Park in London (140 ha) or Central Park in New York (349 ha).

The park character was created by planting extensive stands of trees (approx. 36,000), incorporating 17 ponds, streams and historic buildings, garden monuments and modern themed gravesites in a network of checker boarded plots. The park cemetery integrates many trees from the rampart hedgerows of previous agricultural pasture use.

The general plan of the complex was drawn up on behalf of the city of Hamburg in 1876 by the architect Johann Wilhelm Cordes (1840–1911), who also became cemetery administrator and cemetery director.

Since 1996 the cemetery houses a museum, offers guided tours and is accepted by visitors as a recreation area and landscape garden of silence and tranquility.

The maintenance is carried out in a restrained nature-oriented manner and

2.5 · The Urban Forests and Forest Parks – How the Forest Stayed...

47 **2**

■ **Fig. 2.18**   8 Skogskyrkogården Cemetery in Stockholm. (© Breuste 2012)

provides a habitat for many species of wild animals (for example, deer, squirrels, hedgehogs, martens, foxes, hares and over one hundred species of birds, including tree falcon, great spotted woodpecker, kingfisher, grey goose, green woodpecker, blackcap, robin and eagle owl).

About half of the biomass produced annually (approx. 60 m³ wood chips from tree and shrub cuttings, approx. 5500 m³ leaves, approx. 1400 m³ herbaceous plant components and lawn cuttings) are composted and reused as fertilizer (Leisner and Schoenfeld 1991).

2.5   The Urban Forests and Forest Parks – How the Forest Stayed in the City

Urban forests are often the largest urban natural areas and of great importance for the recreation of the inhabitants close to the city. They are subject to special management that does not focus on the yield of wood and have equipment and care geared to their recreational use.

An **urban forest** is a forest that is located in the immediate catchment area of the city and that fulfils recreational functions for the citizens of the city. The location on municipal city territory or municipal ownership are not necessary in this context. Forest management is geared to recreational use even if economic success is foregone. Urban forests can be near-natural forests or forests originally established for economic reasons. Urban forests are meeting places for urban citizens with "wild" nature and a diverse range of species.

2

In the history of cities so far, hardly any new forests (woodlands) have been planted in and around cities. On the contrary, attempts were made to eliminate existing forests in order to supply the cities with food through agricultural land. This was only done where either the soil hardly promised a profitable agricultural use, the relief conditions (slopes and mountains) or wetlands (moor forests) precluded this. But even there, existing forests were thinned out and used for forest pasture or loosened up. Where forests in urban surroundings were manorial forests, they were partly preserved as hunting grounds, even if the city came closer and closer to them or even enclosed them (Großer Tiergarten Berlin).

With its Grunewald, Köpenicker Stadtforst, Plänterwald, Großer Tiergarten and other forests, Berlin is the most densely forested city in Germany in terms of forest area (■ Table 2.1).

■ Table 2.1   Large urban forests in Germany (Stadtwald 2018)

Urban Forests/Woodlands	Area in ha
Berlin city forests with Grunewald, Köpenicker Forst and other forest areas in and around Berlin	28.500
Baden-Baden, Stadtwald	8526
BrilonStadtforst	7750
Augsburg Stadtwald	7000
Dresdner Heide (Dresden Heath)	6133
Rostocker Heide (Rostock Heath)	6000
Villingen-Schwenningen, Stadtwald	80,000 (5841 designated as "urban forest")
Wiesbaden Stadtwald	5600
Freiburg Stadtwald	5200
Boppard Stadtwald	4360
Frankfurter Stadtwald	5785 (3866 on municipal land)
StadtwaldFürstenwalde	4677
MühlhäuserStadtwald	3093
Weissenburger Stadtwald	2806
Koblenz Stadtwald	2772
Bielefelder Stadtwald	2256
Leipziger Auenwald (Leipzig Floodplain Forest)	2500
Stadtforst Salzwedel	1400

◘ Table 2.1 (continued)

Urban Forests/Woodlands	Area in ha
Lauerholz in Lübeck	960
Steigerwald in Erfurt	800
Dölauer Heide (Dölau Heath) (Halle/Saale)	759
Eilenriede in Hanover	650
Duisburger Stadtwald	600
Fürther Stadtwald	560
Eschweiler Stadtwald	350
Kölner Stadtwald (Cologne municipal forest)	205
Marienhölzung in Flensburg	200
Krefelder Stadtwald	120
Seelhorst in Hanover	100

Baden-Baden (54,000 inhabitants), although the smallest municipality in Baden-Württemberg, has Germany's largest urban forest (8526 ha) and the largest proportion of forest of any municipality (60.8% forest area). This includes shares in the Black Forest National Park, seven nature reserves, predominantly forests in mountainous surroundings under landscape protection. A 40 km long panorama trail leads through the city's forests. It was awarded the title of "most beautiful hiking trail" by the German Tourism Association in 2004. Baden-Baden consciously uses its forest amenities as a quality feature for tourism and living (Eidloth 2013).

> **Definition**
>
> **Waldstadt (Forest City)** is an ambiguous term for urban districts in forest-related locations, for example, in Potsdam, Iserlohn, Karlsruhe and Halle/Saale. The term is intended to express a quality feature of natural amenities that is important for preferred residential locations.
>
> However, a forest city can also be understood as a city with a large or even dominant share of forest in relation to the city area (for example, Baden-Baden with 60.8% forest share or Berlin with 28,500 ha of forest).
>
> A **Forest City** can also be defined as a city that deliberately draws on the benefits and ecosystem services of existing or newly created forest in and around its urban area (for example, Liuzhou Forest City 2017 designed as a planned city in China) (Alleyne 2017; see ▶ Sect. 3.3).

2

An Alluvial Forest Becomes a People's Park – The Vienna Prater

The Vienna Prater is a very extended public forest park in the Danube, covering about 6 km^2, which still consists to a large extent of the floodplain landscape formerly shaped by the Danube. On April 7, 1766 Emperor Joseph II released the original hunting ground for public use. The area was fenced off until 1774. Permission was also granted for the settlement of coffee-boilers and innkeepers, the preconditions for today's Wurstelprater amusement park. The Prater became a centre of entertainment, public festivals and recreation for the metropolis of Vienna and is used by a very large number of visitors, especially on Sundays and public holidays (Haas 2010; Pemmer and Lackner 1974; Schediwy and Baltzarek 1982; Sehnal 2008).

The Berlin Great Tiergarten – Hunting Ground, Vegetable Garden, Now Forest Park

As a hunting ground with enclosed wild animals, the Großer Tiergarten forest had already reached its present size of 210 ha in 1530 and was located outside the gates of the city. At the end of the seventeenth century, the former hunting ground began to be turned into a "pleasure park for the people", in 1742 the fence was removed, and in 1833 and 1838 the landscape architect Peter Joseph Lenné (1789–1866) transformed it into an English park, which it remained until the Second World War. After the war, the trees were almost completely cut down for firewood. Out of about 200,000 trees, only about 700 remained. The free areas were released in 2550 plots for the cultivation of potatoes and vegetables. Parts of the Großer Tiergarten were also used to deposit rubble from destroyed buildings. Reforestation with tree dona-tions from other German cities began in 1949. The waters were silted up, all bridges destroyed, the monuments overturned and damaged. Plans to fill in the Tiergarten's pond and river landscape with rubble were prevented by Reinhold Lingner, the head of Berlin's Main Office for Green Planning. The forest park is an attraction for all Berliners and is used to stage major events (for example, Fan Mile on the Straße des 17. Juni during the 2006 World Cup and the 2008 European Football Championships). Since 1987, the Berlin Marathon has started in the Großer Tiergarten, and part of the route of the Love Parade also ran here from 1996 to 2003. Since the opening of the Tiergarten tunnel in 2006, the north-south traffic runs underground (Wendland 1993; von Krosigk 2001; Twardawa 2006; see ◘ Figs. 2.19 and 2.20).

■ **Fig. 2.19** Großer Tiergarten, Berlin. (© Breuste 2011)

■ **Fig. 2.20** Gardens of the Chehel Sotun Palace in Isfahan, Iran. (© Breuste 2018)

2.6 The Private Garden for Everyone Complements the Public Urban Nature

The manorial garden was part of the first cities and towns since the beginning of urban development. However, it remained reserved for the use of only a very small elite. It was not a kitchen garden, but first a pleasure garden. The transition from (small-scale) garden to park was fluid, and its form and size were also adapted to local conditions and requirements. The garden stood for the "earthly paradise" (see ▶ Sect. 3.6).

2

The urban peasant gardens that were widespread in the Middle Ages were kitchen gardens in the city or front of the city gates. They gave rise to burgher gardens, which were often laid out in front of the city wall due to a lack of space.

For allotment gardens, which are common in towns today, different motives for their establishment were important according to the historical situation. In times of crisis it was always nutritional aspects, in times of stability especially idealistic motives that moved the allotment gardeners.

» "Allotment gardens are important components of the city. They are the last links of the urban dweller to the countryside, where most of today's urban dwellers once came from. The allotment garden association is an important cultural factor, it is a place of learning, recreation and meeting. Allotment garden colonies in the city are green spaces that make the built-up areas habitable" (Schiller-Bütow 1976, p. 1, translated).

The allotment garden and the allotment garden associations came into being with the development of the industrial towns. At the same time, it is part of the preindustrial rural life that was transferred to the cities with the gardens. Many of its accents have changed in the course of development, but its core, the self-designing contact with nature, has remained and is as relevant in modern city life today as it was in the past. In general, several sources of the allotment garden movement can be identified in Germany (see ◘ Fig. 2.21).

The **allotment garden,** also called "Schrebergarte"n, home garden, family garden or land plot, is a leased area in the urban area for use as an orchard, vegetable garden (kitchen garden) and for recreation. Allotment garden estates consist of leasehold plots situated together. Allotment garden tenants usually organise themselves locally, regionally and nationally into allotment garden associations. The use of the gardens is usually subject to an allotment garden regulation and/or an allotment garden law, especially in Europe. The allotment garden plots are usually between 150 and 400 m^2 in size and are often equipped with a small building ("arbour") of about 20–30 m^2 for the storage of equipment or for short stays.

◘ **Fig. 2.21**   Allotment garden in the oldest Salzburg site Thumegg, founded in 1940. (© Breuste 2007)

Aid to the poor and nature education were the roots of the allotment garden movement in Germany and other European countries in the nineteenth century. Although allotment gardening is widespread all over the world, it is hardly anywhere else in the world that it is as developed as in Germany. It has several roots and cannot be traced back to the "allotment garden movement" alone.

Gardens for the poor: The first allotment garden association was founded in 1814 by leasing a piece of church land in Kappeln for gardens of poor. In England, this also became possible for the municipalities from 1819 onwards. The poor gardens are the direct predecessors of today's allotment gardens. Bourgeois and aristocratic circles wanted to help themselves generously. In Kiel, Flensburg, Königsberg, Frankfurt a. M. and Leipzig allotment gardens for the poor were established after 1830. The poor gardens were not legally protected and often only existed for a few decades. They were gradually bought up again and used for industrial and construction purposes.

Around the middle of the nineteenth century, industrialisation and urban growth had led to social ills, widespread alcoholism and neglect among parts of the urban population. The **Lebensreform movements** reflected this and hoped to improve urban living conditions by returning to simple, nature-based ways of life, education through gardening and contact with nature. To this end, they propagated the founding of land societies, women's and nature conservation associations, and garden colonies. Garden colonies were to be organized as associations and equipped with playgrounds, light and air baths, and sunbathing halls (Katsch 1994).

Life Reform

The term "Lebensreform" stands for various reform movements as a reaction to developments of modernity that started in the middle of the nineteenth century, especially in Germany and Switzerland. Common features were the criticism of industrialization, materialism and urbanization and an idealization of the state of nature. Representatives propagated a way of life close to nature with ecological agriculture, vegetarian nutrition, rejection of alcohol and tobacco, reform clothing and naturopathy. In physical culture, the aim was to provide people with exercise in fresh air and sunshine, also in the form of naturism (Barlösius 1997).

The **allotment gardens,** which are widespread today, had their origins in efforts to contact nature, education and health promotion. On the initiative of the school director Dr. Ernst Innocenz Hauschild (1808–1866) in Leipzig, the parents of his pupils decided in 1864 to found an association for the propagation of educational issues (Teachers' and Parents' Association), to build up a library for this purpose and to acquire land from the city of Leipzig for a playground and playground for children. The association was named "Schreberverein" in memory and appreciation of the Leipzig physician **Dr. Daniel Gottlob Moritz Schreber (1808–1861)**, who had died three years earlier, and who had already spoken out along these lines and had been closely associated with Hauschild. In 1864 Hauschild inaugurated a playground on a municipal leasehold site at Johannapark as the "Schreberplatz".

2

Three years later, the teacher Karl Gesell (1800–1879) had "children's beds" laid out around this playground, which soon became "family beds". Fencing as a boundary and small huts as weather protection soon became necessary. **In 1869, the first garden ordinances were enacted. This is often given as the birth year of allotment gardening.** In 1876 the Schreberverein moved to a new site on the "Butchers' Meadows" on the old Elster river at the instigation of the town, which wanted to build on the original site. This site still exists today. It was not until ten years after the founding of the first Schreberverein that a second association was founded in Leipzig's southern suburb, and then in the last quarter of the nineteenth century new Schreber and naturopathic associations were founded in rapid succession in Leipzig alone. In 1900 there were already 119 associations with 7741 gardens in Leipzig.

Arbor colonies: Homelessness and inadequate living conditions in the tenement districts, especially in Berlin, combined with rural tradition, gave rise to the desire to improve one's own living conditions by growing fruit and vegetables for one's own use or to replace the lack of living space. In the 1990s of the eighteenth century already 45,000 allotment gardeners ("Laubenpieper") had settled as tenants on undeveloped areas in the urban area of Berlin, often under insufficient hygienic conditions, and founded an important Berlin gardening tradition. In the "wild allotment gardens", spontaneously created marginal settlements, about 40,000 arbor colonists lived permanently in Berlin at the turn of the century as a substitute for housing.

Inspired by the workers' garden project of the Jesuit priest Felix Volpette (1856–1922) in Saint-Étienne, France, the ministerial official in the Reich Insurance Office, Alwin Bielefeldt (1857–1942), founded the first **workers' garden colony** in Berlin in 1901 consisting of 84 workers' gardens under the auspices of the German Red Cross of the Vaterländischer Frauenverein (**Red Cross Gardens**). They were intended "to combat tuberculosis, to strengthen sick and invalid persons and to supplement the often insufficient accident, old-age and invalidity pensions" of poor families and families with many children. The other colonies that developed had playgrounds, lodging rooms, drinking halls, shopping cooperatives, and libraries. The Berlin workers' gardens developed alongside the deciduous colonies and the allotment gardens as a separate form of allotment gardens.

Around 1900 large industrial companies and institutions more often provided their employees with small leased areas for allotment gardening as part of their social benefits. Thus allotment garden associations of the German Reichsbahn, the Deutsche Post, the miners etc. were founded. Especially the railway associations had supraregional importance (in 1909 already 761 associations).

The aim of the efforts for gardens in the city was to improve the inadequate living conditions of the workers through self-sufficiency in food and to improve their health through spending time and physical work in nature. A special target group was children, who were to be encouraged physically and mentally through gardening.

The demand for family gardens, small children's playgrounds, kindergartens and evening places for the young was raised by bourgeois reformers and pedagogues, such as Adelheid Poninska, Countess zu Dohna-Schlodien (1804–1881), as early as 1874 in her book "Die Großstädte in ihrer Wohnungsnot und Grundlagen einer durchgreifenden Abhilfe" ("The big cities in its housing misery and bases of a thorough remedy") (Poninska 1874, published under a pseudonym). Reinhard Baumeister (1833–1917), civil engineer, urban planner and university lecturer, published the first urban planning textbook in 1876, in which the demand for allotment gardens, family arbours, after-work places and children's playgrounds was raised (Baumeister 1879). In 1909 the "Zentralverband deutscher Arbeiter- und Schrebergärten" ("Central Association of German Workers' and Allotment Gardens") (922 associations) was founded. The chairman of the central federation affirmed in 1912:

» "The allotment garden is a valuable as well as simple and not very costly means of promoting the family in economic, health and educational terms" (Katsch 1994, translated).

With the enactment of the Allotment Garden and Small Lease Land Ordinance (Kleingartenordnung) in 1919 as a law of the German Reich, the municipalities were given land rights (protection against dismissal, protection against speculation, compulsory lease) and the task of establishing **Permanent Allotment Garden Estates.** This was the beginning of the social institutionalisation of the "allotment garden movement" (Koller 1988). In 1921 the Reichsverband der Kleingartenvereine Deutschlands (Reich Federation of Allotment Garden Associations in Germany) was founded in Bremen as the organisational framework of the allotment garden movement. From 100,000 allotment gardeners in the founding year the federation grew rapidly to 389,000 members by 1926. From then on there were annual increases of 10,000 until 1930 (Breuste 1996; Katsch and Walz 2011).

Gardening became a mass movement and is still popular today as an attractive way to interact with nature in the city.

Paradise Gardens of the Garden Palaces – The Garden as an Ideal of "Paradisiacal Beautiful" Nature in the Urban Living Environment

One of the seven wonders of the ancient world weres the Hanging Gardens of Semiramis, according to Greek authors a step-like, elaborate garden complex in Babylon. For a long time, the gardens represented the ideal of an earthly paradise in their plant splendour achieved through irrigation.

British Assyriologist Stephanie Dalley argues, with evidence from topographical surveys and historical sources, that the Hanging Gardens may have been part of a palace garden of the Assyrian king Sanherib (c. 745-680 BCE) at Nineveh on the Tigris River, built for Sanherib's wife Tašmetun-Šarrat (Dalley 2013).

The tradition of the garden palace is widespread in the Islamic world between Andalusia in the west (e. g., Alhambra gardens) and India in the east (Fatehpur Sikri gardens). The Chehel Sotun Palace from Safavid times (seventeenth century) in Isfahan (Iran) represents this tradition of the garden of paradise in the midst of a large garden complex (see ◘ Fig. 2.20).

2

Allotment Garden Association "Dr. Schreber" – The First Schreber Association in Germany

The allotment garden estate in Leipzig, established in 1864, is the oldest allotment garden site still existing in Germany. The allotment garden site has 162 plots with an average size of 170 m², a large association meadow with historical playground equipment for children as well as a museum garden and a deciduous garden. The association house, built in 1896 and restored in 1992, houses the German allotment garden museum. Historical gardens with listed arbours from the first half of the twentieth century are also maintained, including garden No. 140, which was managed from 1926 to 1933 by the university professor and Leipzig publisher and bookseller Dr. Heinrich Brockhaus (KGV Dr. Schreber 2018; see ◘ Figs. 2.22, 2.23 and 2.24).

◘ **Fig. 2.22** Memorial plaque for Dr. Schreber and Dr. Hauschild in the allotment garden association estate Dr. Schreber in Leipzig. (© Breuste 2011)

■ **Fig. 2.23** Allotment garden association estate Dr. Schreber, association house and playground, seat of the "German Allotment Garden Museum in Leipzig" (© Breuste 2011)

■ **Fig. 2.24** Almkanal in Salzburg, Austria. (© Breuste 2003)

2.7 How the Waters Became Urban?

Urban water bodies are not a uniform category in the sense of classical water body typology. Their defining characteristic is their urban location and use. The transition between urban waters and waters in the open landscape is therefore fluid. This includes water bodies that were originally natural but have since been heavily modified, as well as artificial water bodies and canals (see ► Sect. 5.4).

> **Urban waters**
>
> Urban waters are limnetic systems, streams and standing waters, located within urban ecosystems and shaped to varying degrees by urban use and design. This can include low-influence (rare) to completely man-made water bodies. Thus, they do not represent a uniform water body type, but are characterized in particular by contact with nature by the surrounding population and recreational use. Urban water bodies are also associated with risks of use that can be reduced or eliminated by appropriate management (Gunkel 1991; Schuhmacher 1998).

A large number of cities were founded along flowing waters, mostly rivers. This secured the vital water supply, opened a transport route for exchange with other areas, made fishing, waste disposal and energy generation possible. Water bodies also protected the cities strategically. Crossing points over rivers (fords or bridges) were important nodes in the trade network (for example, Saarbrücken, Frankfurt or Osnabrück). Flowing waters were one of the most important location factors for towns in the hinterland of the coasts. Watercourse changes in urban areas took place at an early stage and were aimed at improving the usability of watercourses and reducing the dangers they posed. In the medieval European city, the economic aspects (transport of goods, supply and disposal, energy production) of the water bodies were the main focus.

In view of the scarcity of available energy, the regenerating energy of water was essential for the operation of mills (grain, oil, paint mills, etc.) and workshops (hammer mills, saws, manufactories, etc.). Extensive canal constructions were built for this purpose. Water played a decisive role in the production of leather and textiles (Schuhmacher and Thiesmeier 1991).

The urban bathing culture, for example, in the Roman Empire, the Ottoman Empire or Japan, was directly linked to water from flowing waters. For the relatively frequent need to fight fires, the availability of water from wells and rivers was crucial (fire-fighting streams, fire-fighting ponds, "fire lakes", etc.). Water pollution through sewage was already noticeable in the European Middle Ages. Straightened watercourses and a watercourse bed offering as little friction as possible were intended to prevent solid waste from accumulating. Due to recurring problems and to make use of the water, including for irrigation, municipal regulations on its use were issued as early as the sixteenth century (Kaiser 2005).

The development of large and industrial cities in the eighteenth and nineteenth centuries put a strain on urban watercourses and led to unprecedented health problems from water pollution, including pandemics. Despite existing evidence on the spread of disease, unfiltered river water often continued to be fed into water supplies to avoid the high cost of clean, safe drinking water.

To regulate wastewater and improve hygienic conditions in the large cities, work began in the second half of the nineteenth century on the construction of central plants for the discharge of wastewater (alluvial sewers), for example in Vienna in 1850, in Hamburg in 1854 and in Paris in 1856. The high cost of these systems was often criticised as a "pointless luxury", with reference to the self-purification capacity of rivers. For this reason, many European cities in the nineteenth century only had sewage systems, but only a few had sewage treatment plants or trickling filters for the trickling of wastewater, that is, they continued to discharge wastewater into the receiving waters. At the beginning of the twentieth century, at least mechanical treatment plants and trickling fields were increasingly installed. By 1910, some 400 municipalities in Germany were already treating their wastewater, 40% of them using biological processes (trickling fields, soil filtration, meadow irrigation) (Kaiser 2005). Advances in wastewater treatment technology (for example, sludge activation processes in 1914) led to improved treatment performance, although commercial and industrial wastewater remained unaddressed. German rivers, such as the Wupper, Ruhr, Emscher, and Pleiße, turned into open industrial sewers running through cities and remained in this state well into the twentieth century. Only urban hygiene aspects and infectious diseases caused by organic wastewater were discussed. Industrial wastewater was sometimes even regarded as positive in comparison to municipal wastewater because of alleged "neutralizing effects" (Kaiser 2005, p. 51). It was not until the end of the nineteenth century, when agricultural and fishing associations saw their interests threatened by water pollution, that restrictive regulations were enacted for industrial wastewater discharges into flowing waters. The arguments "natural self-purifying power of water bodies" and "dilution effect" were advanced to postpone investments in wastewater treatment. Urban rivers became receiving waters oriented to industrial requirements, straightened, developed and remained polluted. Small streams disappeared underground in a piped sewer system. With this, visible hygienic, olfactory and aesthetic impairments caused by polluted urban waters also disappeared. They also disappeared from the consciousness of urban citizens. With the generational change, the knowledge of the partially underground watercourse system disappeared altogether. Instead of combating water pollution, long-distance water pipelines and deep wells were set up to largely eliminate public health pollution by tapping new water resources. The task of urban watercourses was now no longer to supply water but to dispose of wastewater for municipalities and industry. They retained this task until well into the twentieth century. First steam engines and, from the end of the nineteenth century, electric motors replaced industrial hydropower, which became insignificant as a location factor (Schuhmacher and Thiesmeier 1991).

Flood control became another reason for the modification of urban rivers and streams, especially in the nineteenth and twentieth centuries. By the middle of the twentieth century, the rivers crossing the cities were given an "efficient technical profile" with flood and relief channels to reduce flood peaks (see ◘ Fig. 2.25).

2

■ **Fig. 2.25**   Pleißemühlgraben in Leipzig, Germany. (© Breuste 2003)

In the floodplains, which are protected from flooding, new urban uses, residential development, including new industry (for example, Duisburg, the world's largest inland port), could find space alongside agricultural land. Settlement areas. City and water came into even closer proximity.

» "According to the view of the time, man no longer had to submit to the natural force of the waters, but was now the one who could adapt rivers and streams to the requirements of urban development, industry, flood protection, navigation and energy production. This purely economically oriented view also led to serious changes in water bodies in the decades after the Second World War" (Kaiser 2005, p. 67, translated).

In the twentieth century, Central European urban watercourses developed in a completely new direction. While maintaining the necessary flood protection, the economic functions almost completely disappeared, except for a few transport routes that continued to be used, in favour of leisure and recreation, the enhancement of the cityscape and the function as a habitat for plants and animals, which were now valued more highly (see ■ Table 2.2). However, this does not apply in the same way to rivers in cities on other continents, where economic functions (see ■ Fig. 2.26) or wastewater discharge are still important (see ■ Figs. 2.27 and 2.28).

Table 2.2 Changes in the functions of water bodies and water in central European inland cities due to anthropogenic use and perception (Kaiser 2005, p. 22)

	Before 1750	1750 – 1850	1850 – 1915	1915 – 1950	1950 – 1980	**From 1980**
Protective function	Great	Medium	–	–	–	–
Food, fishery, irrigation	Great	Great	Medium	Low	–	–
Transport route	Great	Great	Medium	Low	Low	Low
Energy supplier	Great	Great	Great	Low	Low	Great
Drinking water supply	Great	Great	Low	Medium	Medium	Medium
Service water supplier	Great	Great	Great	Medium	Medium	Medium
Waste disposal	Great	Great	Great	Great	Medium	Medium
Leisure and recreational use	–	–	–			Great
Upgrading the residential environment	–	–	–	–	–	Great
Habitat for plants and animals	–	–	–	–	–	Medium

Great importance ●	Medium importance ●	Low importance ·	No importance –

Fig. 2.26 Hongkou River in Shanghai, China, as an intensively used transport route. (© Breuste 2006)

Fig. 2.27 Rio Matanza-Riachuelo, the most polluted urban river in Latin America, in Buenos Aires, Argentina. (© Breuste 2006)

■ Fig. 2.28 Historic water town of Tongli in the alluvium of the lower Yangtze River in China, where canals are still important for transportation and not only for tourism. (© Breuste 2006)

The Almkanal in Salzburg, Austria – The Most Important Urban Hydraulic Structure of the High Middle Ages

The Almkanal is a 12 km long canal in the south of the city of Salzburg, which brings the water of the Königseeache, which rises from the Königsee, through the Stiftsarmstollen between Mönchsberg and Festungsberg into the city of Salzburg. Built between 1137 and 1143, it is the oldest medieval water gallery in Central Europe and is still in use today. The oldest part of the Almkanal network (today's Müllner Arm), which runs around the city's mountains, was probably built as early as the eighth century. The Almkanal served to supply the town with water for use, drinking and fire-fighting as well as for the operation of mills. Today the Almkanal has mainly a recreational function and is significant as a cultural monument. Energy is still generated to a small extent (14 turbines, including the oldest electricity power station in the province of Salzburg, a mill and the city's emergency power generator, as well as ponds, cooling and air-conditioning systems, for example, those of the Salzburg Festival Theatre). Under normal conditions, the Almkanal carries about 5.5 m³ of water per second.

The clients were the Salzburg cathedral chapter and the monastery of St. Peter, which was located in the city. It was not until 1335 that Archbishop Friedrich III granted the citizens of the city the right to freely draw water. Until then, the water in the city belonged to the cathedral chapter and the monastery of St. Peter alone. After the construction of a municipal well house on the municipal branch of the Almkanal in 1548, Salzach groundwater could be pumped up with the energy of the Almwasser and used in the city in numerous pipes for fountains, wash houses, baths, horse ponds and basins to store fish. In the second half of the nineteenth century, up to 120 mills were operated on the Almkanal. At that time there were over 400 water rights. The Salzburg Fortress Railway (nicknamed the Tröpferlbahn) was operated from its opening in 1892 until 1959 with ballast water (water reservoir in the mountain station) and was therefore not in operation in winter. Since 1937 a water cooperative unites all users by Almkanal Law ("Almkanalgesetz") obligatory. An "Almmeister" manages the operation of the main chanel of the Almkanal (Klackl 2002; Ebner and Weigl 2014; see ■ Fig. 2.24).

2

Cholera Epidemics – Contaminated Drinking Water Kills Thousands of Urban Dwellers

In 1832, in response to the first cases of cholera in London, under the direction of Edwin Chadwick (1800–1890), it was ordered that sewage and silts from the foul-smelling sewers be flushed into the Thames, from which drinking water was obtained. However, the measure led to the contamination of drinking water and an epidemic with a further 14,000 deaths. The hypothesis that cholera was caused by living organisms in the drinking water only slowly gained acceptance from 1849 onwards. In 1854, John Snow (1813–1858), the pioneer of cholera research, investigated the transmission of cholera via contaminated drinking water and was also able to attribute the severe epidemic in Soho in 1855 to it. After three cholera epidemics had killed 30,000 people, the last of which in 1853/1854 killed over 10,000 people in central London alone, all that was needed was a trigger for now drastic measures. The Great Stink in the summer of 1858 as a result of the unhindered discharge of sewage into the London Thames also reached Parliament as an unbearable effect, which in the same year ordered the construction of the great London sewage network under Joseph Bazalgette (1819–1891) as chief engineer of the London Metropolitan Board of Works and provided 3 million pounds.

It was not until Bazalgette also introduced filtration of drinking water that cholera disappeared from London and never returned (Hamlin 2009).

References

Alleyne A (2017) China unveils plans for world's first pollution-eating 'Forest City'. www.cnn.com/style/article/china-liuzhou-forest-city/index.html. Accessed 11 Jan 2018

Alphand A (1867–1873) Les Promenades de Paris. J. Rothschild, Paris

Amsterdam.org (1999–2018) Kanäle von Amsterdam. https://amsterdam.org/de/kanale-von-amsterdam.php. Accessed 28 Dec 2017

Barlösius E (1997) Naturgemäße Lebensführung. Zur Geschichte der Lebensreform um die Jahrhundertwende. Campus, Frankfurt a. M

Baumeister R (1879) Stadterweiterungen in technischer, baupolizeilicher und wirtschaftlicher Beziehung. Ernst & Korn, Berlin

Bell E (1993) La Antigua Guatemala. La historia de la ciudad y sus monumentos. Elisabeth Bell, Guatemala

Breuste J (1996) Zur Entwicklungsgeschichte der Kleingärten. In: Breuste I, Breuste J, Diaby K, Frühauf M, Sauerwein M, Zierdt M (eds) Hallesche Kleingärten: Nutzung und Schadstoffbelastung als Funktion der sozioökonomischen Stadtstruktur und physisch-geographischer Besonderheiten, vol 3, UFZ-Bericht No. 8/1996 = Stadtökologische Forschungen. UFZ-Umweltforschungszentrum, Leipzig, pp 3–6

Breuste J (2010) Green space, planning and ecology in German cities in the late twentieth century. In: Clark P, Niemi M, Niemelä J (eds) Sport, Recreation and Green Space in the European City, Bd 16. Studia Fennica, Historica, Helsinki, pp 113–124

Breuste J, Pauleit S, Haase D, Sauerwein M (eds) (2016) Stadtökosysteme. Springer, Berlin

Brocklebank RT (2003) Birkenhead: an illustrated history. Breedon Books, Derby

Bruckmann E (2018) Verschönerungsverein Stuttgart www.vsv-stuttgart.de/. Accessed 10 Jan 2018

Burns R, Sanders S, Ades L (2002) New York. Die illustrierte Geschichte von 1609 bis heute. Frederking & Thaler, München

Chadwick E (1843) Report on the sanitary conditions of the labouring population of Great Britain. A supplementary report on the results of a special inquiry into the practice of interment in towns. Made at the request of Her Majesty's principal secretary of state for the Home department. Clowes and Sons, London

Cocking L (2016) A brief history of Alameda Central Park, Mexico City. https://theculturetrip.com/.../mexico/.../a-brief-history-of-alameda-central-park-mexico-city/. Accessed 9 Jan 2018

Corbin A (2005) Pesthauch und Blütenduft. Eine Geschichte des Geruchs ("Le Miasme et la Jonquille. L'odorat et l'imaginaire social XVIIIe–XIXe siècles", 1982). Wagenbach, Berlin

Crompton JL (2016) Evolution of the "parks as lungs" metaphor: is it still relevant? World Leisure J 59(2):105–123. https://doi.org/10.1080/16078055.2016.1211171, www.rpts.tamu.edu/.../Evolution-of-the-parks-as-lungs-metaphor-is-it-still-relevant.pdf. Accessed 13 Dec 2017

Dalley S (2013) The mystery of the Hanging Garden of Babylon: an elusive world wonder traced. Oxford University Press, Oxford

Därr M (2018) Stadtgottessacker Halle (Saale) Denkmalpflegerische Zielstellung-Bestandsaufnahme/dokumentation. www.la-daerr.de/.../stadtgottesacker_halle/stadtgottesacker_halle.html. Accessed 8 Jan 2018

Ebner R, Weigl H (2014) Das Salzburger Wasser. Geschichte der Wasserversorgung der Stadt Salzburg, vol 39, Schriftenreihe des Archivs der Stadt Salzburg. Stadtarchiv und Statistik der Stadt Salzburg, Salzburg

Eidloth V (2013) Baden-Baden, europäische Kurstädte und das Welterbe der UNESCO. Grundzüge einer länderübergreifenden gemeinschaftlichen Bewerbung. Denkmalpflege in Baden-Württemberg 42(3):134–144

Endell A (1908) Die Schönheit der großen Stadt. Strecker & Schröder, Stuttgart

George P (2017) What are the lungs of London? history house, Essex. www.historyhouse.co.uk/articles/lungs_of_london.html. Accessed 13 Dec 2017

Golftoday (2018) Golftoday. www.golftoday.co.uk/clubhouse/coursedir/london.html. Accessed 10 Jan 2018

Greaney M (2013) Liverpool. A landscape history. The History Press, Liverpool

Greiner J, Gelbrich J (1975) Grünflächen der Stadt 2nd impr. ed. Verlag für Bauwesen, Berlin

Gunkel G (1991) Die gewässerökologische Situation in einer urbanen Großsiedlung (Märkisches Viertel, Berlin). In: Schuhmacher H, Thiesmeier B (eds) Urbane Gewässer. Westarp, Essen

Haas I (2010) Der Wiener Prater. Sutton, Erfurt

Hajós G (2005) Parkanlagen in Wien. In: Brunner K, Schneider P (eds) Umwelt Stadt. Geschichte des Natur- und Lebensraumes Wien. Böhlau, Wien, pp 440–449

Hamlin C (2009) Cholera: The biography. Oxford University Press, Oxford

Hansard (1812) June 30. Hyde park. The Parliamentary debates from the year 1803 to the present time. London, Vol. XI. The eleventh Day of April to Fourth Day of July 1808. TC Hansard, London, pp. 1122–1124

Hansard (1848) Public Health Act of 1848. HC Deb 09 March 1853 vol. 124 cc1349-57, London

Hennebo D, Wagner S (1977) Geschichte des Stadtgrüns in England von den frühen Volkswiesen bis zu den öffentlichen Parks im 18. Jahrhundert, vol 3, Geschichte des Stadtgrüns. Patzer, Hannover

Hesse R (1907) Die Parkanlagen der Stadt Magdeburg. Rob. Hesse & Co., Magdeburg

Hoke G (1991) Herrenkrug: die Entwicklung eines Magdeburger Landschaftsparks. Magistrat der Stadt Magdeburg, Magdeburg

Johnson B (2018) The history of Golf. www.historic-uk.com/HistoryUK/HistoryofScotland/The-History-of-Golf/. Accessed 10 Jan 2018

Jones PB (2006) Gunnar Asplund. Phaidon Press, New York

Joyce P (2003) The rule of freedom: Liberalism and the modern city. Verso, London

Kaiser O (2005) Bewertung und Entwicklung von urbanen Fließgewässern. Ph.D Thesis, Faculty Forestry and Environmental Sciences Albert-Ludwigs-University Freiburg, Freiburg i. Brsg

Katsch G, Walz JB (2011) Deutschlands Kleingärtner in drei Jahrhunderten. Zum 90. Jahrestag der Gründung des Reichsverbandes der Kleingartenvereine Deutschlands. Bundesverband Deutscher Gartenfreunde, Leipzig

Katsch G (1994) Ein "Deutsches Museum der Kleingärtnerbewegung" in Leipzig. Vorstand des Fördervereins "Dt. Museum der Kleingärtnerbewegung", Leipzig

2

KGV Dr. Schreber (2018) Kleingartenverein Dr. Schreber http://www.schreber-leipzig.de/. Accessed 11 Jan 2018

Klackl H (2002) Der Almkanal. Seine Nutzung einst und jetzt. self-published, Salzburg

Koller E (1988) Umwelt-, sozial-, wirtschafts- und freizeitgeographische Aspekte von Schrebergärten in Großstädten, dargestellt am Beispiel Regensburgs. Regensburger Beiträge zur Regionalgeographie und Raumplanung (1), Regensburg

Kostof S (1993) Die Anatomie der Stadt. Geschichte städtischer Strukturen, Campus, Frankfurt

Krauskopf K (2003) Natur statt Stadt – Die Verwandlung der Stadt in Natur. In: Rohde M, Schomann R (eds) Historische Gärten heute. Edition Leipzig, Leipzig, pp. 78–83

Landeshauptstadt München, Kulturreferat (ed) (2000) Friedrich Ludwig von Sckell 1750–1823. Gartenkünstler und Stadtplaner in München. Kulturreferat der Landeshauptstadt München, München

Leisner B, Schoenfeld H (1991) Der Ohlsdorf-Führer. Spaziergänge über den größten Friedhof Europas. Christians, Hamburg

Loudon JC (1803) Letter to the editor. Literary J 2(12):739–742

Mader G (2006) Geschichte der Gartenkunst. Streifzüge durch vier Jahrtausende, Ulmer, Stuttgart

Medicus R (2006) Die Hellbrunner Allee und ihre Umgebung – Zur Geschichte der Allee und ihrer Bedeutung. Mitteilungen der Ges für Salzburger Landeskd 146:405–426

Metropolitan Borough of Wirral (2008) The history of Birkenhead Park. https://www.birkenheadpark1847.com. Accessed 26 Mar 2008

Murray JF (1839) The lungs of London. Blackwood's Magazine 46:212–227

Pemmer H, Lackner N (1974) Der Prater. Von den Anfängen bis zur Gegenwart. Neu bearbeitet von Günter Düriegl und Ludwig Sackmauer, 2nd ed Jugend und Volk, Wien

Poninska A (1874) Die Großstädte in ihrer Wohnungsnoth und die Grundlagen einer durchgreifenden Abhilfe. Duncker & Humblot, Leipzig

Raven PH, Evert RF, Eichorn SE (2005) Biology of plants. N.H. Freeman, Basingstoke

Reps JW (1992) The making of Urban America: A history of city planning in the United States, 2nd edn. University Press Princeton, Princeton

Rosenzweig R, Blackmar E (1992) The park and the people. A history of central park. Cornell University Press, Ithaca

Schediwy R, Baltzarek F (1982) Grün in der Großstadt – Geschichte und Zukunft europäischer Parkanlagen unter besonderer Berücksichtigung Wiens. Edition Tusch, Wien

Schiller-Bütow H (1976) Kleingärten in Städten. Patzer, Hannover

Schuhmacher H (1998) Stadtgewässer. In: Sukopp H, Wittig R (eds) Stadtökologie. Ein Fachbuch für Studium und Praxis. Gustav Fischer, Stuttgart, pp 201–217

Schuhmacher H, Thiesmeier B (1991) Urbane Gewässer. Westarp, Essen

Schwarz A (2005a) Der Park in der Metropole. Urbanes Wachstum und städtische Parks im 19. Jahrhundert. Transcript, Bielefeld

Schwarz A (2005b) Ein "Volkspark" für die Demokratie: New York und die Ideen Frederick Law Olmsteds. In: Schwarz A (ed) Der Park in der Metropole. Urbanes Wachstum und städtische Parks im 19. Jahrhundert. Transcript, Bielefeld, pp 107–160

Sckell FL (1807) Denkschrift vom 6. März 1807. In: von Freyberg P (ed) 200 Jahre Englischer Garten München. Offizielle Festschrift, Bayerisches Staatsministerium der Finanzen. Knürr, München, pp 93–113

Scobey DM (2002) Empire city. The making and meaning of the New York city Landscape. Temple University Press, Philadelphia

Sehnal P (2008) Wiens grüne Arena, der Prater. Folio, Wien

Stadtwald (2018). https://de.wikipedia.org/wiki/Stadtwald. Accessed 11 Jan 2018

Taylor DE (2009) The environment and the people in American cities, 1600–1900s. Duke University Press, Durham

Thorén KH (2008) De grønne lungene som forsvant. Om tap av grønnstruktur i byer og tettsteder. In: Berntsen B, Hågvar S (eds) Norsk natur – farvel? En illustrert historie. Unipub, Oslo, pp 223–235

Tietz AA (2004) Der Stadtgottesacker in Halle (Saale). Fliegenkopf, Halle

Twardawa S (2006) Der Tiergarten in Berlin: das Abenteuer liegt um die Ecke. Motzbuch, Berlin

von Freyberg P (2001) Der Englische Garten in München, 2nd edn. Kürr, München von Freyberg P (ed) (1989) 200 Jahre Englischer Garten München. Offizielle Festschrift, herausgeben durch das Bayerische Staatsministerium der Finanzen, pp. 93–113

von Krosigk K (2001) Der Berliner Tiergarten, vol 1 Berliner Ansichten. Braun MS, Berlin

von Schweinitz AF (1993) Der Stadtgottesacker in Halle. Die Gartenkunst 5(1):91–100

Wagner-Rieger R (ed) (1972–1981) Die Wiener Ringstraße. Bild einer Epoche, vol. I–XI. Steiner, Wiesbaden

Wendland F (1993) Der Große Tiergarten in Berlin – Seine Geschichte und Entwicklung in fünf Jahrhunderten. Gebrüder Mann, Berlin

Wiede J (2015) Abendländische Gartenkultur. Die Sehnsucht nach Landschaft seit der Antike. Marix, Wiesbaden

Wieland D (2012) Historische Parks und Gärten. Schriftenreihe des Deutschen Nationalkomitees für den Denkmalschutz, vol 45, 2nd edn. SLUB, Bonn

Willeford BR (1979) Das Portrait: Joseph Priestley (1733–1804). Chem unserer Zeit 13:111–117. https://doi.org/10.1002/ciuz.19790130403

Wilner F (2017) Amsterdam. London, New York: Welt-Städte (3/4). 1800–1880: Schock der Moderne. Erstsendung 16.12.2017. Arte France, Iliade Productions, Les Films de L'Odyssée

Worde N, van Heyningen E, Bickford-Smith V (1998) Cape Town. The making of a city. David Philipp Publishers, Cape TownZedler JH (1743) Grosses vollständiges Universal Lexicon aller Wissenschafften und Künste, vol 38, Halle. https://www.zedler-lexikon.de/. Accessed 23 Dec 2018

How Urban Nature Exists in the Context of Nature and Culture?

Contents

3.1   The Relationship Between City and Natural Environment – 71

3.2   Example Forest City: Integration into the Natural Environment – 77

3.3   Example Desert City: Turning Away from the Natural Environment – 78

3.4   Example Mountain City: Dealing with Extreme Natural Environment – 82

© Springer-Verlag GmbH Germany, part of Springer Nature 2022
J. Breuste, *The Green City*,
https://doi.org/10.1007/978-3-662-63976-4_3

3.5   Forest and Wilderness
        in the City: Places for Religion
        and Rituals – 89

3.6   Gardens and Parks:
        Designed Nature for the Urban
        Recreational Landscape – 93

        References – 99

Urban nature emerged in adaptation to, but also sometimes in opposition to, the nature of the urban environment. With the export of the idea of "beautiful" urban nature from humid Europe, public urban design with gardens, shrubs and trees was also transferred to other, even arid zonobiomes. There, too, green gardens and park oases were preferred and created and maintained at great expense. The idea of Persian (paradise) gardens spread to India, but remained an isolated urban nature for elites only, as did Chinese and Japanese gardens.

Desert cities already traditionally represent urban nature in their oasis kitchen gardens and decorative gardens of the palaces. Mountain cities also try to integrate urban nature under extreme climatic and orographic conditions. The forest cities largely integrate themselves into their traditional natural environment, the forest, which often continues to exist in the towns in various forms. Forest as primary wilderness is here almost everywhere accepted wilderness.

New urban wildernesses that emerge after the abandonment of uses have largely not (yet) been accepted. Forests and wildernesses are often places of religion and rituals in cities as well.

Gardens and parks, with their either individually active or contemplatively aesthetic uses, are indisputably the preferred urban nature.

3.1 The Relationship Between City and Natural Environment

The Green City is a cultural product in a natural environment. The type of cultural coexistence in cities, including the relationship to nature (preference and repression) **(cultural space)** and the natural features of the localisation space (especially vegetation, water balance, relief and climate) determine the expression of the concrete urban nature. The Green City is thus an individual reality. However, it is comparable in different cultural areas and **zonobiomes** and even shows certain comparable elements in a global view due to cultural similarities.

Definition

Civilizations (Kulturerdteile) are coherent large areas of common, relatively uniform culturally formative human life forms on the basis of their natural environment. They are based on the individual origin of culture, on the special combination of landscape-shaping natural and cultural elements, on the independent, spiritual and social order and the context of the historical process (Kolb 1962; Huntington 1996; Newig 2014).

In general, 10 cultural civilizations (with transitional spaces between them) are distinguished:

1. Angloamerica
2. Australia/Oceania
3. Europe

3

4. Latin America
5. Orient
6. East Asia
7. Russia
8. Sub-Saharan Africa
9. South Asia
10. Southeast Asia

Definition

A **zonobiome** is defined as a zonal plant formation including the animals living in it, that is, a large climatically uniform habitat within the geo-biosphere. Nine zonobiomes are considered globally:

1. Evergreen tropical rainforest (**ZB I**),
2. Tropical seasonal forest, seasonally dry forest, scrub, or savanna (**ZB II**),
3. Hot deserts (**ZB III**),
4. Sclerophyllous, frost-sensitive shrublands and woodlands (**ZB IV**),
5. Temperate evergreen forest (laurel forests) (**ZB V**),
6. Frost-resistant, deciduous, temperate forests (**ZB VI**),
7. Grasslands (steppes) and temperate deserts (**ZB VII**),
8. Taiga (**ZB VIII**) and
9. Tundra (**ZB IX**) (Walter and Breckle 1999).

Culture and nature play an equally important role at the beginning of urban development. It is almost always culture that dominates. The dominant role of culture in the formation of cities gives rise to two situations in which cities are founded:

1. **Autochthonous cities** (locally developing cities): Cities emerge in a pre-existing agrarian cultural landscape through the emergence of performance advantages for an environment or through the founding act of an authority. The indigenous population becomes townspeople (cultural development). Examples are early cities in the cultural centres of Egypt, the Near East, China or urban developments in the Roman Empire or Medieval Germany. Cities are localized in a "known" natural environment.

2. **Allochthonous cities** (colonist cities): Cities emerge as colonist cities. The initial urban population, at least the elite, is imported and determines the urban development. In this case, the culture of the elites is established in a new location and possibly encounters already existing other culture(s) (cultural expansion). Examples are: Greek colonization cities in the Mediterranean from the seventh century BCE, German colonization of the East in the twelfth century, European colonist cities in America in the fifteenth century, in Australia and Oceania in the eighteenth century, in Africa since the fifteenth century, Russian colonist cities in Siberia in the seventeenth century. The cities are located in a new natural and cultural environment.

This situation changes in the course of urban development due to the influx of non-local population into the cities, cultural integration of possibly existing indigenous, agricultural population and generally progressive urban development. The cultural expansion, however, spreads a culture that is foreign in itself into new, often unknown natural areas and leads to new conflicts there with the "new" nature found. The original disintegration into the natural environment slowly changes into an integration of the city into the nature (often also into the culture) of its surroundings. This can consist in adaptation, but also displacement of this nature. Cities, even when they develop on site in the existing culture, have always developed in a confrontation with the nature they find. In the process, attempts are made to dominate nature, to reshape it, to displace it, or to eliminate it altogether (which never succeeds). Nature is seen as the enemy of the city, since it develops dynamically, without cultural decision, according to its own laws (see ▶ Sect. 7.4). With the knowledge of the laws of nature, attempts were made to use them for urban development or at least to evade their negative processes (for example, floods, coastal floods, mudslides and avalanches, etc.) or to control these processes.

Every city is first located in a natural environment. This natural environment is in constant change due to human use. The original natural environment becomes a new natural environment (see ▢ Fig. 3.1).

》 "In the first place, the concept of the city loses its general and systematic validity. Instead, the geographical and landscape conditions play the decisive role. An already existing and continuing city is always determined by the natural environment in which man begins to re-establish himself" (Benevolo 1999, p. 39, translated).

The "location factor nature" such as **traffic favorability** (for example, location on rivers, fords, and the coast), **security** (for example, location on easily defensible hills, river terraces outside floodplains), and **health** (for example, location away from marshes and swamp forests) is first of critical importance in locating a town (Ullman 1941).

The "European City" or the "Western City", as Benevolo (1999) calls it, is the concept of a closed city, as a counterpart to the open country. Benevolo (1999, p. 95) refers to the years between 1050 and 1350 as **"the time of the urbanization of Europe",** when this concept of the city was Europeanized. With the paradigm of the separation of city and nature, European urban landscapes of core cities and surrounding cultivated land emerged over centuries, which even today are cultural heritage worthy of our protection and preservation (Breuste 2006, 2011, 2012).

》 **"Here the city, there the country** – this contrast will for a long time shape the picture of the world and be equally determinative of consciousness and social reality. The city is an enclosed district or a collection of such enclosures, where the art of economizing over medium and smaller distances flourishes – this is what we have since called "architecture", while the far older art of taking possession of and cultivating the unlimited land is gradually being forgotten" (Benevolo 1999, p. 20, translated).

》 "The goal is not the expansion of the city into the landscape, but the formation of a core, the concentration, the encapsulation from the landscape. Now, for the first time, the real strength of Western civilization comes into play, and some of the **most impressive European urban landscapes** emerge" (Benevolo 1999, p. 74, translated).

3

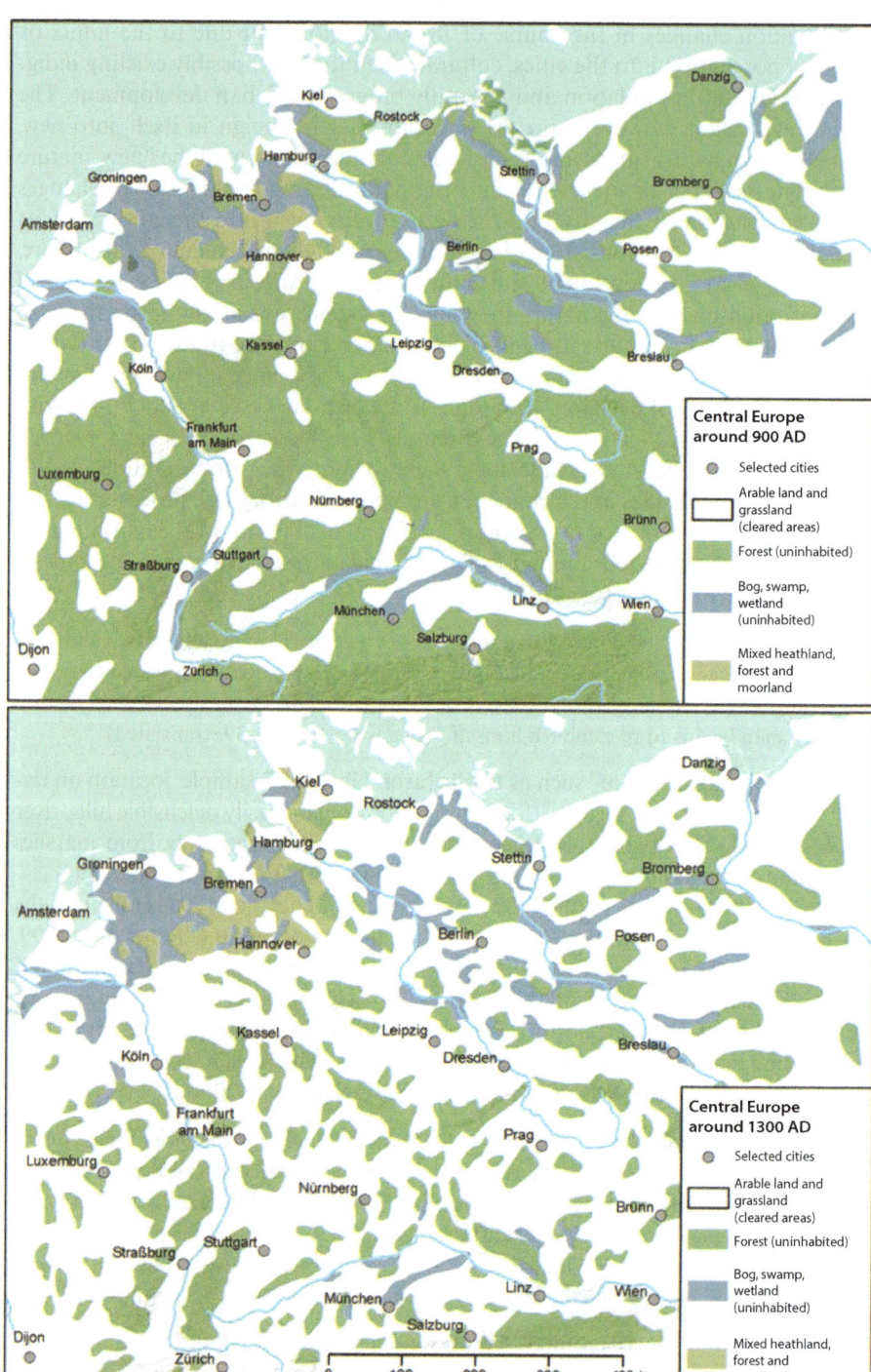

■ **Fig. 3.1**   Forest expansion, clearing and resettlement – The land expansion in Germany (c. 900–1300). (© design: Breuste, drawing: W. Gruber, after Rudolf and Oswalt 2009, p. 89)

It is only in the period after the Second World War that city and country (land-scape) come together again in modern landscape planning (Breuste 2012). In many places, "embedding in the landscape", in a natural environment that is viewed positively, is a goal, for example, realised in Scandinavian cities since the 1940s and 1950s.

The natural environment is still relevant today, most visibly in extreme "hostile to settlement" natural conditions such as arctic natural areas, tropical rainforests and high mountains. Here, the "enemy role" of nature as an "opponent" of the city becomes most obvious. But also in forest landscapes, steppes or on coasts the city and its culture always had to fight "against nature" to be able to exist. It therefore, makes sense to include this confrontation of differently culturally shaped cities on the one hand with the surrounding different local nature on the other hand in the consideration of the Green City. It still influences, even after seemingly "won" confrontation, the relationship of city dwellers to "their" nature in the surroundings of the cities and to the integration of certain, positively accepted nature in the cities.

Landscape: Cultural landscape – Natural landscape

"Landscape" refers to a space on the earth's surface that can be distinguished from other areas by features that can be recorded by natural science and can be perceived holistically. In the cultural landscape, the physiognomic shape of the landscape and its ecological process sequences are profoundly influenced by human use. Together with the opposing term "natural landscape", a landscape shaped exclusively by natural processes, a complementary pair of terms (dichotomy) is created (Breuste 2006, see for example, Wöbse 1999).

In the course of its history, the city has tried to detach itself more and more from the natural environment. It strives for an artificial, controlled living environment for the new, urban way of life of its inhabitants. This can best be created in the forms of living and in the interaction of all now urban structures in a closed structure. This (core) city, deliberately sets itself apart from the countryside, the surrounding landscape. The core of the urban is thus the denaturation, the independence from nature and its processes. This never quite happened, because nature also belonged and still belongs to urban life as a garden for the production of at least one, ever-decreasing part of the food needed in the city, and as an aesthetic object of culturally perceived beauty.

Urban structure

"Urban structure" refers to a real model of the differentiation of a city. It is composed of the elements of development, use, population, social situation, economy, but also nature. This results in a spatial mosaic of structural elements. The urban nature of different kinds or types is such a structural element of the urban structure (Wickop et al. 1998).

But it is always a controlled, reduced and artificial nature that is allowed to be part of the city. The city is never independent of nature, because it needs food, which can only be produced "outside" the city in type and quantity. Nor can the city escape the large-scale regional processes of nature. Cities are not safe from heat waves, pandemics, pest calamities and floods.

Many cities are created in the struggle against nature and its processual effects. A "balance" with nature is usually not sought. This idea only emerges when, with the development of the industrial city, it becomes clear that the removal and discharge of pollutants, "light, air and sun" are necessary for health. Nature is brought back into the cities in a partial and controlled way for these beneficial effects. It is supposed to help compensate for the disadvantages accepted with urban development, at least when this does not lead to too high a burden economically and technical solutions alone are not sufficient.

Urban life, it is recognized, is not a healthy life. Rural life, it is idealized, promotes health. Turning away from the urban is impossible; turning towards the integration of the rural into the urban appears as a possible beneficial solution. Healthy cities, it is generally accepted, are cities with a health-promoting natural endowment. Which nature, to what extent, where it should be placed and by whom it should be used, is the result of a social negotiation process. This process is currently underway all over the world. It is not about a "back to nature", but about letting nature play a role in urban life. What that role will depend on culture and ecological and regional conditions (see ▶ Sect. 8.8).

With the emergence of the idea of recreation, first the urban environments, and then the cities themselves, should be places for recreation. They were extremely poorly equipped for this in the course of industrialization, but can be improved with this goal in mind. Recreation is projected almost entirely onto nature and landscape. Nature is now to be given a place in the city as a space for recreation. In addition, there are further demands on nature to solve problems (see ▶ Chap. 5).

Nature can by no means be a panacea for all major urban problems (*nature-based solutions*). The sometimes idealised expectations of nature as a "saviour" are often far too optimistic. Nor is society by any means completely open to a more clearly nature-determined urban environment for its living environments (see ▶ Sect. 7.4). But the fact that the discussion has only just begun and that exemplary solutions can be presented is a good start. The end result will not be a nature-dominated city, but a differently structured city that also comes to terms with its natural environment in the interior and not only in the urban environment (see ▶ Sect. 7.1).

Cities have always had to find their location "in nature". In which nature depended not only on the basic natural conditions but also on the preceding cultural development. European cities developed predominantly as regional centres and places of trade in a landscape that had already been used agriculturally. They positioned themselves as **"city in rural area"**. In the forest lands of the taiga of Europe, Asia or America, things were different. Here cities emerged as civilizing outposts of the imported culture of new settlers who brought both, cultural landscape and cities, with them at the same time. If city ideas and their realizations already existed (for example, Tenochtitlan of the Aztecs, Tikal of the Maya, Cuzco of the Incas, Angkor of the Khmer, Hampi in India, etc.), they were usually ignored and, if they still existed in reality, even destroyed (see ▶ Sect. 3.2).

Malaria: Deadly Disease from the Wetlands Surrounding the Cities of the Tropics and Subtropics

Especially in the tropics, malaria (Italian *mal'aria* = "bad air") was a deadly disease for centuries, spreading in the humid lowlands, swamp and mangrove areas. The cause was thought to be the "unhealthy air" from these areas, as transmission by mosquitoes (genus: *Anopheles*) from these humid breeding areas was long unknown. Since ancient times, the disease was known in wetlands around the Mediterranean Sea.

Because of the disease, the urban population of the colonist cities often decimated rapidly, so that even at the time of localization, more and more attention was paid to settling especially in areas with "good air". For Pedro de Mendoza, who came from the north along the tropi-

cal coasts of South America to the Rio de la Plata in 1536 and founded Buenos Aires, this was the first city on the coast with "good air" (= *buenos aires*). However, malaria was probably first spread to South America by Europeans and the slave trade from Africa.

If the location of towns close to wetlands was unavoidable or irreversible, the wetlands were often drained, for example, on a large scale in the coastal landscape of southern Tuscany, the Maremma, in the 1930s.

The aversion to wetlands in the tropics and subtropics, which is still widespread today, has its roots in the fear of malaria (Sullivan and Krishna 2005; Paganotti et al. 2004).

3.2 Example Forest City: Integration into the Natural Environment

The towns in the forest lands were initially surrounded only by forests and wetlands with little agricultural land, of which many remnant forests still bear witness today. These **"forest towns"** still have more or less original nature (nature of the first kind, see ▶ Sect. 1.3) in their surroundings. It belongs to them and their mostly comparatively short development of only a few hundred years. Such cities surrounded by extensive forests can be found in Europe (for example, Moscow, founded 1147, Stockholm, founded 1252, Vilnius, founded 1323, Helsinki, founded 1550, St. Petersburg, founded 1703) as well as in Asia (for example, Krasnoyarsk, founded 1628, Yakutsk, founded 1632, Irkutsk, founded 1652).

In forest cities, a close relationship between the inhabitants and the surrounding forest is usually maintained over generations. The inhabitants of Helsinki, Stockholm, St. Petersburg or Vilnius allow forest to penetrate right into their residential areas or preserve its remains there as a reference to their natural environment. They are often still familiar with typical forest uses (mushroom and berry picking or hunting). The forest with its inhabitants is welcome nature for them, also in the city.

With the development of tourism, resorts developed in remaining forest areas, often in the mountain and hill country, which became towns and where tourists took their shorter or longer stay especially because of the forests (German terms "Waldluft", "Waldeslust").

3

> **Campos do Jordão: A Forest City for Sao Paulo, Brazil**
>
> Big city dwellers from São Paulo founded a forest town as a tourist destination – Campos do Jordão – in 1874 in the mountains of the Serra da Mantiqueira, 180 km away. The town in the state of São Paulo at 1628 m above sea level has today about 46,500 inhabitants and is a typical forest town. Campos do Jordão is the highest municipality in all of Brazil. The town, a regional tourist destination in the climatically temperate mountains, reminiscent of European resorts (Swiss-style half-timbered houses), except for the vegetation of the native subtropical humid forests surrounding it, with many araucarias. 20 km of hiking trails shoot the attractive forest environment (Campos do Jordão 2018; see ◘ Fig. 3.2).

◘ **Fig. 3.2**   Campos do Jordão, Brazil, view from Morro do Elefante to the Capivari district. (© Breuste 2009)

3.3   Example Desert City: Turning Away from the Natural Environment

If forests were a welcome and accepted environment for forest cities, this does not apply to the nature of dry areas, deserts and steppes, the great grasslands, in America, Asia and Africa, in the vicinity of cities. The "nature of the first kind", the original nature (see ► Sect. 1.3), is usually still present here in its remnants. It is not always appreciated by the city dwellers, many of them allochthonous colonists. In and around the **"cities of dry regions"** it is even fought against to make room for a new nature which the settlers bring with them as a model of the harmonious. This is first the nature of agriculture around the cities and a lush nature of trees, bushes, flowers and lawns in the cities themselves, which has nothing to do with the nature of the environment of the cities. Their urban nature, tree plantations, parks, gardens and agriculture is based almost entirely on irrigation and is sharply distinguished from the nature of the urban environments (see ◘ Fig. 3.3).

■ **Fig. 3.3** Irrigation of roadside greenery in Delhi, India. (© Breuste 2018)

Hot deserts are widespread in six very different cultural regions (Anglo-America, Latin America, Sub-Saharan Africa, the Orient, South Asia, Australia) and, for cultural reasons, different cities are located in them, for example, Tucson, Arizona/USA; Mendoza, Argentina; Windhoek, Namibia; Aswan, Egypt; Yalgoo, Australia. Few scientific comparative studies have been done on their oasis nature, shading and climate regulating tree cover *(urban forest)* (for example, McPerson and Dougherty 1989; Breuste 2013).

Convergent vegetation and a comparable climate create comparable natural conditions and living conditions for settling people. What do the desert cities have in common? They emerge in contrast to the surrounding nature as a cultural counter-image. The city separates itself as much as possible from the surrounding desert nature, creates its own irrigation nature. The city even arises in the struggle against the unloved surrounding nature. With the exception of the large river oasis cities, which already provided living space for ancient cultures, and some fishing towns on the coasts, the desert cities are predominantly colonist cities.

Urban greenery as cultivated and irrigated urban nature is a clear sign of prosperity or even luxury in green deficient cities, such as those in desert environments. A green city in the desert is the antithesis of the desert nature of the wilderness.

Dubai, the capital of the emirate of the same name on the Persian Gulf, is a city in the desert region that has developed an intensively maintained green structure worldwide at great expense for the benefit of its inhabitants. In 2012, 6.7 million visitors used Dubai's public parks (Al Serkal 2013). Seven major parks alone have been newly created since 2012 (Dubai Parks and Resorts 2018).

The Dubai Strategic Plan 2015 *"A green economy for sustainable development"* has firmly established the goal of making Dubai a "healthy city" with a clean and pollution-free environment. Dubai Electricity and Water Authority (DEWA) together with Dubai Municipality have prepared "Green Building Regulations & Specifications" as part of a €2.223 billion (AED 10 billion) "Green Buildings Project" by 2020 to achieve this. However, the "Green Dubai" action does not refer to urban nature (Dubai Statistics Center 2018).

3

Fig. 3.4 Creekside Park in Dubai, UAE. (© Breuste 2006)

Fig. 3.5 Palace and private park of the ruler of the Emirate of Dubai HH Sheikh Mohammed Bin Rashid Al Maktum in the Al Bdai'a district of Dubai, UAE. (© Breuste 2017)

All parks in Dubai are fenced, guarded and intensively maintained. There is a (small) entrance fee. A cultural difference exists in the gender perspective for park use. On special "Ladies Days", one or two days a week, which vary per park, only women or families are allowed to visit most public parks (see ■ Fig. 3.4). Members of the ruling family have intensively maintained extensive private parks (see ■ Fig. 3.5).

Trees in an Oasis City: Mendoza in the Argentinean Dry Region

In the west of Argentina, on the edge of the Andes at 700 m above sea level, lies the city of Mendoza (population approx. 1 million), founded in 1561 and rebuilt after its destruction by an earthquake in 1861. The arid desert climate, with summer temperatures above 40 °C and less than 200 mm of rainfall a year, produces inhospitable heat and drought. The vegetation of the Monte shrub steppe, the natural vegetation formation of the region, cannot provide shade. An urban forest of street trees and large parks is intended to provide this. It was planted as early as the late nineteenth century and irrigated by a system of ditches *(acequias)* derived from the Rio Mendoza. Both, the tree population and the irrigation system, are not in the best condition today, but are essential for the quality of life of the city's inhabitants. In 1907 and 1986 legal measures were taken for its protection and management of this urban forest, but so far only partially implemented. In 1996, a 'Consejo Provincial de Defensa del Arbolado Público' ('Provincial Council for the Defense of Public Trees') was entrusted with the tasks of tree protection.

The urban forest of street and park trees includes an estimated 400,000 trees, all at least 50–100 years of age with some significant reductions in vitality. The Gran Mendoza urban core alone has 48,811 trees (Carrieri 2004).

Representative surveys of the street tree population showed the following results:

- Approximately 70% of the irrigation canals have inadequate management.
- More than two thirds of the trees suffer from acute water stress.
- The existing open tree grates are almost always undersized due to

sealing and their soil is extremely compacted.

- Mechanical damage to the roots and trunk of trees occurs frequently.
- Damage caused by improper pruning by contractors outside the trade.
- The tree population is highly appreciated by the residents as a source of shade.
- The residents hardly notice tree damage and the problems of irrigation, and are only prepared to help with tree care to a limited extent (they consider this to be an exclusively public task).
- The responsible authorities hardly take the degradation of the tree condition seriously.

The tree population is composed exclusively of 75 non-native tree species, including five species from Asia (50%), Europe (26%) and North America (10%), which account for 86% of the population (*Morus alba., Fraxinus excelsior, Fraxinus americana, Platanus acerifolia, Melia azedarch*). The only native tree of the Monte shrub steppe, the algarrobal *(Prosopis flexuosa,* Spanish *algarrobo)*, is not represented at all as a street tree. The beautiful, large-crowned and shady growth of exotic trees is the reason for the choice of exotic trees, although these in particular have high water requirements (Breuste 2013; Faggi and Breuste 2016; see ◘ Fig. 3.6).

3

◻ Fig. 3.6 Acequia irrigation of street trees in the oasis city of Mendoza, Argentina. (© Breuste 1995)

3.4 Example Mountain City: Dealing with Extreme Natural Environment

The mountain towns mostly developed late after the settlement of the lowlands. People only moved to the mountains when the lowlands offered inhospitable agricultural settlement conditions, such as in parts of the Andes in Latin America. Mountain towns are often associated with the extraction of mineral resources. The term "mountain city" is commonly used for this. This almost always led to extensive transformation of the surrounding natural environment, alteration of the water network by dams, canals and ponds, deposition of dead rock and extensive infrastructure works. In most cases, considerable deforestation also took place in their surroundings in order to meet the great demand for wood as a source of

Fig. 3.7 "Mountain city" of Ouro Preto, at 1200 m a.s.l., Brazil. (© Breuste 1995)

energy and timber in the past. Mountain cities are localized in orobiomes (see ■ Fig. 3.7). Especially in tropical humid areas, orobiomes have often been chosen as more favourable and healthier places for the establishment of cities (for example, Latin America, India, Indonesia).

Orobiome

Orobiomes are mountain habitats within a climatic zone, a zonobiome, characterized by the division into altitudinal zones. Compared to lowlands, they are exposed to more extreme natural conditions (cool, damp climate, wind, vegetation adapted to it up to above the timberline and low-yielding, shallow soils) (Walter and Breckle 1999).

The mountain cities usually have reduced possibilities of expansion due to the relief. They inevitably adapt to the surface forms and leave the extremes such as gorges, steep slopes and peaks uninhabited, often with little or no use. Here one finds the remnants of "nature of the first kind" (see ▶ Sect. 1.3) as representatives of extreme sites. In contrast, the more favourable, more level areas have mostly already been used for colonisation (see ■ Fig. 3.8).

A number of large cities, including several cities with over a million inhabitants, are also located in orobiomes (see ■ Tables 3.1 and 3.2). Modern mountain towns developed in the extreme habitats of the mountains from the nineteenth century in Europe, first as summer resorts, then increasingly in the twentieth century as winter resorts for tourism, then also outside Europe.

These places were created only because of the attractiveness of the nature surrounding them, either for summer tourism, but mostly for winter tourism. The natural environment of tourist towns was often dramatically transformed for the recreational needs of winter tourists (lifts, slopes, relief design, water management, snowmaking, deforestation, infrastructure). Cities with tens of thousands of inhabitants, some of them only seasonal, and their urban structures emerged from only small mountain villages. For this purpose, access infrastructures were built and energy, water, food and material are continuously transported to high-altitude places.

3

■ **Fig. 3.8** Guatemala City at 1533 m N. N. on a volcanic plateau dissected by deep erosional canyons. (© Breuste 2010)

■ **Table 3.1** Examples of large cities and tourist towns in the mountains above 1000 m MSL altitude

Big cities in the mountains			Tourist mountain towns	
	MSL (altitude in m a.s.l.)	**Population in 1000 inhabitants**		**MSL (altitude in m a.s.l.)**
Cali, Colombia	1018	2401	Chamonix, France	1035
Kaxgar, China	1270	341	Metsovo, Greece	1160
Salt Lake City, USA	1288	1124	Ischgl, Austria	1377
Ulaanbaatar, Mongolia	1350	1380	Lech, Austria	1444
Kathmandu, Nepal	1400	1003	Campos do Jordão, Brazil	1628
Guatemala City	1533	2918	Obertauern, Austria	1664
Guadalajara, Mexico	1566	1495	Ifrane, Morocco	1664
Maseru, Lesotho	1600	220	Val d'Isere, France	1785
Campos do Jordão, Brazil	1628	51	St. Moritz, Switzerland	1822
Windhoek, Namibia	1655	326	Obergurgl, Austria	1907
Almaty, Kyrgyzstan	Until 1700	1508	Tochal, Iran	1910
Srinagar, India	1730	1193	Mussoorie, India	2006

◘ **Table 3.1** (continued)

Big cities in the mountains			Tourist mountain towns	
	MSL (altitude in m a.s.l.)	Population in 1000 inhabitants		MSL (altitude in m a.s.l.)
Kabul	1791	4635	Nainital, India	2084
Erzurum, Turkey	1950	762	Shimla, India	2276
Sanaa, Yemen	2250	1708	Aspen, USA	2438
Addis Ababa, Ethiopia	2355	3385	Valle Nevado, Chile	3000
Mexico City	2310	8851	St. Moritz, Switzerland	1822
Bogota, Colombia	2640	8081	Obergurgl, Austria	1907
Quito, Ecuador	2850	1619		
Shangri-La (until 2001 Zhongdian), China	3160	130		
Cusco, Peru	3399	349		
La Paz, Bolivia	3640	757		
Lhasa, China	3650	902		
Potosi, Bolivia	4067	175		
El Alto, Bolivia	4150	842		

◘ **Table 3.2** Examples of historic mining towns in Europe

Name	MSL (altitude in m a.s.l.)	Mining since	promotion of
Falun, Sweden	110	1641	Copper Ore
Hallein, Austria	450	1198	Salt
Johanngeorgenstadt, Erzgebirge, Dtschl.	892	1654	Tin and silver ore, later uranium ore
Altenberg, Erzgebirge, Dtl.	400 to over 800	1451	Tin Ore
Freiberg, Ore Mountains, Dtl.	400–500	1162/1170	Silver Ore
Clausthal-Zellerfeld, Harz, Dtl.	560	1268/1529	Silver, lead, copper, later zinc ore

(continued)

3

◘ Table 3.2 (continued)

Name	MSL (altitude in m a.s.l.)	Mining since	promotion of
Banska Stiavnica (Schemnitz), Slovakia	600	1217	Gold and silver ore
Banská Bystrica (dt. Neusohl), Slowakei	362	1255	Gold and silver ore, later copper ore
Kremnica (German: Kremnitz), Slovakia	550	1328	Gold and silver ore
Kutná Hora (engl. Kuttenberg), Czech Republic	254	1152	Silver Ore
Schwaz, Austria	545	1420	Silver and copper ore
Schladming, Austria	745	1322	Silver, lead, copper, later also cobalt and nickel ores
Eisenerz, Austria	736	1230	Iron ores
Libiąż, Poland	310	1906	Hard coal
Ronchamp, France	353	Middle eighteenth century	Hard coal
Tarnowskie Góry (dt. Tarnowitz); Polen	320	1526	Lead, silver and zinc ores

Mountain resorts are particularly unsustainable settlements from the point of view of resource efficiency. From the point of view of the "green city", they are often not "green cities" either, despite extensive natural surroundings, because there is no understanding of the care of the nature of the orobiomes, but nature is adapted to economic needs in an extreme way and overused. Nevertheless, "skiing in the midst of untouched nature" (for example, Tourist-Info Ruhpolding 2018) is also advertised in places where there is no question of untouched nature. This already becomes most obvious when looking at the natural environment there in summer (Alpenverein Österreich 2017; Ringler 2016/2017).

The "nature of the first type" that still exists (see ▶ Sect. 1.3) is not a representative image of nature before settlement. It is often reduced to extreme habitats, for example, in bogs, on steep slopes and peaks or on coasts and in gorges. These are the habitats that could only be used with difficulty or not at all in the course of cultivation. This has to be taken into account when it comes to nature acceptance by urban populations today. As remnant wildernesses, people turned away from them for a long time, until in some areas their religious or romantic transfiguration gave them new values and made them the destination of visitors again.

The White Ski Towns of the ALPS Are Not Green Towns

Alpine winter tourism is one of the most important economic sectors in Austria. However, it is also one of the biggest drivers of the destruction of mountain nature. Winter sports related nature change around ski towns provides high nature risks and ecological degradation. This concerns especially French, Swiss and Austrian ski towns because of their altitude.

A new study by the Austrian Alpine Association on the ecological footprint of ski resorts in the Alps shows the dramatically ecologically unfavourable situation of ski towns. An intervention index (impact points) is based on land consumption, clearing, grading, erosion areas, snowmaking and other characteristics.

The 10 places with the most negative points in Austria are:
1. Sölden, Tyrol (120)
2. Ischgl, Tyrol (105)
3. Obergurgl-Hochgurgl, Tyrol (95)
4. Leogang-Saalbach Hinterglemm, Salzburg (85)
5. Kaprun-Kitzsteinhorn, Salzburg (80)
6. Schmittenhöhe, Salzburg (80)
7. Innerfragant, Carinthia (88)
8. Kleinkirchheim/St. Oswald, Carinthia (84)
9. Wet field, Carinthia (64)
10. Schladming Ski Area, Styria (95)

The size and location of mountain ski towns are closely related to their environmental footprint.

Gurgl in the Ötztal Alps in Tyrol, Austria, surrounded by peaks up to 3500 m high and several glaciers, is located in the Gurgler Valley at an altitude of 1770 to 2154 m a.s.l. (Hochgurgl), Obergurgl is considered the highest village in Austria at 1907 m above sea level. With 564 inhabitants, the village does not seem to be a town in the true sense of the word. Tourism in the twentieth century transformed the former village into a town and fundamentally changed the surrounding cultural and natural landscape of an entire valley. Today, with over 4000 guest beds and around 85,000 guests annually, Gurgl is a white ski town with an average winter population of almost 5000 residents and at least as many day visitors. Obergurgl and Hochgurgl have created around them an extensive ski area with 24 lifts and 110 kilometers of ski slopes, a natural environment, exclusively for winter sports, with considerable natural risks, which the ski town has increased even further. Until the nineteenth century, the frequent eruptions of the Gurgler Eissee Lake caused devastation throughout the Gurgler valley. The town was also frequently affected by avalanches, which led to the destruction of buildings and claimed lives. In January 1951, all buildings of Untergurgl and Angern were destroyed by avalanches, killing seven residents (Alpenverein Österreich 2017; WWF Austria 2018; see ◘ Fig. 3.9).

3

Fig. 3.9   Obergurgl, at 1900 m a.s.l., Austria. (© Breuste 2008)

Potosí: The Silver City Creates a New Mountain Landscape Since the Sixteenth Century

Potosí (population 175,000), founded in 1545, lies between 3976 m and 4070 m above sea level, at the foot of Cerro Rico. Its silver wealth made the city one of the largest in the world in the early seventeenth century, with 150,000 inhabitants, although only about 13,500 people mined silver underground. It still depends on its silver and tin deposits.

The barren, cold and humid environment of the alpine Puna grassland at 4000 m a.s.l. still allows hardly any agriculture. Most of the population lives from trade and transport of food and other goods, such as construction and firewood, black powder, coca, tourism and the transport of silver over long distances (Cipolla 1998; see ▪ Fig. 3.10).

■ **Fig. 3.10**   Mountain town of Potosi, at 4067 m a.s.l., Bolivia. (© Breuste 2015)

3.5 Forest and Wilderness in the City: Places for Religion and Rituals

Religiously revered nature (for example, dwelling places of deities, ancestor worship, place of initiation rites) is found as sacred trees, mountains, stones and forests in various ethnic groups and religions of script less cultures, but not only in these. Mostly these are natural places, formerly somewhat distant from settlements, towards which settlements have already expanded. However, it also happens that these natural sites were deliberately created in settlements, including towns. This is the case, for example, with the shrine forests of Shintoism (Japan) and the initiation bush land of the Xhosa (South Africa) Bauer (2005).

> **Urban Nature as an Object of Religious Worship: Shrine Forests in Japanese Cities**
>
> In the Japanese Shinto religion (Shintoism, Shintō), kami, or deities, inhabit plants, trees, stones, water, and other natural objects. They have their abode in dense forests and there in so-called *himorogi* (thickets). Therefore, there is no forestry intervention in these forests. Certain trees, natural or planted, also function as natural shrines. Shrines are erected to show respect and reverence to these deities. Sacred forests surround these shrines and are strictly protected. Such shrines and their protective forests are also found within Japanese cities. The Sagano Bamboo Forest in Kyoto, like other bamboo forests, has protected Tenryu-ji Temple as a symbol of strength since the fourteenth century.
>
> The Meiji Shrine (Japanese 明治神宮 Meiji-jingū) in Tokyo is dedicated to

90   **Chapter 3 ·** How Urban Nature Exists in the Context of Nature and Culture?

3

the souls of Meiji-tennō and his wife Shōken-kōtaigo. The planted evergreen forest of about 120,000 trees, about 365 different species, was created by donations of trees from all over the country and is now about 100 years old. It covers an area of 70 hectares. Although it was created primarily for religious reasons and continues to be a religious ceremonial site, the forest is now also a popular and highly frequented recreational area.

The number of Shintō shrines in Japan is estimated at more than 150,000, many of them in cities. Since 1897, shrines of particular religious significance have been listed as national treasures, and since 1951 they have been registered in a Cultural Property Protection Act. The list includes 39 entries. Most of these shrines are now located in or on the outskirts of variously sized cities and are surrounded by a more or less large shrine forest. Five shrines are located in Kyoto and six in Nikko, the remainder in 23 in other cities (Creemers 1966; Kato 1988; see ◘ Fig. 3.11).

◘ **Fig. 3.11**   Torii to the forest of Meiji Shrine in Tokyo, Japan. (© Breuste 2010)

Konglin: The Private Forest Cemetery of the Kong Family in Qufu, China

The Kong Family Forest (孔林 "Kong-lin") is the forest cemetery of the Kong family (Confucius and his direct successors) in the city of Qufu in China. It is about 200 hectares in size, walled and has a dense stand of ancient trees. After Confucius' death in 479 BCE, he himself and all subsequent family members were buried here until 1947.

The forest consists of native tree species and is about 2400 years old. The tree population varies between 10,000 and 42,000 trees. Most of the current tree population is probably about 200 years old. The most common tree species are Chinese juniper (*Juniperus chinensis,* cultivar *Sabina chinensis*), gingko tree *(Gingko biloba),* bubble ash *(Koelreuteria paniculata)* and other native species.

Since 1994, the forest cemetery, together with Qufu, has been on the list of UNESCO World Heritage Sites (see ◘ Fig. 3.12).

◘ **Fig. 3.12** Kong family forest in Qufu, China. (© Breuste 2009)

Urban Wilderness for Men Only: Xhosa Urban Initiation Sites in Khayelitsha, Cape Town, South Africa

Khayelitsha is a 38.75 km² district of the megacity of Cape Town in South Africa. With about 400,000 inhabitants, it makes up about 10% of the city's population. It is 97% inhabited by Xhosa, a South African people. For the Xhoas, *Ulwaluko,* the traditional rite of passage for young men to be accepted into the adult community of protection and help through ritual and proving themselves in the wilderness, is a fundamental part of life. The *abakwetha,* the community of those participating in the initiation process (individually *umkwetha*), consists exclusively of mostly young people between the ages of 13 and 16. But the age and duration of initiation are not precisely defined. Circumcision by local healers is part of it. The rituals are also performed in towns where the majority of the Xhosa live, usually for

3

at least three weeks. To participate, a health certificate is now part of the process. Alcohol (in moderation) also plays a role. The rituals, the framework of which is an (over-)life in the wilderness, must thus also be feasible in cities. For this, "initiation wildernesses" are needed in the city as an element of urban nature. Cape Town is a pioneer in the organisation and design of Xhosa initiation in the wildernesses. Each site has a Site Manager who is appointed by a committee of the community responsible for the site, confirmed and employed by the South African Heritage Resource Agency, the Department of Arts and Culture. He ensures access control, registration of *umkwetha* (initiation participants), compliance with the rules for the use of the controlled wilderness and communication with the neighbourhood residents. *Ikhankathas* (guardians and supporters) and the *ingcibi* (the traditional operator/healer) assist him, and for the first eight days of stay, one member of each *umkwetha* family also assists. There are two initiation seasons a year (November to January and April to June) in which, in order to avoid permanent vegetation damage, half of an area is used at a time. On the area, the *umkwetha* build shelter huts *(amabhoma)* from branches and shrubs with the help of their families and the *Ikhankathas*. A fee is charged for the use of drinking water and showers, which are compulsory for hygienic reasons. Toilets are mostly missing. Holes next to the huts serve for this purpose. The trees and bushes of the initiation wilderness" may not be used as firewood or for building huts *("Bring your own branches")*, fire must be controlled, the area cleaned.

The Langa site, initiation in the wilderness is the first of its kind in Cape Town and in the entire Western Cape province. The area has been used informally since the 1930s and was formally established between 2005 and 2008. The area is 8.2 ha in size and is used by approximately 240 *umkwetha* annually. The area is now formally dedicated and managed by the City of Cape Town.

This is a dune site in what is known as the Cape flats, the extensive alluvial sand flatland to the east of the town centre. The site is fenced off. Langa Site is an example, in terms of infrastructure and security, of all the other sites now established and yet to be established. As initiations otherwise take place uncontrolled and under unsafe health conditions all over and around towns in the Western and Eastern Cape and different areas are used for this purpose, there are always subsequent health problems and even deaths of the *umkwethas*. The enclosure of the areas with bushes as a final visual protection is recommended, but mostly missing.

For dune stabilisation, Port Jackson willows *(Acacia saligna)*, evergreen undemanding, drought-tolerant, nitrogen-fixing, fast-growing and fire-resistant woody plants up to 10 m in height, were introduced from south-western Australia as early as 1848. While suitable, they also proved to be a rapidly spreading invasive species that is now being controlled, particularly in the Western Cape. However, the City of Cape Town's Nature Conservation Department has excluded the Langa Site from its control measures (replacement planting of native Natal figs, *Ficus natalensis*). On the initiation sites, the Port Jackson willows provide shade. It can be assumed that they and their young growth are also partly used as firewood (Chen et al. 2010; see ◻ Fig. 3.13).

◘ Fig. 3.13   Langa initiation site in the Khayelitsha district, Cape Town, South Africa. (© Breuste 2005)

3.6 Gardens and Parks: Designed Nature for the Urban Recreational Landscape

▪ **Rural landscape is transforming into a recreational area**

Productive agriculture is what made most cities possible in the first place. Agriculture in and around cities, animal husbandry and crop cultivation can be found in medieval European cities as well as in Mesopotamian, Persian, Indian and Chinese cities, in the cities of the Inca, Maya and Aztecs. Only this "nature" was economically useful and therefore, as long as it remained so, integrated into the urban structure. Agriculture was always only an accompanying economic function, because cities had their economic focus in crafts, trade, later industry and service, not in agriculture.

However, "rural nature" as an economic relict function of rapidly growing European cities or as a conscious economic support measure of urban citizens (*Ackerbürger*) can still be found today in fields, meadows, pastures, orchards, gardens, vegetable plots and grave land of all kinds and dimensions in many cities. The significance of these natural remnants has often changed, fallow land is not uncommon, additional uses such as recreation (for example, litter meadows, orchards) are added (see ◘ Fig. 3.14).

Rural nature becomes part of recreational parks if management does not prevent this and attractive infrastructure allows access (see ◘ Fig. 3.15).

3

Fig. 3.14 Orchard meadow below Strahov Monastery in Prague, Czech Republic, part of the gardens of Petřín Hill. (© Breuste 2017)

Fig. 3.15 Leopoldskron park landscape with farms in Salzburg, Austria. (© Breuste 2013)

■ **Globalized urban nature: From "nature painting" to artificial, extravagant and interchangeable park nature**

With the worldwide spread of European culture in the nineteenth century, European park and garden culture also spread (see ▶ Sect. 3.2). Since by this time the English garden as a landscape garden had already displaced the French park with its geometric forms in new designs almost everywhere on the continent since the second half of the eighteenth century, it was the English park whose formal language spread. It included modelled landscapes with green lawns, groups of bushes, small woods and individual trees, watercourses, ponds, bridges, pavilions, vistas and visual axes. These landscape architectural forms determined the taste in the design of public parks worldwide at the latest from the design of Central Park in New York, where they were recomposed for the first time on a large scale in a city park (see ▶ Sect. 2.3). The public park designed in this way first spread to North and South America and the British colonies. As a result, the idea of the public park also experienced a global expansion and, with these forms, penetrated other cultures in which it had not previously existed, for example, China, India and Arabia. In the process, the individual and the locally symbolic were often lost. Above all, however, the idea of recreating, of painting, nature as it could be found locally outside the cities, idealized in the park as in a painting, was lost. The *"global picturesque"* became a *"'signature' in Western landscape architecture style,"* a globalization of urban nature aesthetics and horticultural landscape architecture (Ignateva 2010). In the nineteenth century, the implementation of plants from all over the world into English gardens achieved an unprecedented richness of color and form of public park nature, changing this goal and foregrounding sensory impressions through the appeal of exoticism. *"Victorian Gardenesque"* became *"Global Gardenesque"*, urban garden nature, third kind nature (see ▶ Sect. 1.3), urban park and urban greening were globalized (Schenker 2007; Roehr 2007; Ignateva 2010).

» "Similar to picturesque, global gardenesque is quite a simplified version of Victorian time which has completely lost its meaning and innovative character. Today, it is just a symbol of 'pretty', 'colourful' and 'beautiful' urban homogeneous 'global' landscape" (Ignateva 2010, p. 123).

This was also associated with the spread of the design element "lawn", which today plays a dominant role in parks and gardens worldwide. It is even used as a vegetation element where its natural prerequisites are not given and it must be kept alive by intensive care and irrigation, for example, in cities in arid zones.

Lawn: The most species-poor and artificial form of vegetation is popular as a design element in urban nature

Lawns are an anthropogenic form of ground-covering vegetation in residential areas. Lawns are sown, installed and maintained as an aesthetic design element in parks and gardens and/or for walking on, in some cases in parks and gardens, but above all in sports facilities (for example, football stadiums, golf courses, tennis courts).

Lawns consist of grasses that penetrate the humic topsoil like felt through roots and runners, are often not very sensitive to treading and are not used for agricultural purposes. They are maintained in species composition and structure by regular mowing of the grasses, as the selection pressure of mowing favours plants that cope well with high radiation intensity. Extreme species poverty is accepted in order to achieve the goal of a uniformly closed green carpet ("English lawn"). Such "cultivated" lawns have less than a dozen grass species per square meter. In the case of a flower-rich lawn, this means up to 60 plant species, which form a valuable habitat for a large number of insects. Lawns are of little importance as habitats, but enjoy great popularity as horticultural design tools, especially when species poverty is high and cover is uniform. Converting them, at least in part, into species-rich meadows is often the goal of design measures (Hard 1985; Ignateva 2018).

3

■ **Fig. 3.16**   Lawn at Creekside Park Dubai, UAE. (© Breuste 2006)

This is the case, for example, in parks in the southwest of the USA, in the United Arab Emirates, Oman or Saudi Arabia. Often, more than 50% of park areas are lawns with shrubs, clumps of bushes and individual trees, or overall open lawns that are often not desired to be walked on, a nature to be looked at, not used (Ignateva 2017).

» "The lawn which plays an essential role in picturesque and gardenesque styles is now one of the most powerful symbols of Western culture" (Ignateva 2010, p. 123).

The picturesque, which was originally intended to emphasize naturalness (Zuylen 1995), has changed to the artificial and extravagant. This also includes the abandonment of native species in favour of colour and form effects. Globalization, which is about the exchange of cultures, goods and ideas, also affects the exchange of nature. Cities with their nature are its exchange centers and best examples (Short and Kim 1999). The reference back to native nature under the name of "biodiversity" as a design element is propagated (Ignateva 2018; see ▶ Sect. 6.1; see ■ Fig. 3.16).

- **The private garden remains popular and is spreading worldwide as an urban natural element**

The controlled and useful nature as ordered, cultivated and fruit-bearing nature has always been ideal, reflected in diverse ideas of paradise. The ideal nature was not the uncontrolled and danger-mounting wilderness, but the garden in the broadest sense (Mitchell 2001). It too was associated early on with ideas of beauty; flowering plants, singing birds, and the like were among them. But first and foremost the garden was wrested from wild nature and was something useful and for that reason alone beautiful. To live in a garden appears to many religions and cultures as an ideal state (see ▶ Sect. 2.6).

Only the wealthier among the city dwellers could allow themselves to become completely independent of the utilitarian function of the garden and make a place in the garden of idealized natural compositions of flora and fauna, often these from far away. Their ideal was a philosophically based harmony with nature, the like of which had never existed in man's struggle for survival in his cultural history.

This is exemplified by the gardens of Roman villas as well as, for example, those of the Tang dynasty in China, which students, poets and officials created for themselves as a place to stay and work in "harmoniously" designed nature.

Very special gardens were created only for teaching and study purposes (for example, botanical gardens), medicinal purposes (for example, herb gardens, monastery gardens) or for courtly hunting amusements (for example, zoological gardens). Their design depended on the purpose. However, they were always designed nature, without the purpose of food production (Mader 2006).

The model of the economic garden was thus contrasted with the model of the harmonious garden, which stood alone for emotional refreshment. Such gardens were long reserved for the wealthy and only became fashionable in Europe for broader middle classes with the Enlightenment and the hope of purification and education through spending time in them.

Transitional or mixed forms occur between all garden forms. In principle, however, these garden forms still exist today and have spread worldwide with Western culture.

Chinese Gardens Are Different: They Take the Natural Landscape as Their Model, Miniaturising and Idealising It

The early gardens of Chinese antiquity were laid out around real mountains and bodies of water, embedding them within themselves. In the Wei and Jin dynasties (220–265 and 265–420, respectively) and the Northern and Southern dynasties (420–581), natural landscapes began to be recreated. By the Ming (1368–1644) and Qing (1644–1912) dynasties, recreating nature became an art and the most important element of Chinese horticulture in private gardens. The gardens select elements from nature to create impressions with their shapes and colors. They miniaturize and compile. Animate and inanimate nature are worked with equally, stone, water and vegetation, according to the naturally occurring vegetation of the region. In this way, the gardens of the North differ from those of the South. But the designed landscape is always an inhabited one, with pavilions, galleries, halls, walls, with the aim of successive impressions during the walk, planned down to the last detail. The art, in which silent contemplation and the appeal to all the senses are central, lies in the execution of detail. The most famous private gardens of the Ming and Qing dynasties were mainly located in the prosperous Jiangnan region, on the lower Yangtze River and in Beijing, which was characterized by a long humanist tradition. Each garden is a synthesis of landscape, flowers, trees and buildings. Of the more than 170 private gardens in Suzhou at the beginning of the twentieth century, 60 are still completely preserved.

The imperial gardens of the Ming and Qing dynasties took historical experience from private gardens and combined it with elements of ancient and contemporary gardens from home and abroad. They had the advantage of having larger areas and design possibilities. They were also a symbol of power and were intended to be majestically imposing. Replicas of existing landscapes (for example, Emperor Qianlong's summer residence in Chengde) can be found in them, as can the materialization of Buddhist religious ideas (for example, the summer palace in Beijing).

The gardens belong to cultivated privacy or imperial display. They are not parks for the inhabitants of the cities (Qingxi 2003).

Yipu is a small private garden from the Ming Dynasty in Suzhou. It was inscribed on the UNESCO World Heritage List in 2000 (see ■ Fig. 3.17).

■ **Fig. 3.17** Yipu Garden in Suzhou, China, UNESCO World Heritage Site. (© Breuste 2016)

References

Al Serkal MM (2013) Ladies day at Dubai parks prove popular. www.gulfnews.com/…/ladies-day-at-dubai-parks-prove-popular-1.1175924. Accessed 16 Jan 2018

Alpenverein Österreich (2017) Blick unter die Schneedecke: Wie der Wintertourismus alpine Landschaften zerstört. Neue Studie mit Ländervergleich des ökologischen Fußabdrucks unserer Skigebiete. https://www.alpenverein.at/…/2017_03_14_der-oekologische-fussabdruck-unserer-skigebiete.php. Accessed 20 Jan 2018

Bauer W (2005) Heilige Haine – Heilige Wälder. Ein kulturgeschichtlicher Reiseführer. Neue Erde, Saarbrücken

Benevolo L (1999) Die Stadt in der europäischen Geschichte. Beck, München

Breuste J (2006) Mitteleuropäische Kulturlandschaft im Spannungsfeld zwischen Bewahren und Gestalten. Sauteria, Schriftenreihe f. System. Botanik. Floristik und Geobotanik 14:9–27

Breuste J (2011) Stadt in der Landschaft? Landschaft in der Stadt? Der suburbane Raum in ökologischer Perspektiv. In: Stiftung Natur und Umwelt Rheinland-Pfalz (ed). Stadtlandschaft – die Kulturlandschaft von Morgen? 9:6–17

Breuste J (2012) Der suburbane Raum in ökologischer Perspektive – Potenziale und Herausforderungen. In: Scheck W, Kühn M, Leibenath M, Tzschaschel S (eds) Suburbane Räume als Kulturlandschaften (= Forschungs- und Sitzungsberichte der ARL Nr 236, Hannover). Akademie für Raumforschung und Landesplanung, Hannover, pp 148–166

Breuste J (2013) Investigations of the urban street tree forest of Mendoza, Argentina. Urban Ecosyst 16(4):801–818

Campos do Jordão (2018) https://de.wikipedia.org/wiki/Campos_do_Jordão. Accessed 11 Jan 2018

Carrieri SA (2004) Diagnóstico y propuesta sobre la problemática del arbolado urbano en Mendoza. Cátedra de Espacios Verdes, Facultad de Ciencias Agrarias, Universidad de Cuyo, Mendoza

Chen Q, Connolly MS, Quiroga LJ, Stewart A (2010) Initiation site development in Khayelitsha, Cape Town: addressing the challenges of urban initiation while preserving tradition and culture. Worcester Polytechnic Institute. wp.wpi.edu/…/Initiation-Site-Development-Proposal-compressed.pdf. Accessed 25 Dec 2017

Cipolla CM (1998) Die Odyssee des spanischen Silbers. Conquistadores, Piraten, Kaufleute. Wagenbach, Berlin

Creemers WHM (1966) Shrine Shinto after World War II. Brill, Leiden

Dubai Parks and Resorts (2018) https://www.dubaiparksandresorts.com/en. Accessed 16 Jan 2018

Dubai Statistics Center (2018) Green Dubai. www.dubai.ae/en/Lists/…/DispForm.aspx?ID=33&category… Accessed 16 Jan 2018

Faggi A, Breuste J (2016) Mendoza metropolitan y sus estrategias de adaptación al cambio climático. In: Nail S (ed) Cambio climatico: Lecciones de y para Ciudades de America Latina. Universidad Externado de Colombia, Bogota, pp 277–294

Hard G (1985) Städtische Rasen, hermeneutisch betrachtet. In: Backé B, Seger M (eds) Festschrift Elisabeth Lichtenberger. Klagenfurter Geographische Schriften 6:29–52

Huntington SP (1996) The Clash of Civilizations and the Remaking of World Order. Simon & Schuster, New York

Ignateva M (2010) Design and future of urban biodiversity. In: Müller N, Werner P, Kelcey J (eds) Urban biodiversity and design – implementing the convention on biological diversity in towns and cities. Wiley-Blackwell, Oxford, pp 118–144

Ignateva M (2017) Manual. Lawn alternatives in Sweden from theory to practice. Swedish University of Agricultural Sciences, Uppsala

Ignateva M (2018) Biodiversity-friendly design in cities and towns. Towards a global biodiversinesque style. In: Ossola A, Niemelä J (eds) Urban biodiversity. From research to practice. Routledge, Milton Park, pp 216–235

Kato G (1988) A historical study of the religious development of Shintō. Greenwood, New York

Kolb A (1962) Die Geographie und die Kulturerdteile. In: Leidlmair A (ed) Hermann von Wissmann-Festschrift. Geographisches Institut der Universität Tübingen, Tübingen, p 46

Mader M (2006) Geschichte der Gartenkunst. Streifzüge durch vier Jahrtausende, Ulmer, Stuttgart

3

McPerson EG, Dougherty E (1989) Selecting trees for shade in the Southwest. J Arbocult 15:35–43

Mitchell JA (2001) The wildest place on earth: Italian gardens and the invention of wilderness. Counterpoint, Washington, DC

Newig J (2014) Was versteht man unter Kulturerdteilen? https://www.kulturerdteile.de. Accessed 11 Jan 2018

Paganotti GM, Palladino C, Coluzzi M (2004) Der Ursprung der Malaria. Spektrum der Wissenschaft 3:82–89

Qingxi L (2003) Chinas Klassische Gärten. Culturw China Series, Beijing

Ringler A (2016/2017) Skigebiete der Alpen: landschaftsökologische Bilanz, Perspektiven für die Renaturierung. Jahrbuch des Vereins zum Schutz der Bergwelt 81–82:29–154

Roehr D (2007) Influence of Western landscape architecture on current design in China. In: Stewart G, Ignatieva M, Bowring J, Egoz S, Melnichuk I (eds) Globalisation of landscape architecture: issues for education and practice. Petersburg's State Polytechnic University Publishing House, St. Petersburg, pp 166–170

Rudolf HU, Oswalt V (eds) (2009) Atlas Weltgeschichte. Klett, Stuttgart

Schenker H (2007) Melodramatic landscapes: nineteenth-century urban parks. In: Stewart G, Ignatieva M, Bowring J, Egoz S, Melnichuk I (eds) Globalisation of landscape architecture: issues for education and practice. Petersburg's State Polytechnic University Publishing House, St. Petersburg, p 36

Short JR, Kim YH (1999) Globalization and the city. Addison Wesley Longman, Edinburgh Gate

Sullivan D, Krishna S (eds) (2005) Malaria: drugs, disease and post-genomic biology. Springer, Berlin

Tourist-Info Ruhpolding (2018) Ruhpolding. https://www.ruhpolding.de/…/winterurlaub-chiemgauer-alpen-bayern-ski-alpin.html. Accessed 20 Jan 2018

Ullman E (1941) A theory of location for cities. Am J Soc 46(6):853–864

Walter H, Breckle S-W (1999) Vegetation und Klimazonen, 7th edn. Ulmer, Stuttgart

Wickop E, Böhm P, Eitner K, Breuste J (1998) Qualitätszielkonzept für Stadtstrukturtypen am Beispiel der Stadt Leipzig: Entwicklung einer Methodik zur Operationalisierung einer nachhaltigen Stadtentwicklung auf der Ebene von Stadtstrukturen (= UFZ/Bericht 14/98). UFZ Leipzig, Leipzig

Wöbse HH (1999) „Kulturlandschaft" und „historische Kulturlandschaft". Informationen zur Raumentwicklung 5(6):269–278

WWF Österreich (2018) Blick unter die Schneedecke. Wie der Wintertourismus alpine Landschaften zerstört. https://www.wwf.at/…/blick-unter-die-schneedecke-wie-der-wintertourismus-alpine-la… Accessed 27 Dec 2018

Zuylen G (1995) The garden. Vision of paradise. Thames & Hudson, London

What Services Urban Nature Provides?

Contents

4.1   What Ecosystem Services
        Do We Expect From Urban
        Nature? – 102

4.2   Which Urban Natures Provide
        What Kind of Ecosystem
        Services? – 117

4.3   How Can Urban Ecosystem
        Services Be Valued? – 124

        References – 127

4

Up to now, urban citizens have felt what urban nature actually provides and have stayed where these demanded services were offered optimally in comparison. However, these services were usually not measurable. The particularly beautiful park, the floodplain with many bird species, the place where children can play safely in the shallow water or the shady avenue were only known and used locally as categories.

How much does which urban nature achieve? This question is often at the centre of planning discussions when it comes to making changes and proving justifications for existing nature. It is not only a question of aesthetic and urban design values, but also of the qualification of concrete services such as recreation, climate regulation, influence on the water balance and biodiversity. Often, cities do not have concrete and up-to-date data on these issues. Much of this data can only be obtained through complex and lengthy scientific studies, user surveys or measurements, for example, on temperature changes in vegetation. It is not uncommon for there to be a lack of time, resources and expertise to make such decisions.

The concept of *ecosystem services*, often based on indicators that allow a rough quantification of individual services (e.g., scaling from 1 = very good to 5 = non-existent services), can provide an overview of urban nature services at different scales (local to regional). It is important that the assessment is location-specific and relates to concrete services. For this purpose, non-monetary and monetary methods will be developed, tested and discussed.

It should also not go unnoticed that, in contrast to "untouched" nature outside urban areas, these services are generally not provided free of charge, because the design of urban nature costs land and management. Use and benefit can be compared with each other.

Scientific research is also increasingly concerned with determining the performance of specific types of urban nature and developing methods for this purpose. This is not only a matter of providing input for planning decisions, but also of urban development as a concept-oriented towards performance-oriented urban nature, and of comparing the performance characteristics of comparable urban nature and optimising them. It is also a question of what services and in what quantity are expected from urban nature at a specific location, in a specific neighbourhood, in a specific city. To meet the demand for services better and better is also the endeavour of urban planning.

4.1 What Ecosystem Services Do We Expect From Urban Nature?

The concept of "natural capital" (Schumacher 1973) was developed in *environmental economics* in the 1970s. It refers to "natural" ecosystems (Daily 1997; Costanza et al. 1997; Haber 2013). It was introduced into the international environmental discussion in the 1990s. It was not until 1999 that it was applied to (designed) urban ecosystems (Bolund and Hunhammar 1999) and popularized by the Millennium Ecosystem Assessment (MA 2005). Today, it is widely used everywhere, including in used ecosystems and in cities (Grunewald and Bastian 2013a, b; Haase et al. 2014; TEEB 2011; Naturkapital Deutschland – TEEB DE 2016).

103

4

4.1 · What Ecosystem Services Do We Expect From Urban Nature?

Urban Ecosystem Services – The Benefits of Urban Nature for City Dwellers

Urban ecosystem services refer to services provided by urban nature and used by people. They are based on ecological functions from which people actively or passively derive a direct benefit for their well-being (De Groot et al. 2002; Fisher et al. 2009). As urban nature is usually nature in the process of being designed and maintained, its services are associated with costs.

» "Final ecosystem services are components of nature, directly enjoyed, consumed, or used to yield human well-being." (Boyd and Banzhaf 2007, p. 619)

The concept of urban ecosystem services aims to analyse, measure (non-monetary and monetary) and evaluate the benefits of urban nature for city dwellers and make them the basis for design action. "Naturkapital Deutschland – TEEB DE" (Naturkapital Deutschland – TEEB DE 2016) is the German follow-up study to the international TEEB study (The Economics of Ecosystems and Biodiversity, TEEB 2011). It highlights the connection between the "services of nature", the creation of value in the economy and human well-being. An economic perspective is intended to make the potential and "services of nature" more tangible and clear. This should make it easier to include nature as a service provider in private and public decision-making processes.

The third report of the project "Ecosystem Services in the City – Protecting Health and Enhancing Quality of Life" addresses the connections between nature's services, human health and well-being. The permanent protection and promotion of natural capital in urban areas should contribute to health and well-being, economic development, social prosperity and the preservation of our natural livelihoods (Naturkapital Deutschland – TEEB DE 2016, p. 7).

The terminology on ecosystem services is not yet consistent (Bastian et al. 2012a, b). De Groot et al. (2002) state "ecological functions" as the basis of ecosystem services.

Bastian et al. (2012a) include ecological functionality (structures, components and processes) in *"ecosystem properties"* and see this as the basis of ecosystem services. Haase et al. (2014) and Naturkapital Deutschland – TEEB DE (2016) summarize basic properties (e.g. habitat supply, carbon and nitrogen cycling, decomposition, primary production) as "ecosystem functions" that characterize ecosystems *(service providing units)* in certain ways.

According to Boyd and Banzhaf (2007), some "ecosystem services" mentioned by Daily (1997) or the Millennium Ecosystem Assessment Report (MA 2005) are not services related to the user, but "ecosystem processes or functions".

Ecosystem services can be captured by indicators. Bastian et al. (2012a) add the category *"potentials"* between the properties and services of ecosystems, which assesses natural assets from the user perspective in order to compare the *service capacity of* ecosystems with actual services, including risks, carrying capacity and stress resistance/resilience.

Natural components of urban ecosystems *(providers* or *generators)* provide the services *(ecosystem benefit) that* are used by people/residents of a city or urban region *(benefiter)* (De Groot et al. 2010; Grunewald and Bastian 2010; Bastian et al. 2013; Bastian 2016; Breuste et al. 2016; O'Brien et al. 2017) (see ◘ Fig. 4.1).

4

The economic perspective should allow urban citizens to better benefit from urban nature:

» "An economic perspective can help to raise awareness of urban nature concerns; it can demonstrate to society what it means to lose urban nature or to preserve it; it can encourage more systematic recording of all the advantages and disadvantages of a decision; and it can provide more space for opportunities for citizen participation in decision-making processes." (Lienhoop and Hansjürgens 2010; Naturkapital Deutschland – TEEB DE 2016, p. 13, translated)

There can be favourable (synergies) or competing (*trade-offs*) interactions between different ecosystem services. When land is built on, it is no longer available for other purposes. Soil compaction and sealing, for example, reduce or prevent the soil's filtering capacity and groundwater recharge. Which services should be preserved where is subject to a consideration of the benefits of the services. Users of the benefits are the urban residents on whom these benefits have a very direct effect (more recreational areas rather than buildings) or an indirect effect (flood protection through infiltration). Urban decision-makers should consider ecosystem services when using urban nature directly or indirectly.

To this end, the TEEB approach includes the following steps:
1. Identify and acknowledge (acceptance and appreciation).
2. Record and evaluate (clarify the value).
3. Taking values into account in decisions (valorisation through the creation of instruments and measures).

Monetary and non-monetary methods are used.

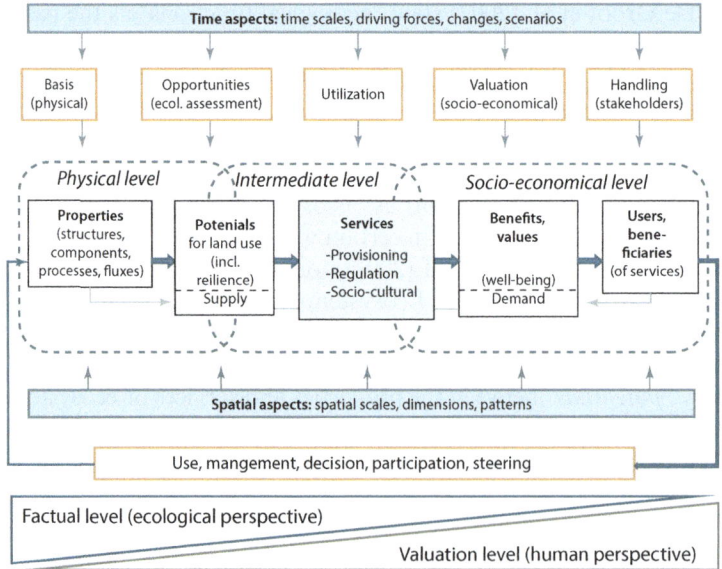

■ **Fig. 4.1** Conceptual framework for the analysis of ecosystem services with special consideration of spatial and temporal aspects. (© Bastian et al. 2013)

105

4

4.1 · What Ecosystem Services Do We Expect From Urban Nature?

Overall, the aim is to analyse and evaluate the type, performance and scope of ecosystem services for the contributions to *human well-being* perceived actively or passively by humans, and to incorporate these into decision-making.

If ecosystem services are about "natural" ecosystems (Daily 1997; Costanza et al. 1997; Haber 2013) in the sense of at least self-regulating natural systems, this cannot apply to cities. Here, city dwellers are not merely recipients of services provided by the city-nature system "for free," so to speak. Urban dwellers can design ecosystems in their cities in ways that benefit them most. Design of urban ecosystems is already taking place everywhere, but usually without sufficient consideration of nature-based services for the well-being of urban dwellers. The services are also not permanent and unchangeable, but can be controlled in type and extent. They are synergistic or competitive with each other. Land take by urban nature is associated with land price foregone for other possible uses. Further costs arise for the care and maintenance of the concrete urban nature in the desired condition. The efficiency of urban ecosystem services can be determined within certain limits by design, maintenance and unconscious actions and can thus be reduced or increased (e.g. management of green spaces, street trees) (Langemeyer et al. 2018).

In the urban context, ecosystem services are thus located at the interface between urban nature and society. Ecological functions of urban nature only become ecosystem services through their benefits for individual people, various social groups or society as a whole. Differences in meaning can occur in this process (Naturkapital Deutschland – TEEB DE 2016). The resulting benefit has a perceived and valued value (e.g. recreation in a city park) or is consumed without any particular awareness of value (clean air). Often, ecosystem services are only recognized as a value and benefit when they have already been reduced or disappeared (e.g. lack of greenery, tree felling in streets, air pollution) (see ■ Fig. 4.2).

The Millennium Ecosystem Assessment (MA 2005), initiated by the United Nations, uses three categories of ecosystem services that have direct benefits for humans: **Provisioning services, Regulating services, and Cultural services**. Supporting or **basic services** include processes such as soil formation, nutrient cycling, and photosynthesis. Habitat functions for animals and plants are also mentioned here (Mace et al. 2012). Naturkapital Deutschland – TEEB DE (2016) states **biodiversity** and basic services of ecosystems as an indispensable basis for provisioning, regulating and cultural services, but does not see a direct link to human well-being. While all other services certainly cannot be provided without basic services, the assessment on biodiversity is not clear. Haber (2013) calls it "misleading to claim that biodiversity is the basis of human life" (p. 32). There are also many findings that prove that even without special biodiversity, ecosystem services can beneficially emerge in cities (e.g. even an intensively maintained tree stand of non-native woody plants of only one species yields a climate-regulating effect at the site). It is undoubted that at least biological components must be present in order to provide ecosystem services, as stated by Naturkapital Deutschland – TEEB DE (2016). However, whether this needs to be described as "single or multiple components of biodiversity" (Naturkapital Deutschland – TEEB DE 2016, p. 26) may be doubted. ► Chapter 6 is devoted in detail to biodiversity (see ■ Figs. 4.2 and 4.3). Basic services, which are merely the basis for ecosystem services, will not be systematically addressed here.

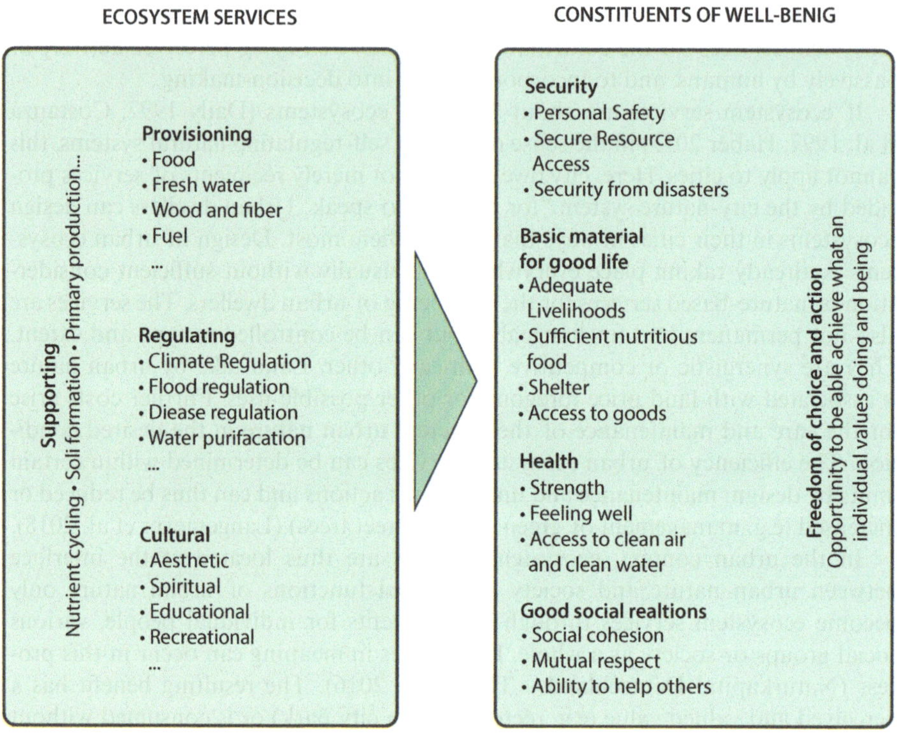

4

ECOSYSTEM SERVICES

CONSTITUENTS OF WELL-BENIG

Supporting · Primary production...

Nutrient cycling · Soli formation · Primary production...

Provisioning
- Food
- Fresh water
- Wood and fiber
- Fuel
- ...

Regulating
- Climate Regulation
- Flood regulation
- Diease regulation
- Water purifacation
- ...

Cultural
- Aesthetic
- Spiritual
- Educational
- Recreational
- ...

Security
- Personal Safety
- Secure Resource Access
- Security from disasters

Basic material for good life
- Adequate Livelihoods
- Sufficient nutritious food
- Shelter
- Access to goods

Health
- Strength
- Feeling well
- Access to clean air and clean water

Good social realtions
- Social cohesion
- Mutual respect
- Ability to help others

Freedom of choice and action
Opportunity to be able achieve what an individual values doing and being

■ **Fig. 4.2**   Millennium Ecosystem Assessment approach to ecosystem services and their importance for human well-being. (MA 2005; TEEB DE 2016; BfN 2012)

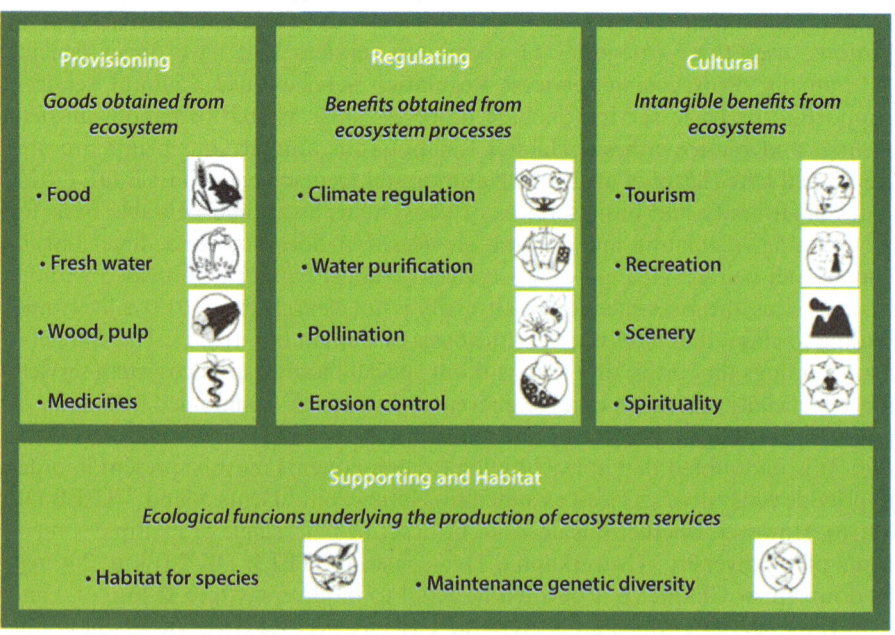

Provisioning

Goods obtained from ecosystem

- Food
- Fresh water
- Wood, pulp
- Medicines

Regulating

Benefits obtained from ecosystem processes

- Climate regulation
- Water purification
- Pollination
- Erosion control

Cultural

Intangible benefits from ecosystems

- Tourism
- Recreation
- Scenery
- Spirituality

Supporting and Habitat

Ecological funcions underlying the production of ecosystem services

- Habitat for species
- Maintenance genetic diversity

■ **Fig. 4.3**   Urban ecosystem services TEEB (2011). (Source: Haase (2016))

107 **4**

4.1 · What Ecosystem Services Do We Expect From Urban Nature?

The categories of provisioning, regulating and cultural services can be further differentiated and subdivided (Naturkapital Deutschland – TEEB DE 2016; Haase 2016). Only services that are provided to or used by city dwellers are considered. Ecosystem services that are provided in the urban area but are not used there (e.g. food production for other regions, carbon sequestration to improve the global climate balance, etc.) should be considered separately.

The city not only consumes services from ecosystems on its own territory, but also benefits from services provided by ecosystems in the surrounding area (e.g. regional climatic air exchange, drinking water supply) or allows other regions to benefit from ecosystem services produced on its territory (e.g. out-of-town visitors to natural areas in the city, food production for a regional or supraregional market).

Provisioning services: This group of services includes the supply of food (e.g. from croplands, greenhouses and urban gardens), raw materials (e.g. timber from urban forests, building materials from quarries and drinking water from local sources).

Regulating services: These services, which have a more indirect effect on humans, relate to the reduction of thermal and air pollution, the reduction of water pollution, the reduction of flood hazards through water retention potential, erosion protection, the filtering effect of soils on the quality of groundwater and carbon sequestration to reduce the greenhouse effect (Elmqvist et al. 2013; Fisher et al. 2009; Grunewald and Bastian 2015).

Cultural services: Cultural services are provided in particular by green spaces with physical and mental recreational functions. They also include gaining emotional experience of nature, knowledge of nature, a sense of home and spiritual or aesthetic significance.

The undesirable effects of urban nature on individuals, groups or society are referred to as "disservices" (impairments) (Lyytimäki and Sipilä 2009; von Döhren and Haase 2015). These include damage to built structures due to plant growth, hazards due to obstructed views or breakage of trees in the street space, and health impairments due to plants and animals (allergies, disease transmission). "Green gentrification" (Wolch et al. 2014) refers to undesirable displacement processes of the residential population by improving green amenities, increasing the attractiveness of the residential location, the residential value and the rent and property prices.

Negative effects caused by natural processes, in part also from outside into the cities (e.g. floods, landslides, mudflows, etc.) are risks emanating from nature, which are always better calculated and managed, but can never be completely ruled out. Even if the causes lie in natural processes from outside, these processes are regulated in the cities, again with the inclusion of services provided by nature (e.g. retention areas, compensation areas) (see ► Sect. 8.8) (see ◘ Fig. 4.3).

The study "Naturkapitel Deutschland – TEB D" (Naturkapital Deutschland – TEEB DE 2016) lists the services of urban nature that are considered important in even more detail under the categories (not those of urban ecosystem services):
1. Urban nature promotes good living conditions.
2. Urban nature promotes health.
3. Urban nature promotes social cohesion.
4. Stadtnatur makes it possible to experience nature and environmental education in the city.
5. Urban supplies.
6. Urban nature as a location factor.

"Used and demanded by humans" and "based on ecological functions" (see definition above) should be the criteria for a concept of ecosystem services. Other services, such as effects as a location factor or effects on the real estate market, are not included here. This does not mean that they should not be considered, but just outside an "ecosystem concept". Health services, which can probably be described as one of the most important services, cannot easily be placed in a "regulatory services" category either. Thermal and air-hygienic services certainly belong there, but perhaps also the effects on mental health. To assign health to cultural services alone would not do justice to the complexity of the subject. This still opens up a wide field of investigation.

Undoubtedly, **social activities** take place **in urban nature** (and elsewhere) that promote social cohesion. This includes communication between social groups, responsible cooperation, neighbourhood building, integration of migrants and social fringe groups, youth work, and much more. However, this does not primarily take place in urban nature, often not even primarily in open space, and above all it is not based on "ecological functions". Urban nature is often only a backdrop for these activities. Sometimes urban nature also becomes a direct object to promote social cohesion, for example, in community interaction directly related to nature (e.g. community gardening). Overall, however, it is the public space that every city needs for precisely this social cohesion in various forms, sizes and distributions. It is a reflection of society. Wolf D. Prix (born 1942), co-founder of the architects' cooperative Coop Himmelb(l)au, assesses this task of public urban space as follows: "Public space is the freedom to do what one wants" (ZDF 2018), and urban nature undoubtedly offers a multitude of suggestions and opportunities for this.

The Millennium Ecosystem Assessment (MA 2005) assigns ecosystem services to three performance classes:

- **Provisioning services:**
  - Provision of food.
  - Provision of raw materials.
  - Provision of drinking water.

109   **4**

4.1 · What Ecosystem Services Do We Expect From Urban Nature?

■ **Regulating services:**

Regulatory services include participation in achieving a healthy living environment. The general protection of resources is not in the foreground here.

— Reduction of the air temperature.
— Reduction of air pollution.
— Reduction of noise pollution.
— Reduction of soil and groundwater pollution.
— Reducing the contribution to climate change.

■ **Cultural services:**

— Physical and mental recuperation.
— Emotional nature experience.
— Acquisition of natural knowledge.
— Spiritual or aesthetic appreciation.

Gómez-Baggethun and Barton (2013) distinguish, without reference to the above three classes of performance, the following services of urban ecosystems that they have studied in more detail, without ranking them in importance:

— Food production.
— Regulation of the water cycle and runoff reduction.
— Temperature regulation.
— Noise reduction.
— Air purification.
— Mitigation of environmental extremes.
— Waste treatment.
— Climate regulation.
— Pollination and seed dispersal.
— Animal observation.
— Disservices.

Only those services provided by urban nature that are consciously demanded or unconsciously benefiting the people living in the city should be recorded as ecosystem services. The fact that services can be provided outside of cities by urban nature and that services provided by nature outside of cities can be used in cities is not considered here (exchange of nature-based services). A supra-regional "performance account" in this respect would, of course, always show cities with enormous performance import surpluses, if only through the performance "provision of food".

The "export services" of carbon sequestration as a contribution to supra-regional, national or global climate protection is also opposed to a disproportionately larger "export" of pollutants that drive climate change. Even though "export services" are now quite analyzable and calculable, and these negative exports from cities need to be reduced, they are not direct services to the people in the city. The CO_2-neutral city is a supra-regional and global task. However, ecosystem services are about direct services to people in the city. When it comes to reducing cities' contribution to climate change, the focus should not be on *end-of-pipe* carbon sequestration, with its minimal capacity compared to pollutant emissions, but on reducing pollutant emissions at the source of emissions. This is about many orders of magnitude more in terms of performance, but then it is also no longer about urban nature as a performer.

Overall, urban ecosystems should not be considered primarily in terms of their performance in reducing the negative effects of human activity, or in terms of their capacity to absorb even more pollutants (*end-of-pipe*). The notion of ecosystems as "absorbers" and "degraders by natural processes" of these waste products should be used only cautiously and always with a view to local material flows. The idea of using water bodies as cesspools has been abandoned for more than a hundred years. Nevertheless, the "cleaning effect" without ecosystem damage (!) can certainly be considered in a limited, primarily local environment, but just not as a regional urban strategy.

Wittig et al. (1995) have already formulated five principles for dealing with urban nature, which serve to preserve its performance characteristics:

1. Reduction in the use of open space for settlement purposes (land efficiency).
2. Promotion of local and regional material flows (transport efficiency).
3. Economical use of non-renewable raw materials and energy sources (resource efficiency).
4. Reduction of the release of pollutants/emissions into the environment (risk reduction, environmental protection – not pollutant binding!).
5. Protection of all life environments (air, soil, surface waters and groundwater) and of structured and differentiated nature (biodiversity).

The vision of the vibrant, safe, sustainable and healthy city has become the universally aspired goal (Gehl 2015). It allows only one view of urban ecosystem services: that from the perspective of humans as beneficiaries. The result is to identify where exactly, with which ecosystem services, to whom (individuals, social groups or society), which benefits arise from urban nature. What are the conscious *demands* and what is the *supply of* nature?

The attractiveness of urban nature is based on a broad spectrum of urban nature types, usability and a great diversity of structures.

A review of international publications on cultural ecosystem services *of* urban and suburban urban nature identified seven urban nature *types (types of green infrastructure)* in 132 studies (2003–2017). Almost half of the studies refer to forest areas, followed by other green spaces and parks (O'Brien et al. 2017). Ecosystem services of water areas (6 out of 132 studies) and urban protected areas (5 out of 132 studies) are investigated only to a small extent (O'Brien et al. 2017) (see ◘ Fig. 4.4).

111

4

4.1 · What Ecosystem Services Do We Expect From Urban Nature?

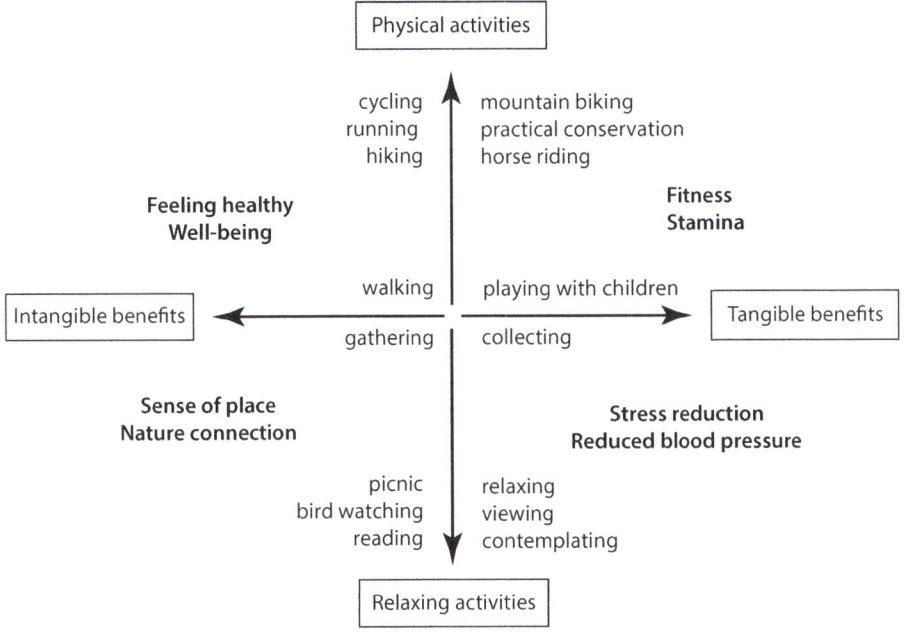

Physical activities

cycling · mountain biking
running · practical conservation
hiking · horse riding

**Feeling healthy
Well-being**

**Fitness
Stamina**

walking · playing with children

Intangible benefits ← → Tangible benefits

gathering · collecting

**Sense of place
Nature connection**

**Stress reduction
Reduced blood pressure**

picnic · relaxing
bird watching · viewing
reading · contemplating

Relaxing activities

Fig. 4.4 Range of practices and benefits of urban and suburban green infrastructure, based on 132 case studies. (O'Brien et al. 2017, p. 243)

Use of the Services Offered by Urban Green Spaces

Supply and demand determine the value of a product in a world of commodities. Urban nature is such a product and can also be judged by this scale.

For this purpose, Hegetschweiler et al. (2017) examined 40 European studies published in the "Web of Science" between October 2014 and May 2015 on the topic of **urban green spaces and user behaviour**. The majority of the studies dealt with situations in Northern and Central Europe and were dedicated to urban forests and forest parks. Two-thirds of the studies used typical socio-demographic data, surveys, some combined with visual analysis, to determine demand, perceptions, access and preferences. Spiritual benefits, educational or research benefits, or the contribution of green space to cultural heritage or local identity were not mentioned at all. Most studies, in order to assess the provision of ecosystem services, use the information on the size and shape of green spaces, amenities (e.g. for sports, children's play or recreation), access to water and vegetation structures. Accessibility of green spaces is a crucial factor for the usability of the offer. One-third of the studies deal with biodiversity aspects, related to the level of maintenance of vegetation.

Aesthetic aspects, structural diversity, accessibility and usability are generally dominant factors in the use of the range of cultural ecosystem services provided by urban green spaces. Different user groups (e.g. due to age or lifestyle) also have differentiated usage requirements (*demands*). Impairments (e.g. noise, pollution, overuse) are hardly inhibiting factors for the use of areas with few alternatives. Health and relaxation are the most important motivations of the users. It can be seen that the quality of the services offered by green spaces (size, location, equipment, inter-

nal diversity) clearly determine their use, number of users, frequency of use and satisfaction with the offer, and that less specific (e.g. sport) than general benefits of use ("general recreation") are claimed. This also reveals further, hitherto unused performance potentials of green spaces.

As the visual quality and diversity of structures have a significant influence on the perception, preference and use of green spaces, this is now being further investigated, for example, through the use of photographs or images to be interpreted (e.g. Sugimoto 2011, 2013; Richards and Friess 2015). The aim is to find out what exactly constitutes urban nature's attractiveness and to better match the supply of green spaces to demand (see ◘ Fig. 4.5).

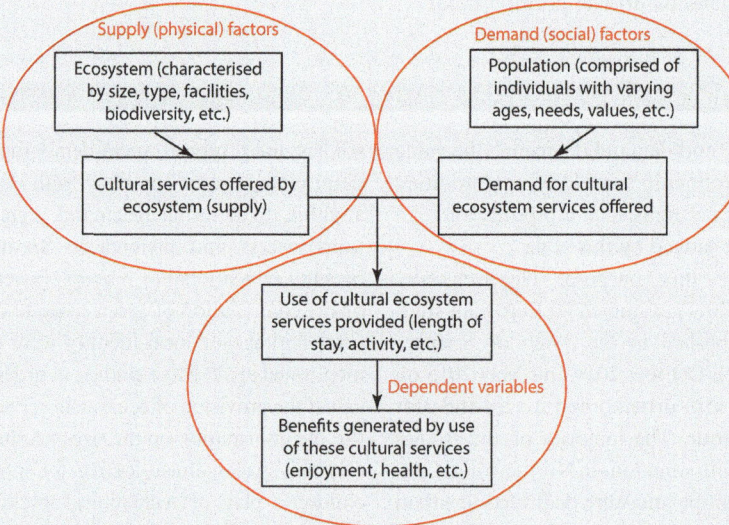

◘ **Fig. 4.5** Supply and demand determine the use of cultural ecosystem services and their benefits. (Hegetschweiler et al. 2017, p. 49)

113   **4**

4.1 · What Ecosystem Services Do We Expect From Urban Nature?

Urban Green Spaces in Focus – Studies Confirm They Are Important for Health and Well-Being

An analysis of the publication on human-environment interactions of **urban green spaces** in the ISI Web of Science© and Scopus© (Kabisch et al. 2015) for 2000–2013 shows that the number of studies published annually increased tenfold during this period. Most of the 219 studies evaluated are from Europe and the USA, with increasing numbers from China. There are very few studies from Africa, Latin America, Russia and South and Southeast Asia.

Kabisch et al. (2015) explain the increase in studies by the fact that urban nature and green spaces in the city are receiving more and more attention in terms of their importance for health and well-being, including in research. Mainly, surveys and qualitative analyses (key interviews) were conducted. The largest number of studies dealt with perceptions of urban green spaces by different user groups. Another group of studies was dedicated to direct or indirect health effects for residents and visitors. Other papers examine environmental justice, access and supply of urban green space for different groups of residents. Other publications examine the development of urban green spaces, their planning and management (see ◘ Fig. 4.6 and ◘ Table 4.1).

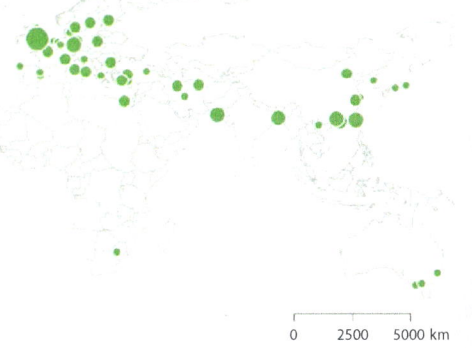

◘ **Fig. 4.6** Regional distribution of the 219 case studies on human-environment interactions in urban green spaces. (Kabisch et al. 2015, p. 28, design: J. Breuste, drawing: W. Gruber)

4

■ **Table 4.1**   Urban ecosystem services and quality of life indicators in the dimensions of sustainability. (Breuste et al. 2013a; Haase 2016 after MA 2005 and Santos and Martins 2007)

Sustainability dimension	Urban ecosystem services	Components of urban quality of life concept
Ecology	Air filtration Climate regulation Noise reduction Rainwater drainage Water supply Wastewater treatment Food production	Health (clean air, protection against respiratory diseases, death from heat and cold) Security Drinking water Food
Social	Landscape Recreation Cultural values Environmental education	The beauty of the surroundings Recreation, stress relief Intellectual enrichment Communication Residential
Economics	Food production Tourism Recreational function	Income protection Investments

Results of Interdisciplinary Research on Urban Biodiversity and Ecosystem Services

Kremer et al. (2016) summarize seven points as a result of a broad European project on Urban Biodiversity and Ecosystem Services (BiodivERsA URBES, ▶ http://urbesproject.org/):

1. **Land use and land cover data** are widely used as readily available **indicators** of urban ecosystem services. However, they are of limited value for comparative studies and cannot replace empirical field studies.
2. Understanding **supply-demand relationships** for urban ecosystem ser-

vices requires cross-disciplinary and cross-scale considerations.

3. Urban ecosystem services **are** mediated **by non-ecological elements, such** as physical built infrastructure, technology, social practices, and the cultural context in which people experience human-environment relationships.
4. Urban nature enables **contact with nature** in cities. This is what enables people's values and relationships with nature to develop in the first place.

4.1 · What Ecosystem Services Do We Expect From Urban Nature?

115

4

5. The **relationships between urban biodiversity and urban ecosystem services** are unclear, poorly documented and require further empirical findings.

6. Incorporating the **concept of urban ecosystem services into practice** requires overcoming outdated disciplinary barriers, bridging existing gaps between science, policy and governance, and linking the concept to planning frameworks and their tools.

7. **Comparisons between cities** are fundamental for understanding the drivers of structure, function and processes of urban ecosystems and for differentiating between dynamics that are locally particular and those that are widespread in an urban context.

(Kremer et al. 2016).

Urban Ecosystem Services – What Is Internationally Being Studied and Where?

While global ecosystem services are already dealt with in detail in many studies, such studies that specifically assess urban ecosystem services are still largely lacking. Haase et al. (2014) carried out an extensive evaluation of the 393 publications on this topic available internationally in the ISI Web of Science. In total, 217 publications in a variety of journals from different disciplines dealt with original analyses and/or assessments of the supply and demand of urban ecosystem services and their application in land use management. Most of the research was conducted in Europe, the USA and China. Since 1973, the number of papers published annually has increased sharply, especially since 2000, with **50% of studies addressing regulating services**, 20% supporting or basic services not considered at all in Naturkapital Deutschland – TEEB DE (2016), 15% cultural and 11% provisioning services. ▢ Figure 4.7 shows which ecosystem services are most studied in cities. While the large number of studies on climate and air quality regulation is understandable, the number of studies on carbon storage, a service on which the well-being of city dwellers does not directly depend, is surprising. More than 60% of studies are devoted to only one particular service. The interaction of benefits (synergies and competitions) is hardly studied, but is of great importance for practical benefit development. Most studies deal with the scale of the city as a whole or even the city region. The local scale of performers is much less studied. Only 5 out of 217 studies deal with this scale (*neighbourhood and site-level*), although this is exactly the perception scale of city residents (Breuste et al. 2013b). Only two studies (Imhoff et al. 2004; Haase 2009) address long-term analyses. More than three-quarters of the studies (78.1%) do not include specific interest and user groups (*stakeholders*). About half of the studies (48.9%) do not address the development of tools for ecosystem services valuation. Deficiencies in research appear to be:

- Process understanding, especially of spatio-temporal scales.
- Link with economic aspects and quality of life.
- Application of tools for participatory and multi-criteria evaluation.
- Involvement of citizens and stakeholders with different points of view.

4

- Under-representation of large parts of the world (especially Africa, parts of Asia and Latin America) in internationally accessible studies.

However, it should be taken into account that there are a large number of studies locally and regionally that are written in national languages and do not necessarily use the keyword "Urban Ecosystem Service", but certainly produce results in this regard and are not covered by the ISI Web of Science. These valuable results do not "make it" into the international literature and are not noticed outside their home countries (see ◘ Figs. 4.7 and 4.8).

Number of studies:
- 5
- 10
- 20

0   2500   5000 km

◘ **Fig. 4.7** Geographical distribution of the 217 publications on urban ecosystem services examined. (Data from Haase et al. 2014, p. 417, design: J. Breuste, drawing: W. Gruber)

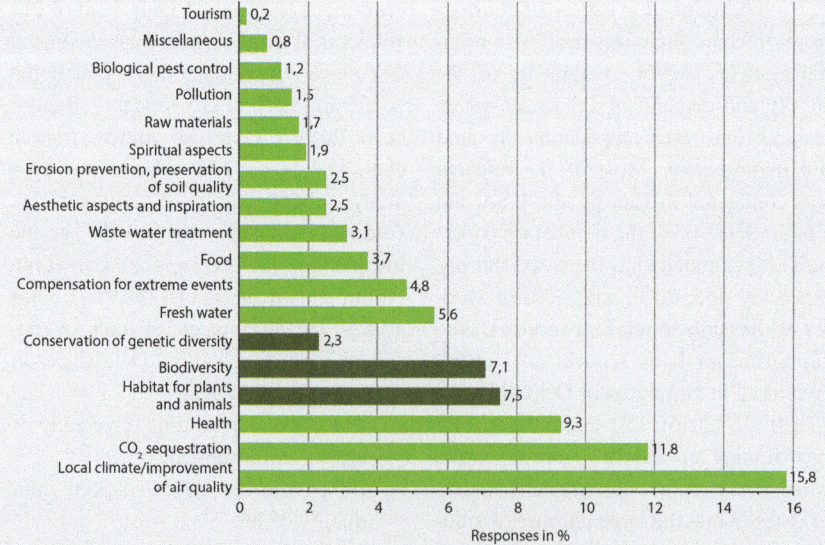

◘ **Fig. 4.8** Ecosystem services covered in the 217 studies on urban ecosystem services examined. (Haase et al. 2014, p. 418, design: J. Breuste, drawing: W. Gruber)

117

4

4.2 · Which Urban Natures Provide What Kind of Ecosystem Services?

4.2 Which Urban Natures Provide What Kind of Ecosystem Services?

Boyd and Banzhaf (2007) argue for units of account for ecosystem services so that ecosystem services as public goods can be linked to *units of account* and comparability becomes possible.

Already at the beginning of the debate on urban ecosystem services was the identification of service provider ecosystems to which services can be assigned and which can serve as reference units. The example of Bolund and Hunhammar (1999) for Stockholm initially only showed which urban nature provides services at all (see ◘ Table 4.2).

In the last 20 years, analyses of the relationships between performer ecosystems and benefits have been carried out in more detail, especially in relation to actual performer ecosystems. The analysis by Niemelä et al. (2010) in Helsinki is an example of this (see ◘ Table 4.3).

As service providers of urban ecosystem services (*service providing units*, Haase et al. 2014, p. 418) are named:

− Individual plant species (e.g. trees, individual animal and plant groups).
− Vegetation structures are composed of plant communities (e.g. forest, parks, water bodies).
− General (not subdivided) vegetation and biodiversity.

Overall, this means that an approach to structuring urban nature-based on vegetation and water body structures is practically applicable for balancing ecosystem services. This approach is therefore also pursued here. It goes beyond the classifica-

◘ **Table 4.2** Urban nature as a provider of urban ecosystem services (service provider ecosystems) using the example of Stockholm

	Street tree	Lawns/ parks	Urban forest	Farm- land	Wet- land	River/ creek	Lake/ sea
Air purification	x	x	x	x	x		
Microclimate regulation	x	x	x	x	x	x	x
Noise reduction	x	x	x	x	x		
Rainwater drainage		x	x	x	x		
Wastewater treatment					x		

After Bolund and Hunhammar (1999), modified

◘ Table 4.3   Relationship between urban ecosystem services and their performers

Group	Ecosystem services	Power generating unit
Utilities	Wood production	Different tree species
	Food: game, berries, mushrooms	Different species in terrestrial, freshwater and marine ecosystems
	Freshwater, soil	Groundwater discharge, suspension and storage
Regulation services	Microclimate regulation on street and city scale, change in heating costs	Vegetation
	Gas cycle, O_2 production, CO_2 storage	Vegetation, especially forests, trees
	Habitat supply	Biodiversity
	Air purification	Vegetation cover, soil microorganisms
	Noise attenuation in built-up areas and along transport routes	Protected green areas, dense/natural forests, soil surfaces
	Rainwater storage, attenuation of heavy rainfall peaks	Vegetation cover, (sealed surfaces), soils
	Rainwater infiltration	Wetlands (vegetation, microorganisms)
	Pollination, maintenance of plant communities, food production	Insects, birds, mammals
	Humus production and maintenance of nutrient content	Waste, invertebrates, microorganisms
Cultural services	Recreation	Biodiversity, especially in parks, forests and aquatic ecosystems
	Psycho-physical and social health benefits	Forest nature
	Knowledge creation, research and education	Biodiversity

After Niemelä et al. (2010, pp. 3229–3230), modified

tion into four types of nature and differentiates them further. The following urban nature types are presented in more detail in terms of their service potential and its acceptance and use by city dwellers (the proportion (of 100%) of studies on this according to Haase et al. (2014) is shown in brackets):

1. Urban parks (4.4%).
2. Urban forests (18.8%).
3. Gardens (1.2%).
4. Wildernesses (3.3%).
5. Urban waters (8.5%).

119

4

4.2 · Which Urban Natures Provide What Kind of Ecosystem Services?

◘ Tables 4.2 and 4.3 also reflect the selection of urban ecosystem services considered essential. These will be followed up in the following chapters in relation to the spatial structure of the service providers for ecosystem services in the city.

Internationally, a globally representative analysis by Haase et al. (2014) (see excursus in ▶ Sect. 4.1) shows that studies on urban ecosystem services focus most on forests (18.8% of studies), land use patterns (15.6% of studies) and green infrastructure as a whole (11.7% of studies). This does not reflect the real and differentiated importance of these performers. The category "land use patterns" also shows that data available here only on a city-wide or regional scale, which do not always meet the data requirements in terms of typification and spatial resolution, are often used for interpretations on ecosystem services. The scale level of individual areas that actually provide measurable services is often poorly studied. It appears too costly for a smaller spatial scale as in data collection. Water bodies, gardens and urban wildernesses, although significant performers are significantly underrepresented in international studies. However, this may be different in national publications, for example, in Central Europe discussing urban wildlands (e.g. Breuste and Astner 2018) (see ◘ Fig. 4.9).

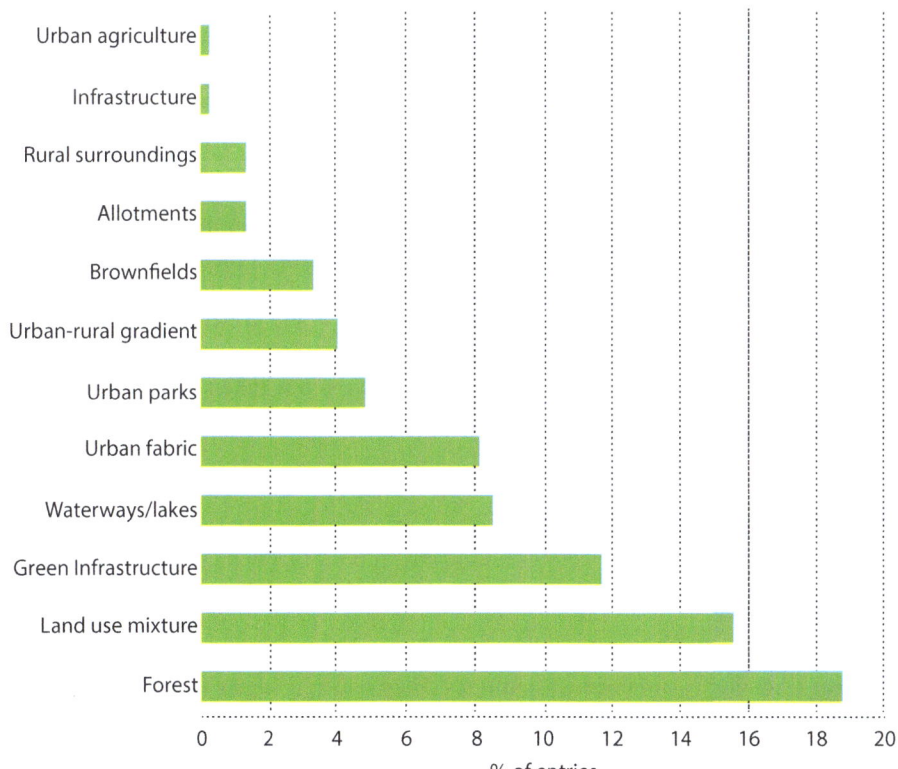

◘ **Fig. 4.9** *Service providing units* of urban ecosystem services considered in 217 studies on urban ecosystem services. (Haase et al. 2014, p. 418, design: J. Breuste, drawing: W. Gruber)

From the above-mentioned broad spectrum of urban ecosystem services, it is always necessary to select those that are particularly significant for the areas under consideration in order to use them for practical planning decision-making.

Stiftung DIE GRÜNE STADT, which is particularly concerned with the practical application of existing knowledge in planning in Germany, focuses on four main areas of application:

1. Urban climate and air pollution control,
2. Species diversity,
3. Health,
4. Costs, and
5. Urban Planning (THE GREEN CITY Foundation 2018).

At the heart of all decisions is the *benefit to human well-being* (De Groot et al. 2002), **healthy living in the city**, and the contribution of urban nature to it. This includes:

1. Promotion of healthy climatic and air-hygienic conditions,
2. Reduction of noise pollution,
3. Physical and mental recovery, healthy living environment (physical and mental health),
4. Local healthy food production, and
5. Promotion of emotional experience of nature and cognitive knowledge of nature.

Urban nature can be divided into four broad categories (see ▶ Sect. 1.3) (Kowarik 1992) (see ◘ Table 4.4).

The classification in ◘ Table 4.4 and ◘ Fig. 4.10 allows an overview of the spatial structure of urban nature. At the scale of a city's ecosystems, a substantial part of urban nature can be captured in "urban structure types" in which nature dominates. This allows good compatibility with the land use categories commonly used in urban planning and accounting of services provided by these units as service providers. However, the urban nature that is present in urban structure types of built-up areas, for example, trees, gardens, etc., is not taken into account (see ▶ Box 4.1 and ◘ Fig. 4.11).

Urban Structure Types

Urban structure types are urban spaces with comparable, typical, clearly physiognomically distinguishable features and configurations of buildings, open spaces and vegetation. They are largely homogeneous with regard to their type, density and area proportions and the various characteristics of the open spaces (sealed areas, vegetation types and woodland). Urban structure types thus summarise areas of similar environmental conditions. They exhibit comparable habitat functions and ecosystem services in their real representatives and are interfaces between science and urban planning (Breuste and Endlicher 2017).

121   **4**

4.2 · Which Urban Natures Provide What Kind of Ecosystem Services?

■ **Table 4.4** Ecosystem services of the four urban natures approach. (Breuste et al. 2013a using Kowarik 1992; Bolund and Hunhammar 1999, modified) (see ■ Fig. 4.10)

Nature category	Vegetation structure type	Provided ecosystem service	Potential ecosystem service
(a) Remnants of original nature	Forest Wetland	Wood production, recreation, microclimate regulation Wastewater treatment, stormwater runoff reduction, habitat for plants and animals.	Experience of nature, education
(b) Nature of the agricultural landscape	Meadows, pastures, grassland, arable land	Food production, microclimate regulation, reduction of rainwater runoff	Recreation, habitat for plants and animals, nature experience
(c) Urban gardens and parks	Flower beds, lawns, bushes, hedges, tree groups, etc. Roadside greenery, residential greenery Gardens, allotments, parks, urban trees	Cultural services, aesthetics Air purification, microclimate regulation, stormwater runoff reduction Recreation, microclimate regulation, air purification	Habitat for plants and animals, stormwater runoff reduction Recreation, habitat for plants and animals Habitat for plants and animals Nature experience, education.
(d) Novel urban ecosystems	Spontaneous vegetation Urban wilderness	Habitat for plants and animals, microclimate regulation	Habitat for plants and animals, nature experience, education, recreation

Urban nature also appears in a large-scale view as individual plants, individual animals, etc. Of course, this view is also useful, for example, when it comes to the concrete local protection of urban nature (species and biotope protection). In the following treatment of the provision of services, however, only the level of urban structure types dominated by urban nature is used, except for individual trees. Individual urban nature types are dealt with in ► Chap. 5.

The **promotion of social cohesion** and of **economic success** through urban nature was not included separately in the consideration. However, social aspects are not only not excluded, but are a core component of life in the city. In this respect, social references can be found for all the above-mentioned services. **Urban nature as a stimulus for the economy (locational advantage)** is not dealt with specifically here. This would require a separate, certainly extensive publication.

The concept of the three categories of urban ecosystem services was deliberately used only to a limited extent in the rest of the presentation, but without completely disregarding it.

4

		Biomass summer	Biomass winter	Total biomass	Temperature regulation capacity	Air pollution reduction capacity	Noise pollution reduction capacity	Scenery
LAWN	 Frequently mowed Evergreen Intensive maintenance	1	1	**1**	1	1	0	0
ORNAMEN- TAL PLANTS	Intensive soil management, Open vegetation cover, Fully open in winter	2	0	**1**	2	2	1	1
MEADOWS	 2-3 times mowing per year Evergreen	3	1	**2**	3	3	1	0
BUSHES	Intensive green in vegetation periode Low maintenance degree	4	2	**3**	5	6	4	6
DENSE BUSHES & HEDGES	 Dense vegetation structure Low maintenance degree	6	2	**4**	6	6	5	4
PARK VEGETATION	 Soil covered by vegettion Bushes & trees/single & patches High degree of maintenance	13	3	**8**	12	10	8	6
CONIFEROUS FOREST	 Evergreen, soil low covered by vegetation Less bush layer	12	8	**10**	10	8	16	16
DECIDOUS FOREST	 Tree vegetation layer Ubderstorey layers depending from its density	18	6	**12**	16	14	14	14
MULTI- LAYERED GREEN	 Trees with two sublayers High degree of biomass in summer	20	12	**16**	18	20	20	20

■ **Fig. 4.10** Ecosystem services of vegetation structures. (Modified using Grzimek 1965; cited in Greiner and Gelbrich 1975, Fig. 18, p. 26/27, 20-part scale between 0 = no performance and 20 = very high performance)

Disservices, synergy and trade-off effects as well as **risks** due to urban nature are also included in the presentation below. On **biodiversity** or the "habitat service" (see ▶ Chap. 6). The general reduction of soil and groundwater pollution and the reduction of **the contribution of cities to climate change (carbon storage)** were excluded from consideration as supra-regional aspects. The production of **drinking water and raw materials** in urban areas was also not considered. Although this is still a common practice in many cities around the world, it is gradually being

123 **4**

4.2 · Which Urban Natures Provide What Kind of Ecosystem Services?

Box 4.1 Urban Structure Types (Land Use and Building Structure Types or "Ecological Spatial Units"), Examples (see e.g. Sukopp and Wittig 1993)

Types of built-up areas

- Town centres
- Older block development (until 1918)
- Block and row development of the interwar period (1918–1945)
- More recent post-war row development (1946–1965)
- New commercial and industrial settlements
- Road traffic areas
- Villa development with park gardens

Types of open spaces/green spaces

- Allotment gardens
- Cemeteries
- Parks
- Fallow land
- Urban forests
- Water bodies
- Sports facilities

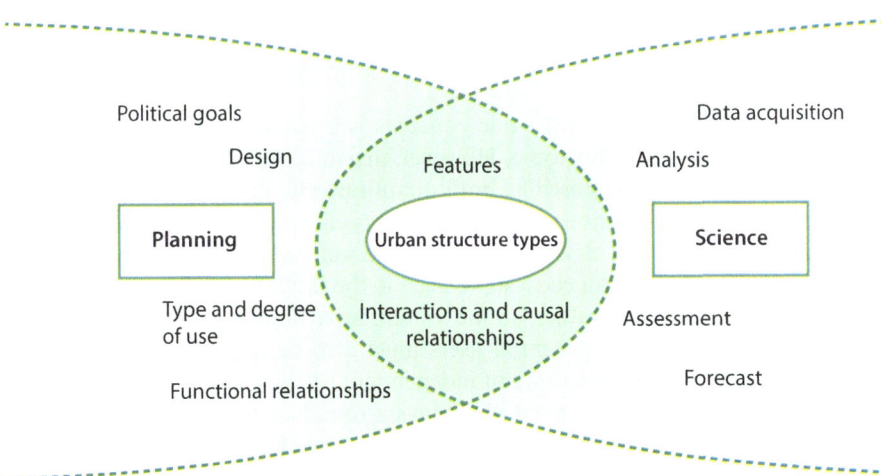

▣ **Fig. 4.11** Urban structure types at the interface between science and planning. (Breuste and Endlicher 2017, using Wickop et al. 1998)

is concerned, in order to secure healthy drinking water supplies and to dedicate land differently. Cities have long since been unable to meet the growing demand for drinking water and raw materials from their own resources on their own territory. A return to this is nowhere strategy. The **spiritual or aesthetic valuations of**

urban nature, although given, are to be assessed as less good in comparison with all other services, and spiritual and aesthetic valuations of nature take place to a much broader extent outside cities. Nevertheless, above all, the theme: **beautiful (green) city** is evident everywhere, because the Green City is supposed to be *useful and beautiful*. However, the theme of **The Beautiful City** would clearly go beyond the scope of this review and also far beyond the contribution that urban nature can make to it (see ▶ Chaps. 7 and 8).

4.3   How Can Urban Ecosystem Services Be Valued?

Overall, it is always about the "value" of ecosystem services of urban nature and behind that the value of the urban nature providing these services, for example, the "value" of a park. In a broader sense, "value" can denote validity, significance or importance for an individual or a community. In a narrower sense, it is an expression of the equivalent of a commodity (expressed in monetary terms).

The valuation of urban ecosystem services a number of promising approaches, methods and tools. These can be roughly divided into two categories:

- Non-monetary approaches (meaning and importance, often difficult to quantify).
- Monetary approaches (nature "worth" expressed in monetary units).

Subdivide.

Gómez-Baggethun and Barton (2013) distinguish economic values, social, cultural and insurance values.

The coexistence of different value concepts is a reality and a value pluralism exists, as is the case with this topic. However, urban ecosystem services should be recorded as completely as possible, both quantitatively and in terms of their significance. Their differentiation is also important (value pluralism), because a "summary value" of different services of one urban ecosystem cannot be compared with those of another urban ecosystem, even if the latter has the same "summary value". The receptors, the valuer, an individual, a particular social group, the real users or the totality of all potential users must also be taken into account. The value of urban nature is therefore not independent of the valuer!

The purpose of the assessment also plays a role, Naturkapital Deutschland – TEEB DE (2016, p. 30) gives the following reasons for this:

- Raising awareness of the importance of nature (attention mechanism).
- Accounting for environmental services, for example, for national accounting purposes (accounting mechanism).
- Communication with stakeholders and/or the public (feedback mechanism).
- Support for priority setting in political decisions (decision-making mechanism).
- Information for the selection and design of instruments, for example, the design of compensation payments or the involvement of stakeholders through the

application of certain assessment procedures (information mechanism) (Naturkapital Deutschland – TEEB DE 2016, see also Lienhoop and Hansjürgens 2010; Gómez-Baggethun et al. 2015).

The spatial scale of the assessment should also be taken into account, for example, building, street, neighbourhood, district, city, region (Gómez-Baggethun and Barton 2013).

The methodological approaches developed so far for the valuation of ecosystem services, especially in cities, are constantly being supplemented and certainly overlap in terms of content.

The majority of the evaluation frameworks developed is based on **indicators** that are more or less already available as data in administrations and statistics or that have to be determined, sometimes with great effort. The latter makes the application of evaluation procedures difficult, as resources, personnel and time are often not available. For example, Andersson-Sköld et al. (2018) develop a five-step valuation framework, including compiling a set of indicators, applying effectiveness factors (to assess the effectiveness of the indicators), estimating impacts, estimating benefits for each ecosystem service, and estimating the total ecosystem service value. They apply this assessment framework to green spaces in Gothenburg, Sweden, along an urban–rural gradient to assess ecosystem services of trees, shrubs, herbs, birds, and bees.

Naturkapital Deutschland – TEEB DE (2016) follows different methodological approaches to capture and assess urban ecosystem services. These are:

- Importance of urban nature from its effects on people's individual health and quality of life (health-related cost variables).
- Participative or deliberative procedures (procedures of participation or "negotiation").
- Quantitative bio-physical and socio-ecological indicators ("ecological assessment", supply-based approach).

The recording and valuation of ecosystem services based on individual preferences includes the valuation of health costs and quality of life as well as the valuation of urban nature in municipal budgeting.

Social-ecological approaches to the assessment and valuation of ecosystem services are currently predominantly based on regulating ecosystem services. Other commonly used approaches to social-ecological valuation focus particularly on the relationship between land use and land use management and the provision of ecosystem services. Also assessed are the bio-physical endowments of a space, especially green spaces, which are compared to the recreational services perceived by users. A disadvantage of such perception-based studies is their high cost and time intensity in preparation and the difficulty of including this in measurement- or model-based analyses of the supply side (Haase et al. 2014) (see ◘ Figs. 4.12 and 4.13).

4

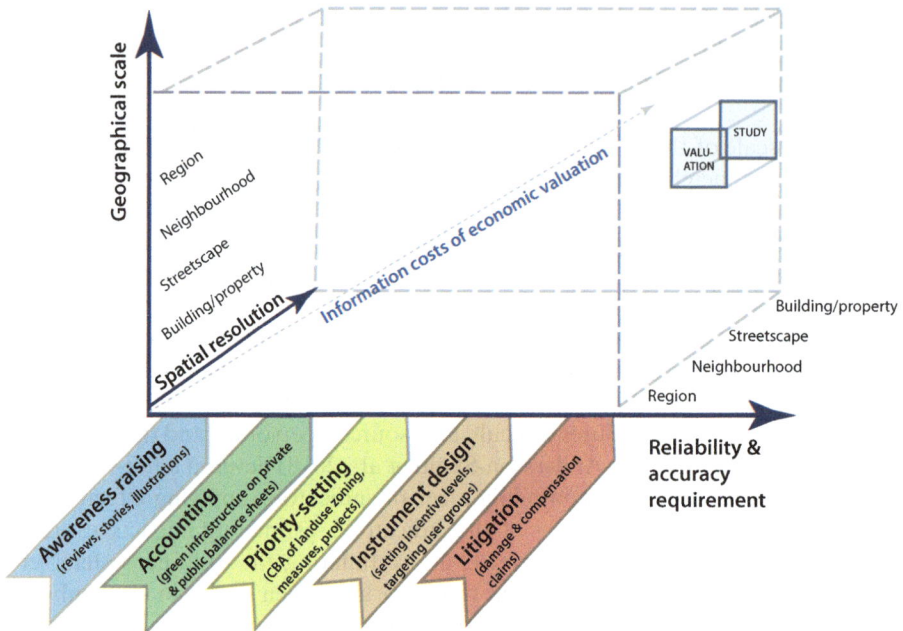

■ **Fig. 4.12** Economic valuation of ecosystem services in different planning contexts. (Gómez-Baggethun and Barton 2013, p. 241)

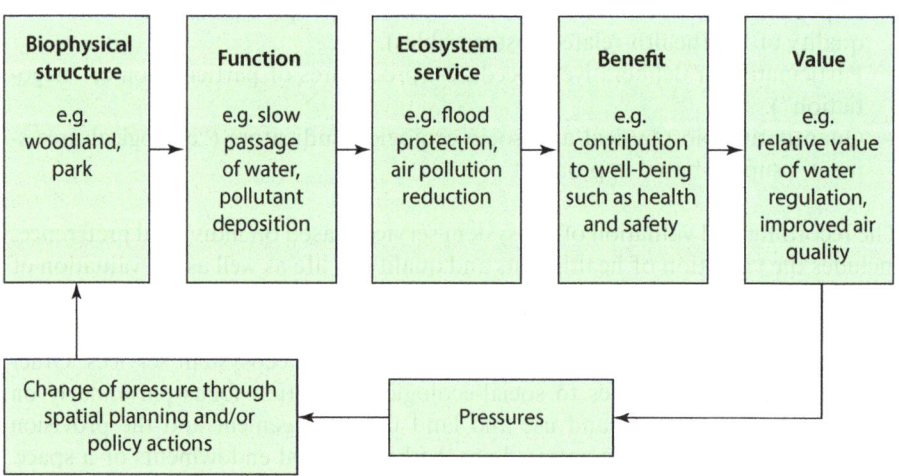

■ **Fig. 4.13** Cascade model of ecosystem valuation for ecosystem services of urban green spaces. (Andersson-Sköld et al. 2018, p. 275)

References

Andersson-Sköld Y, Klingberg J, Gunnarsson B, Cullinane K, Gustafsson I, Hedblom M, Knez I, Lindberg F, Ode Sang A, Pleijel H, Thorsson P, Thorsson S (2018) A framework for assessing urban greenery's effects and valuing its ecosystem services. J Environ Manag 205:274–285

Bastian O (2016) Vom Wert der Natur – das Konzept der Ökosystemdienstleistungen. In: Landesverein Sächsischer Heimatschutz e.V. (eds) Bewahrung der Biologischen Vielfalt – beispiele aus Sachsen. Landesverein Sächsischer Heimatschutz e.V., Dresden, pp 71–79

Bastian O, Haase D, Grunewald K (2012a) Ecosystem properties, potentials and services – the EPPS conceptual framework and an urban application example. Ecol Indic 21:7–16. https://doi.org/10.1016/j.ecolind.2011.03.014

Bastian O, Grunewald K, Syrbe R-U (2012b) Space and time aspects of ecosystem services, using the example of the EU water framework directive. Int J Biodivers Sci Ecosyst Serv Manag 8(1–2):1–12. https://doi.org/10.1080/21513732.2011.631941

Bastian O, Syrbe R-U, Rosenberg M, Rahe D, Grunewald K (2013) The five pillar EPPS framework for quantifying, mapping and managing ecosystem services. Ecosyst Serv 4:15–24

BfN – Bundesamt für Naturschutz (ed) (2012) Daten zur Natur 2012. Bundesamt für Naturschutz, Bonn

Bolund P, Hunhammar S (1999) Ecosystem services in urban areas. Ecol Econ 29:293–301. https://doi.org/10.1016/S0921-8009(99)00013-0

Boyd J, Banzhaf S (2007) What are ecosystem services? The need for standardized environmental accounting units. Ecol Econ 63:616–626

Breuste J, Astner A (2018) Which kind of nature is liked in urban context? A case study of solarCity Linz, Austria. Mitt Österr Geogr Gesell 158:105–129

Breuste J, Endlicher W (2017) Stadtökologie. In: Gebhardt H, Glaser R, Radtke U, Reuber P (eds) Geographie – physische Geographie und Humangeographie, 3rd edn. Elsevier, München, pp 628–638

Breuste J, Haase D, Elmqvist T (2013a) Urban landscapes and ecosystem services. In: Wratten S, Sandhu H, Cullen R, Costanza R (eds) Ecosystem services in agricultural and urban landscapes. Wiley, Chichester, pp 83–104

Breuste J, Qureshi S, Li J (2013b) Scaling down the ecosystem services at local level for urban parks of three megacities. Hercynia N. F. Halle/Saale 46:1–20

Breuste J, Pauleit S, Haase D, Sauerwein M (eds) (2016) Stadtökosysteme. Springer, Berlin

Costanza R, D'Arge R, de Groot R, Farber S, Grasso M, Hannon B, Limburg KE, Naeem S, O'Neill RV, Paruelo J, Raskin RG, Sutton P, van den Belt M (1997) The value of the world's ecosystem services and natural capital. Nature 25:253–260

Daily GC (ed) (1997) Nature's services: societal dependence on natural ecosystems. Island, Washington DC

De Groot RS, Wilson MA, Boumans RMJ (2002) A typology for the classification, description and valuation of ecosystem functions, goods and services. Special issue: the dynamics and value of ecosystem services: integrating economic and ecological perspectives. Ecol Econ 41:393–408

De Groot RS, Fisher B, Christie M (2010) Integrating the ecological and economic dimensions in biodiversity and ecosystem service valuation. In: Kumar P (ed) TEEB – the economics of ecosystems and biodiversity. Ecological and economic foundations. Earthscan, London, pp 9–40

Elmqvist T, Fragkias M, Goodness J, Güneralp B, Marcotullio PJ, McDonald RI, Parnell S, Schewenius M, Sendstad M, Seto KC, Wilkonson C (eds) (2013) Global urbanisation, biodiversity and ecosystem services: challenges and opportunities. A global assessment. Springer, Dordrecht

Fisher B, Turner RK, Morling P (2009) Defining and classifying ecosystem services for decision making. Ecol Econ 68:643–653

Gehl J (2015) Städte für Menschen. Jovis, Berlin

Gómez-Baggethun E, Barton DN (2013) Classifying and valuing ecosystem services for urban planning. Ecol Econ 86:235–245

Gómez-Baggethun E, Martín-López B, Barton D, Braat L, Saarikoski H, Kelemen E, García-Llorente M, van den Bergh J, Arias P, Berry P, Potschin M, Keune H, Dunford R, Schröter-Schlaack C, Harrison P (2015) State-of-the-art report on integrated valuation of ecosystem services. EU FP7 OpenNESS Project Deliverable 4.1, European Commission FP7. http://www.openness-project.eu/sites/default/files/Deliverable%204%201_Integrated-Valuation-Of-Ecosystem-Services.pdf. Accessed 10 Sept 2015

Greiner J, Gelbrich H (1975) Grünflächen der Stadt. Grundlagen für die Planung. Grundsätze, Kennwerte, Probleme, Beispiele. VEB Verlag für Bauwesen, Berlin

Grunewald K, Bastian O (2010) Ökosystemdienstleistungen analysieren – begrifflicher und konzeptioneller Rahmen aus landschaftsökologischer Sicht. GEOÖKO 32:50–82

Grunewald K, Bastian O (eds) (2013a) Ökosystemdienstleistungen. Springer Spektrum, Heidelberg

Grunewald K, Bastian O (2013b) Bewertung von Ökosystemdienstleistungen (ÖSD) im Erzgebirge. Abh Sächs Akad Wiss Leipzig, Mat-Nat Klasse 65(5):7–20

Grunewald K, Bastian O (eds) (2015) Ecosystem services – concept, methods and case studies. Springer, Heidelberg

Grzimek G (1965) Grünplanung Darmstadt. Magistrat, Darmstadt

Haase D (2009) Effects of urbanisation on the water balance: a longterm trajectory. Environ Impact Assess Rev 29:211–219

Haase D (2016) Was leisten Stadtökosysteme für die Menschen in der Stadt? In: Breuste J, Pauleit S, Haase D, Sauerwein M (eds) Stadtökosysteme. Springer, Berlin, pp 129–163

Haase D, Larondelle N, Andersson E, Artmann M, Borgström S, Breuste J, Gomez-Baggethun E, Gren A, Hamstead Z, Hansen R, Kabisch N, Kremer P, Langemeyer J, Lorance RE, McPhearson T, Pauleit S, Qureshi S, Schwarz N, Voigt A, Wurster D, Elmqvist T (2014) A quantitative review of urban ecosystem services assessment: concepts, models and implementation. Ambio 43(4):413–433

Haber W (2013) Arche Noah heute. Sächsische Landesstiftung Natur und Umwelt, Dresden

Hegetschweiler KT, de Vries S, Arnberger A, Bell S, Brennan M, Siter N, Stahl Olafsson A, Voigt A, Hunziker M (2017) Linking demand and supply factors in identifying cultural ecosystem services of urban green infrastructures: a review of European studies. Urban For Urban Green 21:48–59

Imhoff ML, Bounoua L, DeFries R, Lawrence WT, Stutzer D, Tucker CJ, Ricketts T (2004) The consequences of urban land transformation on net primary productivity in the United States. Remote Sens Environ 89:434–443

Kabisch N, Qureshi S, Haase D (2015) Human – environment interactions in urban green spaces – a systematic review of contemporary issues and prospects for future research. Environ Impact Assess Rev 50:25–34

Kowarik I (1992) Das Besondere der städtischen Flora und Vegetation. Natur Stadt – der Beitrag der Landespflege zur Stadtentwicklung. Schriftenreihe des Deutschen Rates für Landespflege 61: 33–47

Kremer P, Hamstead Z, Haase D, McPhearson T, Frantzeskaki N, Andersson E, Kabisch N, Larondelle N, Lorance Rall E, Voigt A, Baró F, Bertram C, Gómez-Baggethun E, Hansen R, Kaczorowska A, Kain J-H, Kronenberg J, Langemeyer J, Pauleit S, Rehdanz K, Schewenius M, van Ham C, Wurster D, Elmqvist T (2016) Key insights for the future of urban ecosystem services research. Ecol Soc 21(2):Art. 29. https://doi.org/10.5751/ES-08445-210229. Accessed 31 Jan 2018

Langemeyer J, Palomo I, Baraibar S, Gómez-Baggethun E (2018) Participatory multi-criteria decision aid: operationalizing an integrated assessment of ecosystem services. Ecosyst Serv 30:49–60

Lienhoop N, Hansjürgens B (2010) Vom Nutzen der ökonomischen Bewertung in der Umweltpolitik. GAIA 19(4):255–259

Lyytimäki J, Sipilä M (2009) Hopping on one leg – the challenge of ecosystem disservices for urban green management. Urban For Urban Green 8:309–315

Mace GM, Norris K, Fitter AH (2012) Biodiversity and ecosystem services: a multilayered relationship. Trends Ecol Evol 27:19–26

Millennium Ecosystem Assessment (MA) (2005) Ecosystems and human well-being: synthesis. World Resources Institute, Washington DC

Naturkapital Deutschland – TEEB DE (2016) Ökosystemleistungen in der Stadt – gesundheit schützen und Lebensqualität erhöhen. In: Kowarik I, Bartz R, Brenck M (eds) Technische Universität Berlin, Helmholtz-Zentrum für Umweltforschung – UFZ. Naturkapital Deutschland – TEEB DE, Berlin

Niemelä J, Saarela SR, Södermann T, Kopperoinen L, Yli-Pelikonen V, Kotze DJ (2010) Using the ecosystem service approach for better planning and conservation of urban green spaces: a Finland case study. Biodivers Conserv 19:3225–3243

O'Brien L, De Vreese R, Kern M, Sievänen T, Stojanova B, Atmis E (2017) Cultural ecosystem benefits of urban and peri-urban greeninfrastructure across different European countries. Urban For Urban Green 24:236–248

Richards DR, Friess DA (2015) A rapid indicator of cultural ecosystem service usage at a fine spatialscale: content analysis of social media photographs. Ecol Indic 53:187–195

Santos LD, Martins I (2007) Monitoring urban quality of life – the porto experience. Soc Indic Res 80:411–425

Schumacher EF (1973) Small is beautiful. A study of economics as if people mattered. Blond & Briggs, London

Stiftung DIE GRÜNE STADT (2018) Urbanes Grün. Für ein besseres Leben in Städten. www.die-gruene-stadt.de. Accessed 3 Feb 2018

Sugimoto K (2011) Analysis of scenic perception and its spatial tendency: using digital cameras, GPS loggers, and GIS. Procedia Soc Behav Sci 21:43–52

Sugimoto K (2013) Quantitative measurement of visitors' reactions to the settings in urban parks: spatial and temporal analysis of photographs. Landsc Urban Plan 110:59–63

Sukopp H, Wittig R (eds) (1993) Stadtökologie. Fischer, Stuttgart

TEEB (2011) TEEB manual for cities: ecosystem services in urban management. http://www.naturkapitalteeb.de/aktuelles.html. Accessed 26 Aug 2014

Von Döhren P, Haase D (2015) Ecosystem disservices research: a review of the state of the art with a focus on cities. Ecol Indic 52:490–497

Wickop E, Böhm P, Eitner K, Breuste J (1998) Qualitätszielkonzept für Stadtstrukturtypen am Beispiel der Stadt Leipzig: entwicklung einer Methodik zur Operationalisierung einer nachhaltigen Stadtentwicklung auf der Ebene von Stadtstrukturen. UFZ, Leipzig

Wittig R, Breuste J, Finke L, Kleyer M, Rebele F, Reidl K, Schulte W, Werner P (1995) Wie soll die aus ökologischer Sicht ideale Stadt aussehen? – Forderungen der Ökologie an die Stadt der Zukunft. Z Ökol Naturschutz 4:157–161

Wolch JR, Byrne J, Newell JP (2014) Urban green space, public health, and environmental justice: the challenge of making cities "just green enough". Landsc Urban Plan 125:234–244

ZDF (eds) (2018) Doku | Terra X – Faszination Erde: Die Weltenveränderer. 43 min. Dauer, Sendedatum: 04.02.2018. https://www.zdf.de/sendung-verpasst. Accessed 5 Feb 2018

What Urban Nature Provides Which Services?

Contents

5.1   The Urban Park – 132

5.1.1   Natural Element Urban Park – 132

5.1.2   Services of Urban Parks – 137

5.2   The Urban Tree Population (Urban Forest) – 147

5.2.1   Nature Elements, Urban Trees and Urban Woodland – 147

5.2.2   Services Provided by Urban Trees – 152

5.3   The Urban Gardens – 174

5.3.1   Natural Element Urban Garden – 174

5.3.2   Urban Gardens as Services Providers – 188

5.4   Urban Waters – 200

5.4.1   Natural Element Urban Waters – 200

5.4.2   Services of Urban Waters – 204

5.4.3   Restoration of Ecosystem Services of Urban Waters – 208

References – 217

© Springer-Verlag GmbH Germany, part of Springer Nature 2022
J. Breuste, *The Green City*,
https://doi.org/10.1007/978-3-662-63976-4_5

Natural elements are found in cities that also occur outside them. What makes their urbanity special, however, is the density of their distribution, the intensity of their use and the variability of their often small-scale structure. The aim in the city is to preserve, re-establish, maintain and develop these natural elements in such a way that they can meet the needs of urban dwellers as far as possible. To this end, the services of urban nature of various kinds should be wisely considered so that they also correspond well to the changing needs of urban dwellers. Many demands on urban nature contradict each other or are only expressed by certain groups of the population. Some cannot be realised at the same time at all (see ▶ Chap. 7). A formative moderation process is therefore required between the services of urban nature and the needs of the population perceiving these services. First of all, there should be clarity about which services can actually be provided by which urban nature and how they are already being used. That is the subject of this chapter. For a better overview, four frequently occurring, typical urban nature types were considered in a selected manner. They are the main service providers, are also widespread nationally and in part worldwide, and are used most intensively: urban parks, urban forests, urban gardens and urban waters.

5.1 The Urban Park

5.1.1 Natural Element Urban Park

Urban parks are among the most frequently and intensively studied types of urban nature worldwide. This is certainly because urban parks are often elements of the urban structure and part of the urban way of life across cultures, often represent the only urban nature that can be used in the city with little public regulation, and are of great importance in the public perception. Nevertheless, they constitute the urban nature that has been late to enter cities for widespread public use (see ▶ Sect. 2.6). Parks that are exclusively for private use on private land are not considered here. In most cities, they are only accessible for use by very small minorities.

> ┌─ **Park** ─────────────
>
> Synonyms: Stadtpark, Parkanlagen (Germany), *metropolitan park, municipal park* (North America), *public park, public open space, municipal gardens* (UK).
>
> Urban parks are horticulturally designed, public urban green space of variable design for recreation of the urban population and beautification of the cityscape. Design elements are vegetation structures (for example, woody plants, open lawns, flowerbeds), infrastructure (for example, network of paths, rest areas), sometimes water areas, in larger parks also buildings, sports areas, children's play areas, cultural or retail facilities, sometimes additionally equipment as toilets, etc. (Schwarz 2005; Konijnendijk et al. 2013).

The attractiveness of urban parks and their high intensity of use is based on the diversity of their features (natural elements and infrastructure) and the related wide range of use options, which appeals to different user interests. Often, city parks are the only alternative for large parts of the population in densely populated inner cities to enjoy nature and escape from everyday stress for a short time.

Urban parks, if they are large, attractive in terms of facilities and located in cities frequented by tourists, are also tourist destinations. At the very least, city tourists visit them as welcome "islands of calm" in the "urban jungle" (for example, Chaudhry and Tewari 2010). Central Park in New York is even listed among the most visited tourist destinations worldwide at number 4 with 40 million visitors annually, Meiji Jingu Shrine in Tokyo (with forest park) at number 7 with 30 million visitors, and Golden Gate Park in San Francisco at number 20 with 13 million visitors (Travel + Leisure 2014).

While parks were often located on the outskirts of cities at the time of their creation and in the nineteenth century (for example, Hyde Park, London, Central Park New York, English Garden Munich), they were soon located in the centre of cities due to rapid urban growth in the nineteenth and twentieth centuries. In the twentieth century, new parks were often created on the now new outskirts of cities as spacious landscape parks, primarily for weekend recreation. The transition to the "open countryside" is often fluid here. The transition between park, woodland park and urban forest is also a fluid one. Whereas in the nineteenth century the park was still a composed, compressed landscape, today it is a section of the landscape, especially on the outskirts of the city.

Public parks are public goods to which all people should have equal access. This equality does not exist in most cities. Parks are not distributed in cities in such a way that equal numbers of people have access to them at equal distances. On the contrary, parks are often located in neighborhoods with more upscale housing standards and incomes of the residential population in the catchment area (Breuste and Rahimi 2015). Causes are the historical development of parks, the general receptiveness of city administrations to the provision of the public good "park", available and affordable land, the morphology of the city and interest policies for certain groups of residents.

Even in social systems that had elevated the equal distribution of public goods to an ideology, there was no equal distribution according to location and access because of the above-mentioned reasons, even if this was striven for as a "standard" of the building industry, for example, in the GDR (TGL 1964). Even according to the criterion "park space per inhabitant", the public good "park" could never be distributed equally (Greiner and Karn 1960; Greiner and Gelbrich 1975, p. 114). Regardless of location, size and access, in order to achieve comparability between cities and to develop targets for meeting demand, planning often works with a "supply value" of square metres of park per inhabitant of a city, and these figures are discussed as a quality criterion for cities. Even though Munich in 2001 had 33.8 m^2/inhabitants. This does not say anything about the actual supply of urban green space, because this depends on the location, size and distribution of the green spaces (Die Welt 2001).

Standards for parks in design, management, and distribution continue to be debated. A somewhat broader consensus is that public parks can be hierarchically classified according to the criteria of "area size", "catchment area" (supply for residents of a surrounding area that can be delimited according to distance from the park) and "area supply in the park per person in the catchment area" (potential users). The exact values to be assessed differ according to culture, tradition and local availability. General cross-cultural demand standards are worthy of discussion and probably cannot be implemented everywhere. Comber et al. 2008 propose Accessible Natural Green Space Standards (ANGSt) for England that may apply to European cities:

- Everyone should be able to reach an urban public park (or other urban natural space) of at least 2 ha in size within 300 m of residence.
- A park of at least 20 ha should be reachable within 2 km from the place of residence.
- A 100 ha green space (park) should be reachable within 5 km from the place of residence.
- A 500 ha green space (park) should be reachable within 10 km from the place of residence.

Are green spaces interchangeable in terms of quality? A park could also be replaced by another green space, for example, an urban forest or a publicly accessible section of landscape on the outskirts of the city, if these standards are geared towards the provision of public green space. Whether these are interchangeable green categories remains an open question, since their range of uses, internal structure, management, etc. are by no means the same.

In its report published in 1995, the European Environment Agency (EEA) already reported, as the result of a study of a large number of European cities, a 15-minute distance to urban green space at which the residential population can be regarded as "provided "(Stanners and Bourdeau 1995).

Grunewald et al. 2016, 2017 calculate the accessibility of public green spaces for the residential population in Germany. In doing so, they use the digital, object-structured vector dataset of the basic landscape model (ATKIS-Basis-DLM), but include as use categories relevant for "recreation" not only parks and green spaces, but also sports, leisure and recreation areas, cemeteries, agricultural land, grassland, orchard meadows, forests, copses, flowing waters and standing waters, regardless of use and development quality. It is assumed that the different nature categories can replace each other equally This cannot be proven by studies at present. In this context, the urban nature category "park" is an important, but by far not the only urban nature category that can be considered for recreation (BBSR 2018). For European cities, Kabisch et al. 2016 calculate accessibility values for urban green spaces, also based on similar use (nature) categories (see ▶ Sect. 8.3).

The location, size and facilities of parks result in an allocation of parks to their "service area" (catchment area), the area from which the majority of users are expected to come. The results of many studies (for example, Belgium/van Herzele and Wiedemann 2003; Sheffield/Barbosa et al. 2007; UK/Comber et al. 2008; London/ Kessel et al. 2009; Denmark/Toftager et al. 2011; Sheikhupura, Pakistan/ Javed et al. 2013; Adana, Turkey/Unal et al. 2016) show that the majority of the urban population in many European cities can reach a green space within a

900–1000 m radius. In cities in other parts of the world, this may even be true for only a minority (for example, Pakistan, Javed et al. 2013; Breuste and Rahimi 2015). The coverage of large parks from a reasonable catchment area decreases with park size, despite increases in the catchment area (for example, Breuste and Rahimi 2015). The importance of local supply with public parks is repeatedly shown by examples in large cities, especially in developing and emerging countries, where entire neighbourhoods are hardly or not at all supplied with green spaces and parks. This is shown, for example, by studies for Istanbul, Turkey (in some cases less than 0.5 m² per inhabitant), for Karachi, Pakistan and Tabriz, Iran (Qureshi et al. 2010; Breuste and Rahimi 2015) (see ◻ Figs. 5.1, 5.2, 5.3, and 5.4).

Internationally, a hierarchical classification of parks is common. The characteristic values used, however, differ considerably from region to region and from country to country, as do the designations. ◻ Table 5.1 attempts to make the designations comparable by standardizing them, but does not use the sometimes divergent designations.

Parks Make Salzburg, Austria, Beautiful – Mönchsberg Hill and Mirabell Gardens

The Mönchsberg hill (150 m above the town) is one of Salzburg's urban hills. With its elongated ridge it shapes the cityscape. As a broken-up forest park with small-scale alternation of forest and meadows and its many vantage points, it is particularly suitable for a **varied walk** (see ◻ Fig. 5.1). It is the most popular urban recreation area and at the same time a destination for city tourists (see ◻ Fig. 5.2).

The Mirabell Gardens, the gardens of Mirabell Palace, are located directly in the much-visited old town of Salzburg. It is considered Salzburg's most beautiful garden (park). Since 1854 it is a public park much visited by locals and tourists because of its historical, ornamental garden design, its **small scale and structural diversity** (see ◻ Fig. 5.3).

Park Categories in Great Britain

A 2002 study by the Institute of Leisure and Amenity Management (ILAM) attempted to bring together the different categories of parking in use in UK cities into one terminology.

1. *Local Park* – up to 1.2 ha, service area 500–1000 m, usually with playground and landscaped green space, with no other infrastructure or amenity features.
2. *Neighbourhood Park* – up to 4 ha, service area 1000–1500 m, landscaped greenery with diverse infrastructure.
3. *District Park* – up to 8 ha, service area 1500–2000 m, variety of landscape features and infrastructure, for example, sports fields, playing fields, children's play area.
4. *Principal/City/ Metropolitan Park* – more than 8 ha, service area entire city, variety of landscape features and infrastructure of particularly high and therefore attractive quality (Dunnett et al. 2002).

5

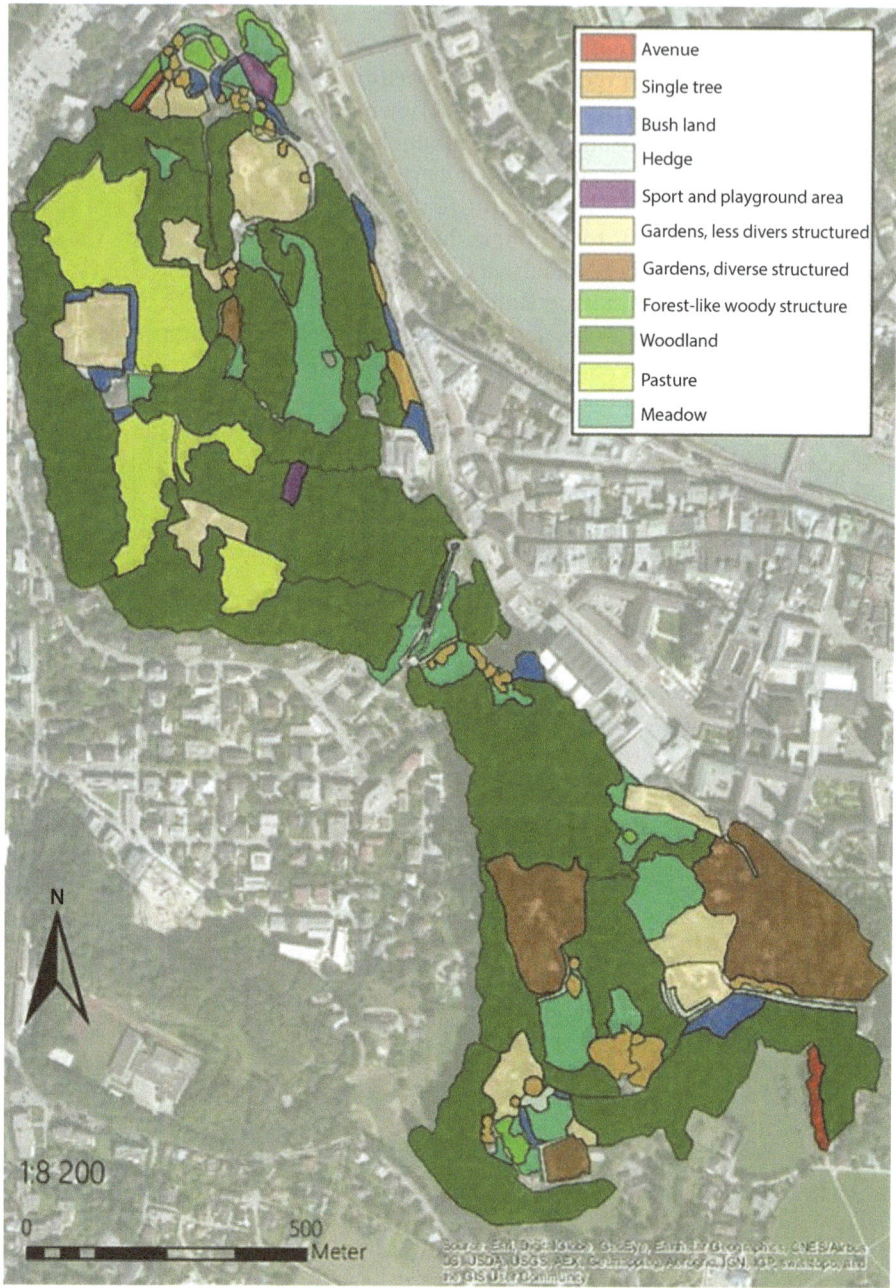

Fig. 5.1 Vegetation structure of the Mönchsberg hill in Salzburg, Austria. (Huber 2016)

Fig. 5.2 Mönchsberg hill in Salzburg, Austria, between the old town in the foreground and Untersberg in the background. (© Breuste 2004)

Fig. 5.3 Mirabell Gardens in Salzburg with Hohensalzburg Fortress in the background. (© Breuste 2004)

5.1.2 Services of Urban Parks

A large number of urban residents are frequent users of urban parks. The park is the "place of desire" of safe, orderly and aesthetic nature under the possible conditions of a city (Tajima 2003). It more or less meets the existing needs of urban dwellers. Landscape architects and planners strive to design parks for this purpose. The needs *(demands)* to go to a park are diverse. They depend on social factors such as age, lifestyle, health status, available free time (in the daily and weekly schedule), living conditions, and attractiveness factors of the parks themselves, such as distance, facilities, attractions, use situation, and other factors (Bonauioto et al. 1999; Carp and Carp 1982; Crow et al. 2006; Tyrväinen et al. 2003; Priego

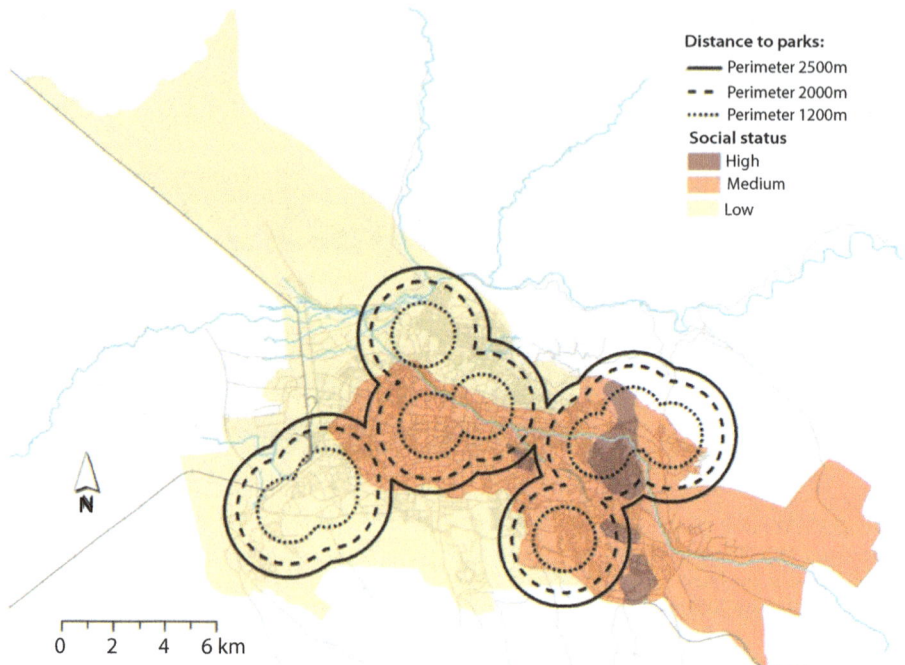

□ Fig. 5.4 Catchment area and social status of the residential environment of urban parks in Tabriz, Iran. (Breuste and Rahimi 2015; Breuste et al. 2016, p. 117)

et al. 2008; Qureshi et al. 2013). Nevertheless, the demand for park use can be summarized in a few categories. Clearly, and across cultural boundaries, the main reason for use is the search for relaxation, recreation, and short-term "time out" from hectic city life, which is perceived as attractive due to its diversity, but also as constantly mentally demanding and stressful (for example, Kaplan et al. 2004; Matsuoka and Kaplan 2008; Schetke et al. 2016).

This is often linked to the search for the "healing power" of nature (for example, "oxygen", fresh air, a healthy environment through nature and an aesthetic counter-image to the built city. This "nature counter-image" presents itself in a similar diversity in a small space as the city itself, only constructed from natural components (Breuste 2004). This search for the ideal, healing, "healthy" and aesthetic nature is a strong driving force for visiting parks (see ► Sect. 2.3). A similarly strong reason, perhaps even stronger, is the search for social contact in a "neutral" setting suitable for this purpose and perceived as pleasant. Those seeking contact often already come to the park together, spend time there, or meet randomly or casually for conversation, shared activities (for example, picnics, games), or just to participate in the social life of other park visitors as observers. Strong reasons for visiting the park include caring for children and walking dogs. Park space is seen as an ideal recreational space for these activities. It is safe (for example, from road traffic), socially neutral, relatively large and diverse (Chiesura 2004; Schipperijn et al. 2010; Milcu et al. 2013; Schetke et al. 2016). Parks try to create "a

▫ **Table 5.1** Examples of hierarchical classification of parks according to the criteria "size in ha", "expected walking distance in m"

Park type		Central Europe (Greiner and Gelbrich 1975)	Great Britain (Dunnett et al. 2002)	China, Nanjing (Liu 2015)	Iran, Mahshahar (Parsanik and Maroofne- zhad 2017)	Iran, Tabris (Breuste and Rahimi 2015)
Local park/ city green	m	400	500–1000		200	200
	ha	1	1.2		0.5	0.5
Residential Park	m	800	1000–1500	500	400–600	200–600
	ha	6–10	4	1–10	1–2	0.5–2
District Park	m	1600	1500–2000	2000	800–1200	600–1200
	ha	30–60	8	10–100	2–4	2–4
City Park	m	3200		5000	1500–2500	1200–2500
	ha	200–400	About 8	Over 100	4–6	4–10
Landscape park/regional park in the outskirts and surrounding areas	m	6500			25–30 min- utes' drive	
	ha	1000–3000			10	

piece of nature" in cities, thus meeting people's need for recreation and relaxation in nature. This is their most noble task. They are the primary places for urban encounters with nature and for learning about nature.

Despite the diverse needs, parks worldwide do not differ significantly in their structure and equipment. They simply try to meet the need for a "natural oasis" in the "urban desert" as best as possible through diverse vegetation structuring, order, cleanliness, safety, use-related infrastructure and attractiveness under the existing framework conditions of the respective city (for example, financing of mainte- nance) (Gälzer 2001; Van Herzele and Wiedemann 2003; Voigt et al. 2014; Parsanik and Maroofnezhad 2017).

As a (core) component of designed urban nature, parks contribute in many ways to the benefit (demanded or not) of park visitors. The park environment can also benefit from the ecosystem services of the park, for example, from aesthetic natural qualities (keywords: natural scenery, "high-quality residential location") or through local climatic air and temperature balancing. There are a large number of studies (for example, Horbert and Kirchgeorg 1980; Horbert 1983; von Stülpnagel 1987; Jauregui 1990; Saito et al. 1990; Eliasson and Upmanis 2000; Puliafito et al.

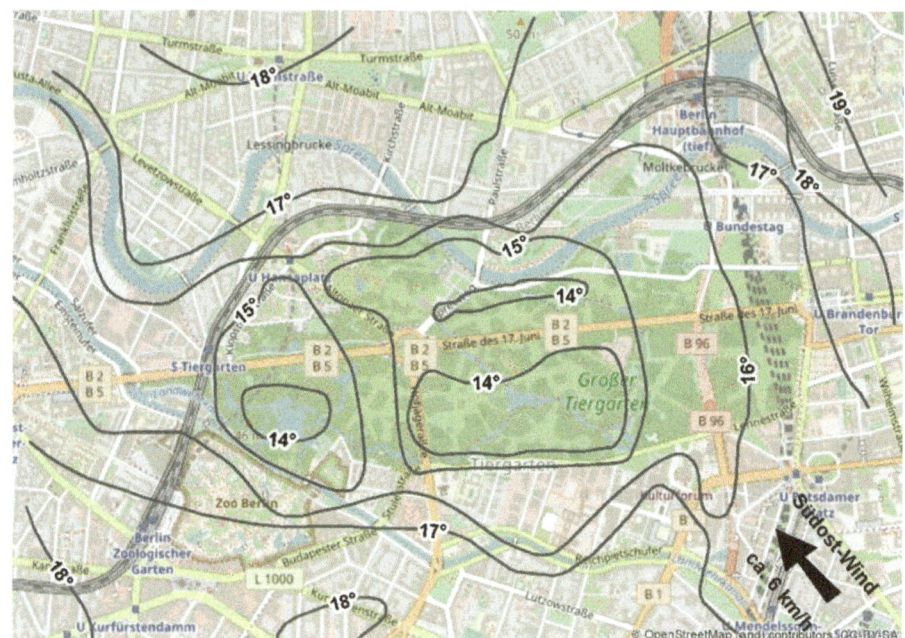

◘ Fig. 5.5 Distribution of air temperature in the Großer Tiergarten forest park and surrounding in Berlin, March 1978. (After Horbert 1983, design: J. Breuste, drawing: W. Gruber)

2013; Carfan et al. 2014; Liu et al. 2017a) on the possible "export" of ecosystem services of parks, especially summer temperature reduction, to a local environment and the necessary physical conditions of this local environment (see ◘ Fig. 5.5). However, it can also be learned from these that these outdoor effects are limited and linked to many often non-existent conditions. Cooling a park environment will only be possible for large parks in a confined environment. In order to enjoy the benefits *(advantages)* of parks, for example, the summer temperature reduction of up to 5 degrees K, parks usually have to be visited (Bernatzky 1958; Fisch 1966; Spronken-Smith 1994; Spronken-Smith and Oke 1998). Only in limited cases the park climate distributes to the neighbouring districts. (see ◘ Figs. 5.5 and 5.6).

Parks provide a variety of other ecosystem services (Swanwick et al. 2003; Haq 2011; Liu et al. 2018). The park as a locality must also be visited for these. Collectively, these services contribute to the **health** (psychological and physical) and **well-being** (physical, mental and social) of visitors through direct or indirect influences (Maas et al. 2006; Newton 2007; Mitchell and Popham 2008; Maas et al. 2009; Coombes et al. 2010; Stigsdotter et al. 2010; Weber and Anderson 2010; Annerstedt et al. 2012; Ward Thompson et al. 2012).

In this context, the frequently used collective term "recreation" is composed of a variety of individual park services (Konijnendijk et al. 2013). Part of these **health services** and beyond are:

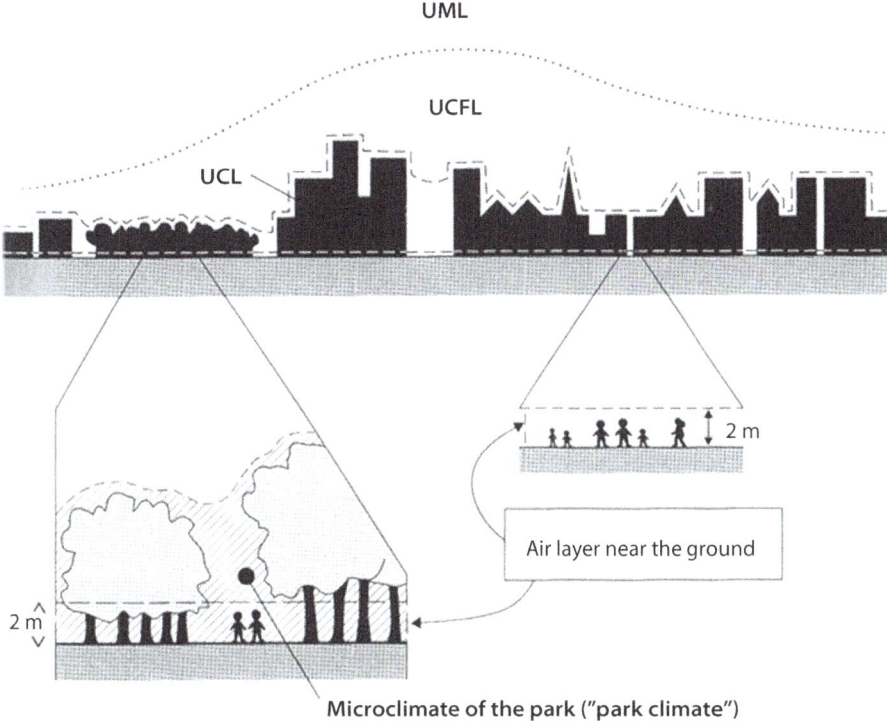

Microclimate of the park ("park climate")

◘ **Fig. 5.6** The urban bio climate (Leser 2007, 2008, changed) UBL = Urban boundary layer, USL = Urban surface layer, UCL = Urban canopy layer, RSL Roughness sublayer

Climatic and physical services:
- Reduction of pollutant levels (PM_{10}, NO_x, CO_x and SO_x)
- Reduction of air temperature
- Increase in humidity
- Regulation of the water balance
- Light and sound variation
- CO_2 sequestration (Konijnendijk et al. 2013; Forman 2014)

These services depend in their quantity on the composition of the physical, especially the vegetation structure of the park.

Other services are often mentioned:
- **Social benefits** (social cohesion, education, cultural mediation, the transmission of historical values (Konijnendijk et al. 2013; Forman 2014)
- **Aesthetic performances** (experience of nature, perception of structure, perception of diversity (Forman 2014)
- Erosion Control
- **Economic benefits** (increasing the value of neighbouring properties, tourism (Konijnendijk et al. 2013; Forman 2014) (see ◘ Figs. 5.7, 5.8, 5.9, 5.10, and 5.11).

5

■ **Fig. 5.7**   Recreation in Parque Cnel. Jordan C. Visocky, Buenos Aires, Argentina. (© Breuste and Wiesinger 2013)

■ **Fig. 5.8**   Recreation in urban nature, James Simon Park, Berlin. (© Breuste 2015)

■ **Fig. 5.9**   James Simon Park, Berlin, as a social meeting space. (© Breuste 2015)

Fig. 5.10 The park as an experience space, Rock Garden, Chandigarh, India. (© Breuste 2016)

Fig. 5.11 The Parque General Las Heras, a green recreational island in Buenos Aires, Argentina. (© Breuste 2010)

The ecosystem service **habitat supply** (habitat for wildlife) is explained separately in ▶ Chap. 6 (Biodiversity) (see ▶ Sect. 6.2).

Despite these proven advantages and the benefits of urban parks for urban dwellers, parks, especially in cities of developing countries, are by no means permanent components of the urban structure. On the contrary, public urban green space is constantly in competition with other uses and often loses out to development and infrastructure projects (Qureshi et al. 2010; Fanan et al. 2011; Haase et al. 2014; Kabisch et al. 2015).

The use of parks changes throughout the day. In Chinese parks, daytime use begins in the early morning, often before 6:00 a.m. with elderly ladies and gentlemen performing their Tai Chi exercises in larger open spaces. Later in the afternoon, retirees gather to play mai jong, chess or cards together in designated areas. Often they bring their birds in cages.

In Bangkok's Lumpini Park, the pattern of use changes every afternoon. Around 5 pm, a thousand to a thousand five hundred runners are in the park at the same time. They use the park, the only safe and quiet environment for them in the far distance, for running training. The other users are now only in the minority until nightfall (see ▪ Figs. 5.12, 5.13, and 5.14).

▪ **Fig. 5.12** Zhongshan Park in Shanghai is one of the much-visited old parks (founded in 1914) of Shanghai, China. (© Breuste 2007)

▪ **Fig. 5.13** Lujiazui Park (10 ha) in Shanghai's largest financial district, Lujiazui, China, is one of the city's newest parks. (© Breuste 2015)

Fig. 5.14 Every evening, Lumphini Park in Bangkok, Thailand, is intensively used by hundreds of joggers. (© Breuste and Astner 2018)

Why Parks Are So Attractive for City Dwellers?

So far, there are mostly only random samples and certainly no nationwide survey on park use. A 2002 survey of ten UK cities (50 local services, 515 respondents) found that 2.25 million people in these cities visit parks 184 million times a year. Extrapolating these results to the urban population of the whole of the UK, 33 million people visit UK parks 2.5 billion times a year.

46% of park users reported using parks more than once a week.

Seven reasons (categories) for park use were given:

- Enjoyment of spending time in nature
- Social activities
- To escape the stress of the city
- Walking (including walking dogs)
- Passive recreation,
- Active recreation (including sports)
- Participation in events (Dunnett et al. 2002).

Similar results can be found in other studies. The survey of 2487 people on park use in Hamburg (Krause et al. 1995) revealed the great importance of parks for the local environment (less than 20 min walking or cycling distance). Longer distances are only accepted if the parks are highly attractive. For the majority of respondents, the park should offer what their immediate living environment lacks: peace, clean air, nature, opportunities for play and action far away from traffic. The most frequent visits are in the afternoon for 2–3 hours. Personal relaxation, but also social interaction is sought. The park is a balancing and contrasting space to "normal life" and the residential environment.

Austrian Pupils Name the Benefits They Get from Urban Parks

In a study conducted by the University of Salzburg (supervised by J. Breuste) on the use of urban parks and contact with nature, 240 schoolchildren aged 14–18 years in the cities of Salzburg and Wels were surveyed in 2017/2018. In open questions, 416 mentions were made on the use of urban parks. One-third of the mention was related to **recreation, relaxation** (33.3%), followed by **activities** (19.2%), park as **social meeting place and place of learning** (15.1%), favourable **climate and environment** (12.0%), beautification of the **cityscape** (9.9%), **contact with nature (9.**1%) and 1.3% other.

This is largely in line with the needs also mentioned by adults in other studies. The main benefit of parks is seen by students as **the possibility of relaxation in a quiet environment,** the temporary escape from the hectic urban life. This is a health-promoting need already expressed by young people.

How Chinese Use and Value the Parks in Shanghai?

Public parks do not yet have a long tradition in China. They are a European import into the urban structure of Chinese cities. The first public park, the relatively small Huangpu Park in Shanghai, was created in 1886 as a "Public Garden" in the British concession (administrative area) of the city. This was followed by Fuxing Park in the French concession in 1909 and further parks based on the European model in the twentieth century.

In a cooperative study of the Technical University Dresden (supervised by J. Breuste) and the East China Normal University Shanghai, 431 park visitors in four Shanghai parks (two new parks: Mengqing Park, Lujiazui Park and two old parks: Changfeng Park, Fuxing Park) were asked about their usage habits, motives for usage and what they value or dislike in the park.

Almost none of the visitors had their own private green space, but more than 2/3 had an apartment larger than 60 m^2, mostly occupied by 3–5 people. Two-thirds were native Shanghainese, more than half with children in the household. Shanghai parks are primarily used by the residential population of their local area. For many, they are the only opportunity for spending time in nature. Half of the visitors, therefore, come to the park on foot, the other half by public transport, bicycle or car. Half of the park users need less than 20 min to get to the park. Parks are used intensively – and not only by retirees. 15.6% of the respondents come to the park daily, another almost 50.5% (!) once or several times a week. Far fewer birds in cages are brought along (traditionally) than dogs. Tai-chi in the morning, walking, playing cards, dancing and meeting others in the park are the most common activities. For this, almost half of the visitors stay in the park for 1–3 h, preferably in the morning or evening. The main reason to visit the park for over 50% of respondents are to enjoy fresh air, be in nature and a quiet environment. 11% come for the morning Tai Chi exercises. All other reasons for use are less significant. Cleanliness, good park maintenance, good air, quietness, emotional experience of nature and pleasant atmosphere are most valued. Water, lawns, woody plants and flowers, that is, a variety of natural features, contribute best to this (Zippel 2016) (see ◼ Figs. 5.12 and 5.13).

The City Administration of San Diego (USA) Describes the Benefits of Its Urban Parks from the Perspective of Planning and Management

- Urban Parks
- Improve and expand the public space
- Are critical elements of place-making
- Serve the densely living urban population in particular
- Complement the urban environment with natural elements
- Are active spaces that support social interaction and contribute to a vibrant city
- Offer a high level of comfort
- Eastern Urban Center (EUC) (2009).

5.2 The Urban Tree Population (Urban Forest)

5.2.1 Nature Elements, Urban Trees and Urban Woodland

Urban woodlands are diverse providers of ecosystem services to urban populations. In many cities, they have largely disappeared due to former agricultural use and later urban expansion. In forest cities (see ▶ Sect. 3.2), they still cover large areas in or around cities. Where they are only present as remnants (for example, Dölauer Heide in Halle/Saale, Auwald in Leipzig, Dresdner Heide in Dresden, etc.), urban woodlands count as particularly valuable ecosystems, as providers of ecosystem services that cannot be offered in this quality by any other urban nature (for example, climate regulation, recreation, the experience of nature, etc.). In addition, they are often considered special urban biodiversity hotspots (see ▶ Sect. 6.2).

A broad, contradictory debate on what should be called urban woodland and what should be called *"urban forest"* and whether or not these are synonymous terms is currently being conducted in the international literature (see especially Randrup et al. 2005). This is due in particular to the different use of terms and conceptual content in Anglophone countries as opposed to, for example, German-language sources.

> **Urban Woodlands**
> - Are characterised by their tree population, which creates its climate and special habitat conditions
> - Are located in cities and their immediate surroundings (urban – per-urban),
> - Are at least 0.3–0.5 ha in size
> - Are publicly or privately owned
> - Are generally accessible to the general public
> - Are providers of ecosystem services (recreation, health and well-being, regulatory services for climate and water balance and timber production) and
> - of biodiversity (habitat function) and
> - provide these primarily for the urban population (Leser 2008; Dietrich 2013).

5

A lower limit in the area size of urban woodlands is given where, due to the small size of the area, special climate and specific habitat characteristics can no longer be developed. Leser 2008 and Dietrich 2013 give 0.5 ha as lower limit, Burkhardt et al. 2008, p. 33, 0.3 ha and 50 m minimum diameter.

Urban woodlands can be publicly or privately owned. Their accessibility is an essential prerequisite for the cultural ecosystem services they provide to urban users (Randrup et al. 2005; Konijnendijk et al. 2006; Gilbert 1989a, b; Dohlen 2006; Burkhardt et al. 2008; Leser 2008).

The English term *"urban woods and woodlands"* and includes *"forest"*, *"wooded land"*, *"natural forest"*, *"plantations"* *"small woods"* and *"orchards"*, regardless of ownership (Randrup et al. 2005).

Urban woodland is thus not a fixed ecosystem unit, but can also be newly created by allowing succession or by planting. This was done, for example, in the "Urbaner Wald" projects supported by the Federal Agency for Nature Conservation in Germany (Burkhardt et al. 2008).

> **Urban Forest – Urban Tree Population**
>
> *Urban forest* refers to all urban trees, regardless of ownership, and considers them as a "resource" and overall provider of ecosystem services from which urban dwellers can benefit. The *urban forest* is thus best interpreted as "urban tree population" or "urban tree stock". It includes woods and woodlands as well as all trees in public and private ownership (street trees, trees in parks, private gardens, cemeteries, brownfields, orchards, etc.) (Dwyer et al. 2000; Randrup et al. 2005; Konijnendijk et al. 2006; Konijnendijk 2008; Pütz et al. 2015; Pütz and Bernasconi 2017) (see ◻ Table 5.2).
>
> » … urban forest includes all trees and their habitat within the city's urban area boundary. This includes trees on both public and private property: along city streets; in parks, open spaces and natural areas; and in the yards and landscaped areas of residences, offices, institutions, and businesses. The urban forest is a shared resource that provides a wide range of benefits and services to the entire community. (Copestake und City of Ottawa 2017).

> » Urban forests are... all urban green spaces with forests, groups of trees, individual trees and other woody plants, including private and public parks, gardens, cemeteries, playgrounds, sports and leisure facilities. (Pütz et al. 2015, p. 232).

□ Table 5.2 Element of the *urban forest*

Item	Description	Forest within the meaning of German forest law	Private ownership
Urban woodland	Woods and woodlands in the urban area, often intensively used for leisure and recreation	Yes/no	Mostly no
Woodland in the per urban area	Forest in the extended use area of cities	Yes	Yes/no
Wooded areas in the settlement area	Wooded areas with forest character	No	Meaning no
Parks	Woodland parks with relatively dense tree population, but also all other parks with copses, groups of trees and individual trees	No	Usually no (exceptions possible)
Urban gardens	Private gardens with fruit trees	No	Yes
Fruit tree plantations, tree nurseries	Usable economic areas	No	Yes
Parkways, street trees, groups of trees, individual trees	Urban trees outside forests and parks on streets and squares and in open spaces	No	No

See also Pütz and Bernasconi (2017)

In some cases, *"urban forest"*is even used as a synonym for *"urban green spaces"* and thus includes not only the tree population but all types of green spaces, including those without trees such as grassland, meadows, pastures, etc. This is explained by reference to the above-mentioned tasks of urban forestry in some countries and the responsible administrations. This is explained by reference to the above-mentioned tasks of *urban forestry* in some countries and the responsible administrations, but not in terms of content (FAO 1998).

> » Urban forest is now a common term that means all of the vegetation and soils of an urban region. (Rowntree et al. 1994, S. 1).

» ...Urban forest includes all trees and their habitat within the city's urban area boundary. This includes trees on both public and private property: along city streets; in parks, open spaces and natural areas; and in the yards and landscaped areas of residences, offices, institutions, and businesses. The urban forest is a shared resource that provides a wide range of benefits and services to the entire community. (Copestake und City of Ottawa 2017).

Urban woodlands predominantly represent forest relicts and are formed through natural succession and planting. Cities located in zonobiomes with a climax forest may have woodland remnants or may become woodland areas through natural succession (see ◘ Fig. 5.15).

Most urban woodlands are planted forests with species composition often different from the site. In this case, the designation "forest" is appropriate. If the combination of woody species is appropriate to the site, it can be referred to as a forest, regardless of whether it came about through original natural development or planting (Kowarik 1995) (see ◘ Table 5.3).

The distinction between urban woodlands and semi-urban woodlands according to distance from the city is less meaningful than according to the degree of anthropogenic use and change, primarily as recreational forests for city dwellers or forest plantation. The character as a recreational forest for urban dwellers natu-

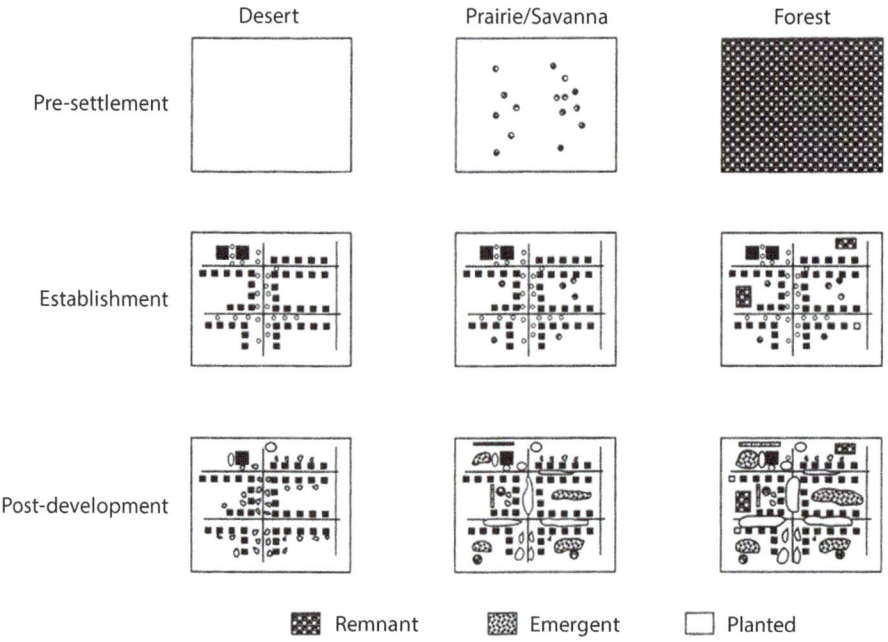

◘ **Fig. 5.15** Evolution of urban tree population *(urban forest)* through woodland relict, natural succession and planting in three ecoregions. (Zipperer et al. 1997, p. 235; Breuste et al. 2016, ► Fig. 4.6, p. 99)

◘ Table 5.3 Differentiation of forests in relation to settlements

Woodland type	Subtype	Spatial position	Function		Urban influ-ence
			Social func-tion	Pro-duc-tion	
Urban wood-lands	Woodlands within semi-urban areas woodlands in sub-urban areas	Isolated in the built-up area between built-up area and open landscape			
Semi-urban wood-lands	Woodlands near cities	Part of the cultivated landscape, close or adjacent to urban areas			
Non-urban woodlands	Forests far from cities	Part of the open (near-natural) landscape, far from urban areas			

After Kowarik 2005, p. 9, modified in Burkhardt et al. 2008, p. 31 in Breuste et al. 2016, p. 104, Table 4.8

rally decreases with increasing distance from the source areas of recreation-seekers, the cities. In individual cases, woodlands that are somewhat further away from the city and have very good infrastructure connections to the city can also "be urban woodlands".

Statistics on the proportion of woodlands in cities usually only include those woodlands that are located in the urban municipal area. In addition, however, all woodlands that fulfill the above-mentioned urban functions for the nearby urban population should be taken into account. This often concerns a radius of several kilometres, depending on the size of the cities, even outside the immediate urban administrative area. Urban woodlands, even as public areas, cannot therefore belong to one municipality alone. This means that the management often has to be coordinated between different municipalities and owners, which is not always an easy undertaking.

The European research project "Urban Forests and Trees" (1997–2002) produced a systematic review of planning, management and use of urban woodlands and urban trees (Konijnendijk et al. 2005). Summary and specific work on the management and redevelopment of urban woodlands in Germany has been carried out in recent years (Kowarik 2005; Kowarik and Körner 2005; Rink and Arndt 2011) (see ◘ Fig. 5.16).

◻ Fig. 5.16 Concept for the development and management of urban forest, Randrup et al. 2005; changed

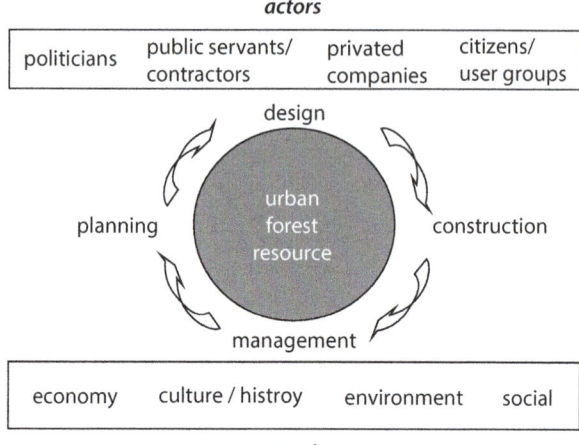

5.2.2 Services Provided by Urban Trees

At least since the late 1960s, there has been an ongoing rethinking of management of urban woodlands away from dominant timber production and toward management of multiple functions and especially recreational ones (Johnson et al. 1990; Kowarik 2005; Burkhardt et al. 2008; Jim 2011; Breuste et al. 2016). Urban forestry, the management of urban forest including the urban woodlands, serves to conserve trees so that they can provide ecosystem services to urban residents, either individually or as a stand.

> **Urban Forestry – Management of the Urban Tree Population**
>
> Urban forestry is the management of urban and peri-urban tree population through interdisciplinary collaboration between different professionals and resorts, involving science, technology, management and governance to optimise the ecosystem services provided by trees to urban dwellers (Jorgensen 1970, 1986; Helms 1998; Konijnendijk et al. 2005, 2006; Konijnendijk 2008; Nath 2012; Pütz et al. 2015; Pütz and Bernasconi 2017; Copestake and City of Ottawa 2017, among others) (see ◻ Figs. 5.17 and 5.18).
>
> » … is a specialized branch of forestry and has as its objectives the cultivation and management of trees for their present and potential contribution to the physiological, sociological and economic well-being of urban society. These contributions include the over-all ameliorating effect of trees on their environment, as well as their recreational and general amenity value. (Jorgensen 1986, S. 177).

■ **Fig. 5.17** Areas of urban forestry. (s. a. Pütz et al. 2015, modified, design: J. Breuste, drawing: W. Gruber)

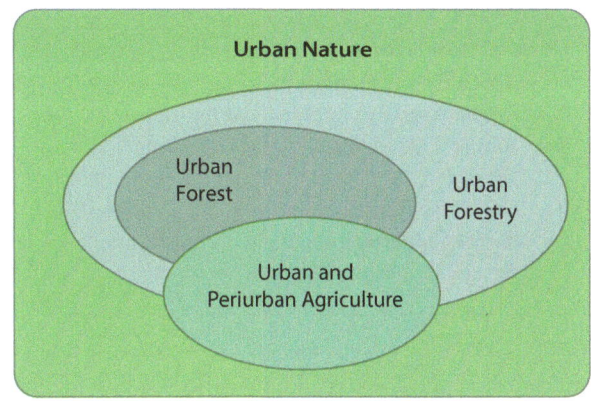

	The urban forest		Urban woods and woodland
	Individual trees		
	Steet and roadside trees	Trees in parks, private yards, cemeteries, on derelict land, fruit trees etc.	(forests and other wooded land, e.g., natural forests and plantations, small woods, orchards, etc.)
Form, function, design, policies and planning			
Technical approaches (e.g. selection of plant material, establishment methods)	← URBAN FORESTRY →		
Management			

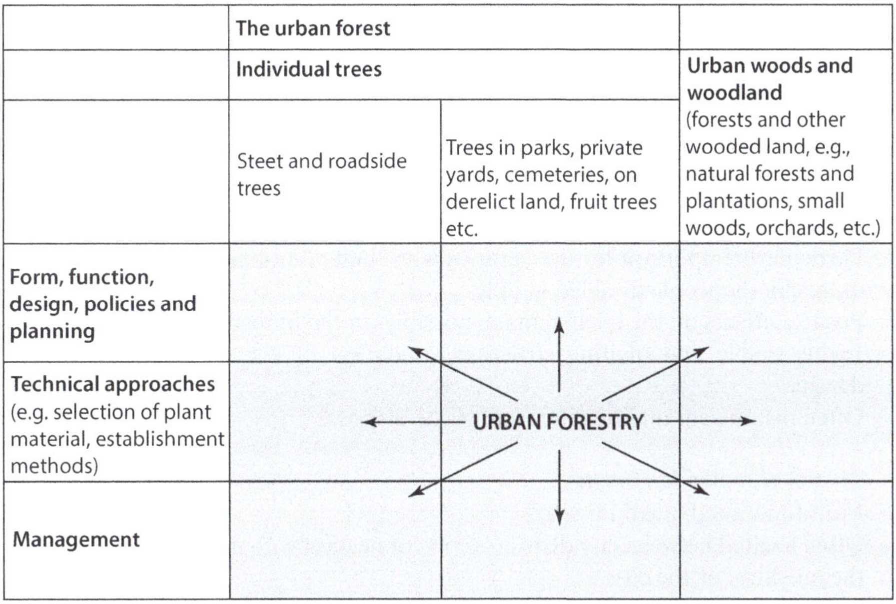

■ **Fig. 5.18** Concept for urban forestry. (Randrup et al. 2005), design: J. Breuste, drawing: W. Gruber

Urban woodlands can be divided into four major groups according to their management and structure, which can be further differentiated according to stratification and structure and by adding other characteristics.

> **Definition**
>
> **Natural woodlands** or their remnants.
>   Nature 1 "old wilderness".
>   **Forest,** strongly characterized by traditional forest management.
>   Nature 2 "traditional cultural landscape".
>   **"Park forest/woodland"**, planted trees in green areas and residential areas.
>   Nature 3 "Functional green".
>   **Succession forests/woodland** on fallow land.
>   Nature 4 "urban wilderness".

Burkhardt et al. (2008) (p. 32, modified) subdivide woody stands and urban wood-lands in Germany according to their functions:

- **Trees and groups of trees near dwellings**
- Daily recovery
- Contact with nature and aesthetic effect
- Wind breakage hazards, shading

- **Neighborhood Woodland/Forest**
- Relatively small woods in the residential area
- Particularly important for user groups with limited mobility, for example, children, elderly people disabled people
- Positive effects on the local climate, possibly on the immediate surroundings
- Light, visible and inviting structure, gradation of the stand-in height and density
- Often insufficient maintenance and waste disposal

- **District Woodlands/Forests**
- Multifunctional, medium-sized
- Often located between city districts or in connection with new building areas on the outskirts of the city
- Use by residents and crossing pedestrians and cyclists
- Information and citizen participation particularly important
- Graduated management related to intensities of use

- **Recreational Woodland/Forests (mostly on the outskirts)**
- Mostly larger than 60 ha
- Different forest structures as mosaic stand possible.
- High diversity and closeness to nature possible.
- Various possibilities to experience nature
- Equipment with paths, meeting points, seating areas, information boards, etc.

- **Productive Woodland/Forest**
- Forest areas outside cities
- Focus on wood production
- With other functions as required (for example, nature conservation, recreation)

In North America in particular, forest administrations are actively involved in the management of urban tree stands through *forest managers*; cities set up their own public service institutions *(urban forest departments)* for this purpose, develop management plans *(urban forest management plans)* and publish on research and management aspects, sometimes in their own regional publications. They thus often have a more important position than the administration of public urban green spaces, and in some cases even take on tasks that in Europe would be the responsibility of the green space administration. This results in a broad field of work, *"urban forestry"*, which does not exist in this form in Europe.

Despite a mostly extensive, city-managed, public tree stock of several thousand trees, this stock can often be significantly exceeded by that of private owners. The extent to which the public management of the *"urban forest"* can also record and manage this stock remains mostly untouched, at least in European cities. Here, there is often not even any knowledge of the private tree stock, it is not recorded by monitoring and is only regulated by general tree protection ordinances that also affect this stock.

It does not seem to make much sense to see the task of urban forests primarily in timber production, although this is also mentioned in some cases:

» Urban forests and woodlands to produce wood products using good silvicultural planning, management and harvesting practice… The prime goal is producing wood products, and usually the secondary goals are protecting and providing clean-water supply, soil erosion protection, and recreational opportunities. (Forman 2014, S. 355).

Overall, economic services are significantly less important for urban woodlands compared to other services. Forest managers even forgo potential profits to be made in favour of environmental and social services (Scholz et al. 2016). This is exactly in line with the concept of urban forestry.

Following Pütz and Bernasconi (2017) (for Switzerland), five challenges can be generalized for future scientific and practical engagement with urban forestry:

- Overcome integrative, cultural, institutional and definitional barriers in order to develop and implement cross-cutting strategies for urban nature (the urban green space).
- Align far-sighted ecosystem management of the urban woodland with the needs of the urban population.
- Establish coherence of tools and programmes for planning, control and monitoring of urban forests and, based on this, develop cross-cutting urban forestry strategies.
- Promote urban forestry socially and politically through dialogical and participatory decision-making processes.
- Secure funding for urban forestry through new partnerships with stakeholders from politics, administration, society and business.

5

Konijnendijk et al. (2005) distinguish three ecosystem service groups of urban trees in a Europe-wide study in 2002:
- **Environmental services**
  - Microclimate moderation
  - Binding of air pollutants and noise reduction
  - Purification effects in the water cycle
  - Biodiversity and habitat functions
  - Landscape
- **Social benefits**
  - Aesthetic value (beauty)
  - Recreation and open space activities
  - Health and well-being
- **Economic performance**
  - Wood production
  - Other economically recovered forest products
  - Heating and building materials

In the 14 European countries studied, Konijnendijk et al. (2005) show that the following services are particularly in focus there:
1. Recreation and open space activities,
2. Scenery,
3. Microclimate moderation,
4. Health and well-being,
5. Biodiversity and habitat function, and
6. Binding of air pollutants and noise reduction.

The Leipzig city administration in Germany reports forest services in the urban context:
- Scenery: urban redevelopment and residential environment: attractive "cityscape", increasing permeability,
- Recreation: additional open space offers,
- Climate moderation: thermal compensation and air pollution control (dust binding), and
- Biodiversity/nature conservation: habitat and biotope network (Dietrich 2013).

■ **Recreation and nature experience**
The main goals of recreation seekers are relaxation, activities in open space and nature. These goals can be optimally achieved in the forest, an ecosystem that is also usually visually separated from the rest of the urban space. Forests have a health-promoting effect due to their special climatic, air-hygienic and psychologically effective properties. Their preservation or new creation should therefore have top priority in the development of green cities.

Urban woodlands have the best prerequisites for the ecosystem service "experience of nature", as they are perceived by children and adults alike as "nature in the true sense", as "natural" green spaces. This is also true when these forests are designed in nature.

Urban dwellers have a fundamentally positive attitude towards nature, even when fears (insecurity) and risks (risk of falling, insect pests, etc.) also have a limiting effect (Hannig 2006; Rink 2004; Breuste and Astner 2018). Roschewitz and Holthausen (2007) state that 96% of the Swiss population visit the forest in summer, 58% even several times a week.

As part of a survey carried out in 1997 on behalf of the Swiss Federal Office for the Environment, Forests and Landscape (BUWAL 1999), 2000 Swiss people were asked about their attitudes to the environment, nature and forests. They provided the following information on their interests in forest use.

- Walking 40.1%
- Recovery 19.1%
- Sports/health 18.2
- Nature experience 9.9
- Gathering 9.9
- Air 8.6%
- Dog walking 7.6%.

Gasser and Kaufmann-Hayoz (2005) stated that 49% of woodland/forest visitors would rather seek solitude and feel disturbed by other people.

In addition to relaxation, rest or activity, the main motivation is the search for an experience of nature (Tyrväinen et al. 2005; Appenzeller-Winterberger and Kaufmann-Hayoz 2005). A study in Zurich urban forests shows different user-related values and preferences. One group prefers natural woodland conditions to intensive maintenance and tidiness. Another group sees the forest as a sporting activity space (cycling, running, horse riding, etc.) (Bernath et al. 2006).

■ **Climate and air hygiene**

The formation of a local climate of its own, a woodland/forest climate, is the prerequisite for many ecosystem services, especially regulatory ones (Baró et al. 2014).

Efficiency of the Climatic and Air-Hygienic Ecosystem Services of the Urban Woodland

The efficiency of the climatic and air-hygienic ecosystem services of the urban woodlands is greatest in large, dense, multi-layered and older deciduous forest stands. To enjoy the benefits of the forest climate, it is usually necessary to visit the forest. In order to benefit from the climatic and air-hygienic effects of urban forests, a precise selection of tree species, their sustainable maintenance and targeted localisation in relation to buildings and residential areas is crucial.

Woodland climate (also forest climate) refers to the climate in the woodlands below the canopy (stand interior climate). Depending on the canopy closure, the

5

forest forms its own (stand) climate, which is clearly different from the atmospheric air layer above it. The canopy is the main energy conversion surface. Canopy density and woodland area influence the formation of specific climatic conditions. The formation of the forest climate also depends on the tree species spectrum and the stand structure and is differentiated vertically (crown climate, trunk climate and the climates of the shrub and herb layer). By reducing radiation, converting radiated energy into latent heat, high transpiration and low vertical and horizontal air exchange, thermal extremes are mitigated both during the day and at night. This temperature difference can be up to 5 K (Nowak 2002a, b), and briefly up to 10 k on hot days (LAF 1995). The more open the forest, the less these climatic characteristics are formed. Loose tree stands and small woods cannot form an optimal woodland climate. The thermal differences cause horizontal pressure differences during low exchange radiation weather conditions, which create a circulation at the woodland edge between internal climate and surrounding atmosphere and have regional climatic ambient effects. During the day, the forest wind is directed out of the forest into open surroundings or, if not obstructed by barriers, into adjacent building quarters, and is only effective at very low wind speeds. Its range is optimally in the tens of meters (Mitscherlich 1981; Kuttler 1998; Tyrväinen et al. 2005). To enjoy the benefits of the woodland climate, for example, on hot summer days, it is usually necessary to visit the forest (see ◘ Figs. 5.21 and 5.22).

Forests are sinks of the greenhouse gas CO_2. However, the CO_2 sequestration amounts to only a fraction of the total urban CO_2 emissions. This limits the possibilities of using urban forests to effectively counteract climate change and CO_2 pollution. CO_2 emissions are most effectively reduced only by reducing the use of fossil fuels, not by urban woodlands/forests (Nowak and Heisler 2010; Nowak et al. 2013).

Due to the large leaf surface and the great roughness, forests have a great binding effect for dust and gas compared to other green spaces. They also reduce noise more effectively than other green spaces (LAF 1995). In a large number of European cities, the most important sources of diverse air pollution are mobile emission sources from road traffic. Some old industrialised cities in Europe (for example, in Eastern and South-Eastern Europe, Russia) and many industrial cities worldwide (for example, in China) are additionally polluted by emissions from industry (Sawidis et al. 2011; EEA 2012).

Absorbed acids and nitrogen compounds enter the forest soil. Forests are "aerosol sinks", reducing, dispersing and diluting dust loads in particular (LAF 1995). However, since they are only deposited, the particles return to the atmosphere or are added to the soil by rain (Nowak 2002a, b). The binding of pollutant gases is very low compared to emissions in cities. Pollutant uptake through stomata is greatest in large old trees (over 70 cm trunk diameter). Forest trees (for example, birch) may also cause undesirable pollen loads with allergic effects (Tyrväinen et al. 2005). The reduction of wind speed can also hinder fresh air transport when forests are unfavourably positioned in air exchange pathways.

Regardless of the fact that carbon sequestration by individual urban trees is vanishingly small compared to forests, woodlands, peatlands, etc., this is also repeatedly calculated and used to justify trees in cities. Nowak et al. 2013 calculated carbon sequestration by the entire *urban forest* based on tree canopy (7.69 kg C m^2, 0.28 kg C m^2) for 28 cities in the USA and identifies 643 million tonnes of sequestered carbon, equivalent to US$50.5 billion.

Studies in the USA confirm this: Trees, properly located, can reduce the energy demand of buildings by up to 25% in summer through wind protection, insulation and shading (Nowak 2002a, b).

» In fact, the main values of urban and peri-urban forests have no market price. These values are termed as non-consumptive use values and include benefits derived for example from a pleasant landscape, clean air, peace and quiet, as well as recreational activities…This category also includes benefits such as reduced wind velocity, balanced microclimate, shading, and erosion control, the economic value of which may be determined through for example reduced costs of heating or cooling or alternative costs of environmental control. (Tyrväinen et al. 2005, S. 101).

Urban woodlands are also valuable in economic terms. Such economic, monetary values are increasingly used as justifications for urban woodlands and trees. A monetary valuation of trees only of urban parks in the USA (Nowak and Heisler 2010) resulted in the following values (see ◨ Table 5.4):

Nowak (2002a, b) states that New York's trees alone absorbed 1821 tonnes of air pollutants in 1994, which would be equivalent to US$ 9.5 million. Similar values are given for other US cities to justify the importance of urban trees. The question of whether US$9.5 million in 1994 has the same value in 2019 points to general problems of monetary valuation.

Beijing's 2.4 million urban trees had absorbed 1261 tons of air pollutants, primarily PM10, in 2002 (Yang et al. 2005). What emissions are simultaneously opposed to these values is not reported. It can be assumed that these emissions are many times higher and thus even trees cannot cope with the basic problem of air pollution. Urban forests are not there to absorb harmful immissions, the emission of which is not prevented for economic-technical reasons, so that the health risk to people is thereby reduced. Emissions must be removed at the source, not end-of-pipe. Instead, urban forests have the task of providing recreation, relaxation and health-promoting activity for urban dwellers.

The sound-absorbing **effect** of leaves, needles, bark and soil reduces noise. The forest's own sounds, which are perceived as pleasant, have an additional noise-reducing effect through selective perception. Dense, multi-layered woody plants or forest stands reduce noise best. Deciduous trees are almost ineffective when leafless. Dense deciduous forests reduce noise by about 0.15 dB(A)/m. To obtain an effective noise reduction of traffic noise (predominantly above 55 dB) by for example, 15 dB(A) to reach a health compatible approx. 40 dB, 100 m of dense decidu-

5

ous forest would have to be placed between the noise source and the recipient. This is hardly possible in a city.

The European Environment Agency states that almost 70 million people in European cities are exposed to excessively high long-term average road noise levels (>55 dB). 44% of the population in Europe's major cities are exposed to noise levels (>50 dB) during sleeping hours, which can have negative effects on health. This cannot be actively counteracted by urban forest. Only those parts of settlements that are separated from main traffic arteries by urban forest, mostly by chance, can benefit from effective noise reduction through the forest. However, these effects can also be deliberately exploited through the targeted localisation of residential buildings (Library of the European Parliament 2013).

In Shanghai, extensive parts of the urban motorway network were surrounded by several decimetre-wide, inaccessible tree plantations in order to bind pollutants from motor traffic and reduce noise. A review of the efficiency of this costly measure has not yet taken place (see ◘ Figs. 5.19, 5.20, 5.21, 5.22, 5.23, and 5.24).

The light inside a woodland has a pleasant effect on the eye and a favourable effect on the psyche due to increased green and infrared components, brightness, colour and shape contrasts and the filtering effect of the canopy. Forests reduce UV light and thus also have a health-promoting effect (Tyrväinen et al. 2005).

Meanwhile, studies on the performance and benefits of individual, very specific urban woodlands are already being conducted to justify their importance (Nowak 2002a, b; Magistrat der Stadt Halle n.d.; Wang et al. 2016) (see ◘ Table 5.6 and ► Sects. 2.5 and 3.2).

The ecosystem service **habitat supply** (habitat for plants and animals, biodiversity) is explained separately (see ► Sect. 6.2).

◘ **Table 5.4** Monetary valuation of park trees in US cities

	All park trees in US cities	**Park trees in Chicago**
General tree value	300 billion	192 million
Reduction of the air temperature	Unknown, expected to be billions of US dollars per year	
Filtration of air pollutants	500 billion	344 million
Reduction of UV radiation	Unknown, but probably significant	
Carbon sequestration	16 billion	11 million
Annual carbon uptake	50 million	32,800

Nowak and Heisler (2010) (in US dollars per year)

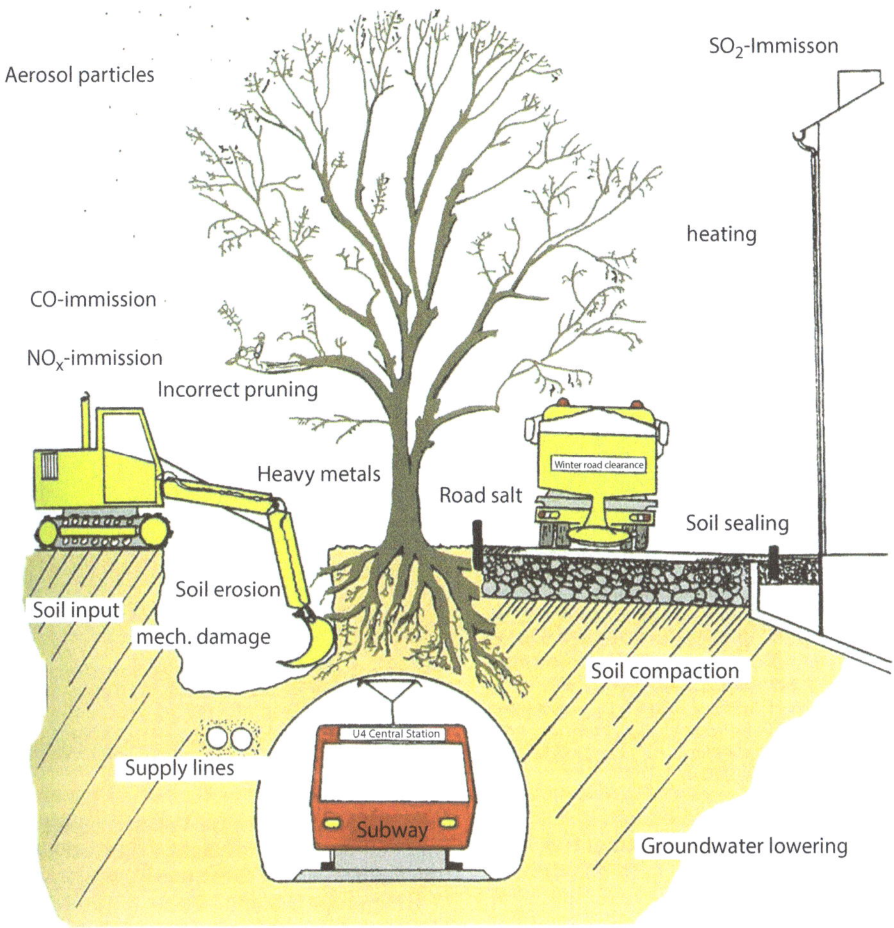

Aerosol particles

SO$_2$-Immisson

heating

CO-immission

NO$_x$-immission

Incorrect pruning

Heavy metals

Road salt

Winter road clearance

Soil sealing

Soil erosion

Soil input

mech. damage

Soil compaction

U4 Central Station

Supply lines

Subway

Groundwater lowering

◘ **Fig. 5.19** Stress factors for urban street trees. (Wittig 2002)

◘ **Fig. 5.20** Old horse chestnut *(Aesculus hypocastanum)* as a street tree in Bagnols-en-Forêt in France. (© Breuste and Faggi 2014)

5

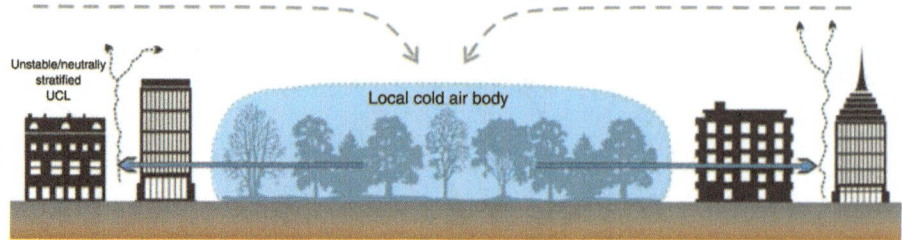

◘ Fig. 5.21 Schematic of the cold air production of a forest park and the air exchange system. (Leser 2008, p. 151, from: Bongardt 2006, design: J. Breuste, drawing: W. Gruber)

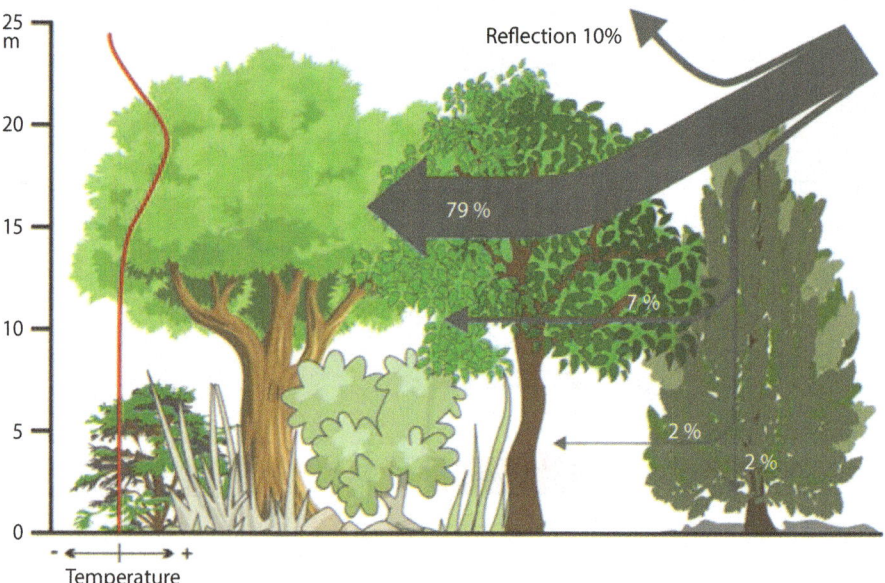

◘ Fig. 5.22 Model of radiation and temperature in a forest. (After Gebhardt et al. 2007 (Design: J. Breuste, Drawing: W. Gruber))

0   250   500 m

Map bases:
Bavarian Survey Administration,
Open street map

―― Road
███ Sealed areas
░░░ Water areas
░░░ Tree cover

◘ **Fig. 5.23** Nymphenburg Park in Munich, surface structures. (After Aevermann and Schmude 2015, design: J. Breuste, drawing: W. Gruber)

◘ **Fig. 5.24** Roadside forest plantings along Shanghai city highways, China. (© Breuste 2004)

Street Trees – Survival Artists on Unfavourable Sites

Street trees are an essential part of the *urban forest,* the tree population of a city. They are in public ownership and are looked after and maintained by city councils.

Trees in the city live under conditions that differ in many respects from those on natural sites. These less favourable site conditions affect the performance characteristics and the life span of the trees. Trees along streets have extremely unfavourable living conditions. Here, they achieve significantly shorter lifetimes and low performance, which are, however, often in particular demand here. An optimisation of the unfavourable site factors would therefore be necessary. This is only possible to a limited extent in competition with other uses in the confined road space, but it can be achieved.

- Particularly unfavorable site conditions are:
- Unpaved tree surroundings that are too small reduce water and gas exchange
- Soil compaction reduces water and gas exchange (oxygen and water deficiency)
- High temperatures due to dense building and sealing increase the water shortage through evaporation
- Lack of rainwater due to channel drainage and sealing
- Limited root space due to underground power systems
- Mechanical damage of the roots due to tasks of the power channels
- Mechanical damage from parked cars and improper tree pruning
- Pollutants from uses in the surrounding area (road salt, vehicle exhaust fumes, dust, gas leaks from underground pipes, etc.)
- Insect pests
- Infections caused by poorly or unattended tree damage.

The main characteristic of street tree sites is drought stress, in addition to pollutant inputs and mechanical damage. The improvement of site conditions must be the first management objective. For these extreme sites, whose unfavourable characteristics will tend to increase in the course of climate change, the best-adapted tree species must be used so that ecosystem services can continue to be provided by these trees in the future (Meyer 1982; Gaston 2010; Roloff 2013; Scholz et al. 2016) (see ▪ Figs. 5.19 and 5.20).

For example, in the "Garden City of India", Bengaluru, the tree population in public spaces continues to decline due to tree felling for road widening. This particularly affects old and large-crowned trees that provide shade. They also make up the bulk of biodiversity, while young, newly planted trees belong to only a few species, can hardly provide ecosystem services, and often do not survive in stressful road traffic. The preservation of large-crowned old trees along roads as high performers is particularly important for ecosystem services. At the same time, new plantings of already larger, viable trees and their special care must be undertaken (Nagendra and Gopal 2010).

Which Nature Is Preferred? – Why Urban Woodlands Are So Important

Which nature is preferred depends on attitudes, mentalities, values, state of health, age and needs. These are not immutable, but vary at certain times in the course of life, in the course of the year and the course of the week and day. Emotional attraction or aversion can be changed by appropriate measures and influencing factors, such as emotional learning, gaining knowledge and experience, group dynamics (children and elders), planning, use of much-used media, and many more.

Only valued nature is also defended and preserved. The degradation and destruction of rejected nature is more often tolerated. The less convenient use of urban forests often takes a back seat to the often more convenient use of well-equipped urban parks.

Value attitudes towards nature are acquired primarily in the city, where three-quarters of all Europeans live, and are transferred from there to the rest of the cultural landscape. It therefore makes sense to learn how to deal with nature, its process flows and its challenges in the city, and thus to learn how to deal with nature. This should begin in childhood, where there is already a broad interest in nature, and later be emotionally encouraged and developed further in a learning manner.

Urban forests and urban woodlands are ideal ways to engage with nature in cities (Rink 2004; Tyrväinen et al. 2005; Baró et al. 2014; Breuste and Astner 2018).

Breuste and Astner (2018) were able to show, using the example of a Linz urban woodland, that a natural woodland ("untidy jungle ") is perceived as less attractive and is used less frequently than a "well-kept, tidy and managed forest" with inviting infrastructure (paths, cleanliness, benches, etc.). If the forest alternative is contrasted with a landscaped, large and attractive green space, this alternative is still used significantly more often than even the woodland. This is irrespective of the fact that even under these circumstances the forest is held in particularly high esteem as "nature in itself", which, however, is not always in demand. The attractiveness and appeal of the forest as the most "natural" nature that can be found in cities is contrasted by the need for cleanliness, order and safety and an urban demand for convenience, which woodlands are naturally less able to provide than other green spaces.

What Is the Nymphenburg Palace Park in Munich Worth to Us?

With an area of 229 hectares, Nymphenburg Palace Park is one of the largest and most important works of garden art in Germany. The park woodland alone is 180 hectares in size, making it one of the largest urban park woodland/forests in Germany. As a work of garden art, it is a listed monument and a landscape conservation area and Natura 2000 site. The park woodland/forest is the result of the transformation of an existing oak-hornbeam forest from 1799 onwards.

A study on the ecosystem services of the park woodland/forest at Nymphenburg Palace in Munich and their evaluation yielded the following results:

5

The monetarily valued ecosystem services of the park forest for the year 2011 (in euros/year) were (Aevermann and Schmude 2015):

- Air pollutant reduction: 392,170
- Reduction of rainwater runoff: 155,824
- Carbon sequestration: 15,080
- Groundwater recharge: 16,821
- **Total: 627,586**

The non-monetizable ecosystem services of recreation, nature perception, and relaxation were not included, demonstrating the difficulty of valuing them in monetary terms.

This compares with administrative and management costs (in euro/year):

- Equipment and machinery: 30,000
- Technical maintenance: 130,000
- Park management: 140,000
- Labour costs: 850,000
- **Total: 1,150,000**

If the park woodland/forest were only valued in monetary terms in this way, the services provided would be offset by costs that are twice as high (Aevermann and Schmude 2015) (see ◘ Fig. 5.23).

New Urban Woodland in Leipzig

The city of Leipzig has 20 km² of woodlands in an area of 300 km². It has set itself the goal of increasing this urban forest stock by 50% to 30 km² and thus to 10% of the city area. To achieve this, it intends to use the potential land (43 km²) available after the abandonment of uses. To this end, it is introducing a new open space category "urban forest". In doing so, Leipzig is using a testing and development project of the Federal Agency for Nature Conservation "Ecological urban renewal through the establishment of urban forest areas on inner-city land in the change of use – a contribution to urban development"(2009–2019). The action in Leipzig is also part of the implementation of the National Strategy on Biological Diversity and the German Strategy for Adaptation to Climate Change.

Particular challenges of designing new urban forests are seen as:

- Linking urban redevelopment and urban nature conservation

- Special legal, planning and technical requirements
- High demands on the design in adaptation to the cityscape
- Access and amenities for urban residents – User requirements
- Mostly smaller areas: Minimum size for ecological stability and own internal woodland climate (from approx. 0.5 ha)
- Tree species selection according to design and use objectives
- Woody plants adapted to the urban climate, climate change and pre-exposed soils.

The pilot project "Urban Woodland Leipzig" (Urbaner Wald Leipzig) is intended to develop recommendations for the development of urban forests that can be transferred nationwide. The project, which is funded by the Federal Agency for Nature Conservation (BfN), aims to establish model forests on inner-city brownfield sites, to develop management approaches for long-term protection and acceptance by the population, and to test their suitability for urban redevelopment.

For this purpose, three sample areas were designed as a new urban forest:

- Schönau Wood (Schönauer Holz), Neue Leipziger Straße, 5.5 ha on a deconstruction area of the large housing estate Grünau
- City Nursery Wood (Stadtgärtnerei-Holz), on the edge of the Leipzig-Ost urban redevelopment area, 3.8 hectares on a former city nursery
- Green train station Plagwitz, 5 ha railway site in the urban redevelopment area Leipzig-West

The 3.8 ha "City Nursery Wood" is the first completed sub-project. The design idea in this case ties in with the previous use "market garden" and includes:

- Reference to the old use – clear division into quarters

- Different woodland structures
- Useful woody plants, flowering shrubs, fruit trees.

Plantings consist of 30–50 cm tall woodland plants that must be fenced for the first 5 years. Different forest patterns are established:

Oak-hornbeam-linden forest, forest type: High forest, single-layered, dense

- Hazel-hawthorn-willow woodland, forest type: low stand of small trees
- Hornbeam-hazel forest, forest type: multi-layered, sparse forest
- Powder-berry grove forest, forest pattern: multi-layered, sparse forest
- Cherry grove forest, forest type: tall, multi-layered, sparse forest
- Walnut grove forest, forest pattern: multi-layered, sparse forest
- Wild fruit grove forest, forest image: sparse stand of wild fruit trees
- (BfN 2010; Rink and Arndt 2011; Dietrich 2013) (see ☐ Fig. 5.25).

☐ **Fig. 5.25** New urban woodland in Leipzig, young planted trees in the City Nursery Wood (Stadtgärtnerei Holz) in Leipzig. (© Breuste 2010)

The 2018–2037 Urban Forest Management Plan (UFMP) for Ottawa, Canada

The Canadian capital Ottawa has about 950,000 inhabitants (2016) and an area of 2790 km². Berlin has about three times as many inhabitants on a third of the area (about ten times the population density).

In June 2017, the Canadian capital decided to sustainably manage, maintain and develop its *urban forest* for the next decades. This should include future challenges and opportunities and management, planning, maintenance, growth and accountability. The plan also includes long-term monitoring of the urban forest and a management plan. Ottawa manages 150,000 street trees alone and tens of thousands of other trees in parks and open spaces. The city also manages 2100 ha of natural forest. In addition, there are 14,950 ha of "greenbelt" managed by the National Capital Commission (NCC), which is both publicly and privately owned and partly used for residential, commercial and institutional purposes. The urban forest is thus a heterogeneous tree stand in all respects, in diverse management and use. Designing a management plan for it is a challenge.

Guiding Principles These guiding principles should be considered whenever actions and decisions related to the urban forest are made:

1. The urban forest must be recognized and managed as a valuable infrastructure asset and a positive investment.

2. The City and its partners should be bold and innovative in urban forest management.

3. Urban forest management is a shared responsibility, and working together is the key to achieving success.

4. Urban forest management encompasses a wide range of actions.

5. Urban forest management must be flexible, adaptive, and information-based.

6. All of Ottawa's residents deserve equitable access to the benefits provided by the urban forest.

7. All trees are valuable, but large trees require special considerations.

8. Increasing diversity builds resilience against climate change and other stressors.

Objectives and Targets The urban forest must be recognized and managed as a valuable infrastructure asset and a positive investment.

– The City and its partners should be bold and innovative in urban forest management.

– Urban forest management is a shared responsibility, and working together is the key to achieving success.

– Urban forest management encompasses a wide range of actions.

(Copestake und City of Ottawa Copestake 2017, pp. 12–13; Urban Forest Innovations et al. 2016).

Urban Trees in the Public Eye – The Jackson Magnolia in Front of the White House in Washington

The magnolia planted on the south side of the White House in 1835 (planted by US President Jackson as an offshoot of a tree on his farm in Tennessee) is a witness to history. Between 1928 and 1998, the tree was also featured on the back of the $20 bill.

In 2017, the tree came under the spotlight of the US public. Due to its poor vitality condition (United States National Arboretum Assessment), it was to be partially topped. The US public expressed dismay and wanted to preserve the tree as a living companion to US history. Former US President

Barack Obama gave away seeds of the tree to Israel and Cuba as a gesture of friendship. In the meantime, young plants grown from the tree's seeds are already growing in a greenhouse, intending to replace the decrepit tree completely soon.

» Mrs. Trump (former US-First Lady –) personally reviewed the reports from the United States National Arboretum and spoke at length with her staff about exploring every option before deciding to remove a portion of the Magnolia tree. (Katz 2017).

Solving Ecological and Social Problems with Urban Forest? – The Example of the Halle-Silberhöhe Forest Town

"Overgrowing the city (by trees)" instead of "over-manage the city" ("Stadt-Verwaldung"instead of "Stadt-Verwaltung") was a declared goal of the artist Joseph Beuys. Beginning with documenta 7 in Kassel in 1982, he planted 7000 oak trees (approximately two-metre-high seedlings) as an art project and "social sculpture". Even with the last planting in 1987 for documenta 8, the project was not over. It is only today that its "forest effect" is now really unfolding (Stiftung 7000 Eichen 2018).

In the course of the deconstruction and conversion of large housing estates in the new federal states, the city of Halle/Saale experimented with a forest concept in its Silberhöhe district (built 1979–1989) on an area of 213 ha. Due to people moving away and after demolition of half of the housing stock (7200 apartments), only about 10,000 inhabit-

ants remained from 40,000 inhabitants in 1989 in 2015. Demolished buildings offered a lot of free space for greenery (Breuste and Wiesinger 2013).

In 2004, an integrated urban development concept with the guiding principle "Forest City Silberhöhe" ("Waldstadt Silberhöhe") was adopted (Geiss et al. 2002; Stadt Halle 2007).

This approach primarily involves the development of large wooded open spaces after building demolition. The residential quality should be increased especially by green spaces and the proximity to the Saale-Elster floodplain landscape (Stadt Halle 2007; Vollrodt et al. 2012).

Almost 50% of the residential area is now green space. 70 ha alone have been reclaimed for new urban forest. Whereas in 1992 each inhabitant had approx. 17 m^2 of public green space at his disposal, in 2014 this was approx. 100 m², that is, almost six times more green space and significantly more than elsewhere in the city or even on average in Germany. The maintenance of these enormous areas for only a few inhabitants as classic park and green spaces is impossible for the city of Halle/Saale for financial reasons alone. So new concepts had to be sought. Development of planted new urban forest seemed to be a suitable concept (Breuste et al. 2016) (see ◘ Fig. 5.26).

The model designates urban woodland as the dominant green space structure on deconstruction sites in the future. The concept ranges from a park-like urban forest in the central green corridor to near-natural reforestation and succession areas (Stadt Halle 2007). A total of 8265 trees were planted in three categories of urban forest:

- 23 ha of woodland areas with forest-like tree density of 203 trees/ha
- 23 ha with park-like tree density of 76 trees/ha and
- 24 ha of short rotation plantations (papeln) (Stadt Halle 2007; Vollrodt et al. 2012).

Vollrodt et al. (2012) refer to a "regenerative woodland park" as "Alternative A" and develop two further scenarios from this, with either further forest or park development as the goal. Which forest the citizens want, what is better, what is worse accepted and used and why, has not been investigated yet. Only then could an ecological project also become a social project and a district

◘ **Fig. 5.26** Areas marked for woodland development up to 2025 in the 2007 urban development concept of Halle-Silberhöhe. (City of Halle 2007) (design: J. Breuste, drawing: W. Gruber, Breuste et al. 2016, p. 238, ▶ Fig. 7.19)

◘ Fig. 5.27 Urban (park) woodland/forest on formerly built-up area in the Silberhöhe district, Halle/Saale. (© Breuste 2007)

gain from woodland with its ecosystem services instead of becoming "forested".

The formulated model "woodland city" hardly became reality and remained a vision for the future. If necessary, the new woodland areas are even used for construction activities (Breuste and Wiesinger 2013; Breuste et al. 2016) (see ◘ Fig. 5.27).

The Vertical Forest – Vertical Densification of Nature

Bosco Verticale (*Vertical Forest*) was the first urban forest project (Barreca e La Varra project) with which Milanese architect Stefano Boeri received international attention (2014 International Skyscraper Award, first place, and Emporis Skyscraper Award, second place).

The two residential towers (110 and 80 m high) of a high-rise complex in Milan were built from 2008 to 2013, completed in 2014 (see ◘ Fig. 5.28).

Efficient use of space through high-rise buildings is combined with the idea of improving biodiversity at the same time. A single-family house development would have required about 7.5 ha of land for the same number of inhabitants. The planted high-rise towers are to be new habitats for insects and birds, stepping stone biotopes between parks, urban trees and other green spaces (biotope network). In addition, the microclimate in the apartments and on the balconies is to be improved, noise and dust bound and generally more contact with nature established.

More than 900 trees (20 species), between 3 and 9 m high when planted, and more than 2000 other plants (80 species) were used on terraces and bal-

5

◘ Fig. 5.28 Bosco Verticale, the vertical forest integrated into buildings, Milan, Italy, Stefano Boeri Architetti 2018

conies in 1.3 m deep concrete troughs. This is compared to a planted forest area of 7000 m². The planting of the façade took almost a year. A management system with three employed gardeners was developed for the care of the plants (for example, by crane from the roof of the house). The necessary irrigation is provided by service water from the houses (Stefano Boeri Architetti 2018).

The "Vertical Forest" is already being copied in Chinese cities (Nanjing, Shanghai and Shenzhen). In Nanjing's Pukou district, the first such structure with planned 600 larger, 500 medium-sized native trees and 2500 shrubs and other plants has been under construction since 2016. Its functions are stated as **biodiversity** regeneration, sequestration of 35 t CO_2 per year and **oxygen production of** 60 kg per day (Gatti 2017). It is also a development impulse and symbol for ecological urbanization in China.

Creating the "Balance with Nature" (Boeri) in Urban Space with the (Horizontal and Vertical) Urban Forest – China Is Planning the World's First Forest City

The Italian architect Stefano Boeri, Architetti Milan/Shanghai and Shanghai Tongyan Architectural and Planning Design Co. Ltd., is planning the *"world's first pollution-eating, Forest City"* for the Chinese government in southern China. Since 2016, the Chinese State Council has been orienting towards *"economic, green and beautiful"*. The planned Liuzhou Forest City identifies itself as a city where people will benefit from the ecosystem services of urban forest in their immediate vicinity. Planned on 138.5–175 ha (figures vary), the "Green Forest City" north of Liuzhou, a city of 1.4 million people and the largest industrial city (engineering and auto industry) in Guangxi province, is expected to have a population of 30,000 and be completed in 2020 (construction began in 2016). It is to be connected to its mother city by a railway. The "Green City" is to combat air pollution as a priority and bind 10,000 t of CO_2 and 57 t of air pollutants annually and produce 900 t of oxygen.

In terms of its dimensions, it is a rather small project by Chinese standards, similar to the German Waldstadt Halle-Silberhöhe, except that here, in addition to the green open spaces, the buildings will also be "forested". More than 70 buildings, residential buildings, hospitals, hotels, schools and office buildings will be covered and surrounded by 40,000 trees.

Stefano Boeri refers to this as *"the first experiment of the urban environment that's really trying to find a balance with nature"*, and *"a place where nature is flowing"*.

Promised as innovations:

- Horizontal and vertical greening of buildings
- Traffic in the city based on electric cars
- More than one million plants of more than 100 species (which is nothing unusual)
- Use of renewable energy (solar energy for building energy and geothermal energy for heating and cooling)
- Largely biological active surfaces (instead of soil sealing)
- Temperature reduction through tree crown shading
- Habitats for displaced wildlife and biodiversity

(Alleyne 2017; Stefano Boeri Architetti 2018).

The City Tree – A Literary Excursion with Bertolt Brecht

» Mr K. and nature

Asked about his relationship with nature, Mr. K. said: "I would like to see some trees now and then, stepping out of the house. Especially since they achieve such a special degree of reality through their different appearance according to the time of day and season. In the cities, too, it confuses us by and by to see always only objects of use, houses and railways that would be empty, unoccupied, meaningless. Our peculiar social order, after all, allows us to count people among such objects of use, and there, at least for me, who am not a carpenter, trees have something reassuringly independent, detached from me, and I even hope that even for carpenters they have something about them that cannot be utilized" (translated by the author).

"Why don't you, if you want to see trees, just drive outdoors sometimes?", he was asked. Mr. K. replied in amazement, "I said I wanted to see them stepping out of the house." (Mr. K. also said, "We must make sparing use of nature. Dwelling in nature without work, one easily falls into a morbid state, something like fever afflicts one.") (Brecht 2018a, lines 214–237, translated by the author).

» **The p**oplar from Karlsplatz

A poplar tree stands at Karlsplatz

in the middle of the rubble city of Berlin,

and when people walk across Karlsplatz, they see her friendly green

In the winter of forty-six

the people froze, and the wood was scarce, and many trees fell,

and it became her last year.

But the poplar there on Karlsplatz still shows us its green leaf today: be thankful, residents of Karlsplatz, that we still have it here (Brecht 2018b, translated by the author).

5.3 The Urban Gardens

5.3.1 Natural Element Urban Garden

Gardens are the last connections of urban dwellers to country life. Gardening is one of the basic activities of man. The cultivation of garden and field crops has always been an accompanying natural use of cities (see ▶ Sect. 2.6). It primarily served to supply the urban population with food. Since this local supply was not sufficient as the urban population grew, gardening and field cultivation in and around towns soon became merely an additional food supply. In times of crisis it was extended to all (still) available land. Thus gardens in their individual or public form belong to the very own natural stock of cities.

When it comes to food production (fruits and vegetables), **urban agriculture** has been widely used as a term since the 1930s (Qinglu Shiro: Agricultural Economic Geography) (Mougeot 2006; Swintion et al. 2007; Barthel and Isendahl 2013; ZALF 2013).

FAO uses the term *"urban and peri urban agriculture"* (FAO 2018).

Urban Agriculture, Urban and Peri Urban Agriculture (UPA)

"Urban agriculture" refers to primary food production in urban areas (city and peri-urban) for the **subsistence needs of the respective urban region.** This includes horticulture (home gardens, allotment gardens, grazing land), arable farming, animal husbandry (poultry, domestic rabbits, urban beekeeping) and aquaculture/aquaponics. Urban agriculture can be practiced in all legal forms (private to communal) and is not bound to any socio-economic intention (self-sufficiency, market production, social exchange).

In practice, however, agriculture is also practised in urban regions that has nothing to do with supplying the respective urban region, but also or exclusively **produces for a supraregional market.** This agriculture is localised in the urban area, but is not linked to this area according to FAO requirements for food security and self-sufficiency.

» Urban and peri urban agriculture (UPA) contributes to food availability, particularly of fresh produce, provides employment and income and can contribute to the food security and nutrition of urban dwellers. (FAO 2018).

Urban and peri-urban agriculture (UPA) is primarily defined in terms of content (food production) and space (urban) (Mougeot 2006; Swintion et al. 2007; ZALF 2013). It fulfils a variety of ecosystem services beyond food production (Artmann and Sartison 2018).

Food is now produced in and around cities in a variety of ways. The main production areas are managed by agricultural enterprises (arable land and grassland, glasshouse crops, etc.). They are predominantly production efficiency and market-oriented. The food market is by no means limited to the urban location. The production does not have to be ecological at all.

In addition to these production sites, which dominate in absolute terms of land area and production volume, food is produced on a smaller scale by individuals and social groups in home, small-scale and community gardens, on rooftops, on house walls, on brownfield sites, in urban marine and freshwater areas, and even in containers on balconies. Artmann and Sartison 2018 call these scattered areas **"edible green infrastructure"** and identify their contribution to food production as their primary purpose. That this, as far as not commercially operated, gives rise to the more horticultural use is beyond doubt.

A review of 166 scientific studies on urban and peri-urban agriculture (UPA) found that it can contribute to ten key societal challenges of urbanisation: addressing climate change, food security, biodiversity and ecosystem services, resource efficiency, urban renewal and regeneration, land management, health, social cohesion and economic growth.

It turns out that even though food production is most often listed as an objective with over 50%, recreation plays another important role in 33% of the studies, environmental education and nature learning in 27%, and nature experience and also climate moderation in 17% (Artmann and Sartison 2018).

According to the FAO definition, horticulture is part of broader urban agriculture (Mougeot 2006). This view of urban gardens focuses on food production as the goal of urban horticulture, but fails to recognize the transformations that have taken place in urban gardens in many developed industrialized countries in recent decades, towards recreational functions and practical activity of gardeners, without primary interests in food production. Urban horticulture and urban agriculture, especially on the scale of small-scale production on small areas, cannot be completely separated by definition.

Urban gardening has multiple goals in addition to food production, and "urban gardening" (Müller 2011) has greatly expanded the range of goals over the past two decades.

Urban Gardens

Gardens are urban natural elements between the private and the public. The gardeners in the private are also the users.

The gardeners for **public urban gardens** create aesthetic nature for other, usually broad user groups, ideally the entire urban population. As a rule, the latter can hardly participate in the design of the public urban gardens and cannot intervene horticulturally after completion. Public urban gardens are public green spaces and parks (for example, Breuste et al. 2016, see ▶ Sect. 2.6).

The **private or communally managed and used gardens** are usually only a few hundred square metres in size, often located close to users as home gardens, allotments or community gardens. In contrast to the large, public urban gardens, they allow nature to be designed according to the wishes and needs of the users. The users are also the designers. They are often visited for cultivation and recreation (Dietrich 2014; Breuste et al. 2016).

Private or communal gardens differ considerably from urban agricultural areas (see ◘ Tables 5.5 and 5.6).

For the majority of the population, gardens were initially kitchen gardens. Only the upper classes (Roman Empire, Egypt, China) afforded ornamental gardens without food production, usually in connection with their residences (see ▶ Sect. 2.6). They were not dependent on a food supply close to their homes. This is still the case today. In addition to decorative and recreational gardens, upscale living can even include private parks, formerly the prerogative of the aristocracy only, often managed by professional gardeners.

In the course of the improvement of living conditions, as is currently the case in Central and Northern Europe, the trend in German gardens, for example, is moving from the kitchen garden to the kitchen-ornamental garden with a recreational function. In Southern Europe, the kitchen garden continues to exist.

Table 5.5 Green space types of urban gardens and urban agriculture

Types of gardens and agriculture	Green space type	Description	Use/perception	Management/Maintenance
Urban gardens	Front yard	Decorative gardens (5–20 m²) in front of residential buildings in the open street space	Private/public	Individual/nursing company
	Home garden	Gardens connected to the private house for ornamental and/or kitchen gardens 150 to over 1000 m²	Private/private	Individual
	Allotment garden	Leasehold plots designed as kitchen gardens with recreational function, 200–400 m²	Private/public perceptible	Individual
	Green between multi-storey residential buildings	Landscaped areas in multi-storey rented housing, in part large areas of several 1000 m²	Semi-public/semi-public	Care companies
	Community garden	Kitchen garden, 100 – several hundred m²	Community/semi-public	Community
Urban agriculture	Farmland	Grain production	Production private or semi-public	Private mechanical
	Grassland	Meadows and pastures	Production private or semi-public	Private mechanical
	Orchard, fruit meadow	Fruit production with high trunk trees	Production private or semi-public	Private
	Plantation	Fruit production with low trunk trees, biofuel production	Production private	Private mechanical
	Horticulture		Production private	Private/individual or mechanical

GreenSurge 2015; Breuste et al. 2016

◼ Table 5.6 Comparison of private or communal gardens – urban agricultural areas

	Private or communal gardens	**Urban agricultural land**
Main objective of use	Varied (food, activity, recreation, social contact)	Food production as an economic objective
Food production	Self-sufficiency Multifarious	To supply the population in the urban region (marketing), specialized
Localization	In the cities (at the house to neighbour-hood) and on the outskirts (allotments) according to land availability, short distance to user/owner	Any according to availability, often remote from owner/operator
Professionalism	Individual as a hobby	Professional as a main or sideline business
Quality target	Aesthetic quality, healthy food	Maximum production of food with good quality
Ecological production methods	Extensive renunciation of pesticides, herbicides and artificial fertilizers	Possible, but not a requirement, production efficiency in the foreground
Labour	Owner/lessee and family members	Owner/lessee and/or wage earner
Ecosystem services	Diverse	Food production (primary)

Urban Gardening – Private and Community Gardening

Gardening is the activity of shaping and maintaining nature (soil, relief, plant cover) as a utilitarian or aesthetic object with freely chosen goals. In private gardens (home gardens and allotment gardens as the main forms), gardeners and users are usually family groups or individuals. In community gardens, a partly socially heterogeneous group of city dwellers with common interests related to gardening is active. Together they are designers and users and have made agreements among themselves to this end.

In both groups, the motivations for gardening can be the strong interest in own nature management, the own production of healthy food and active recreation while gardening. In community gardens there is usually also a motivation to work together (cooperation).

In many cities around the world, urban gardening is not a trend or lifestyle, but a vital part of the economy and necessary for feeding people.

» Urban gardening" is a term that encompasses many forms of gardening in urban areas. The woman who grows herbs on her window sill is as much a part of the urban gardening movement as the man who has tomatoes on his balcony or the collective who have turned an abandoned lot into a thriving community vegetable garden, though collective projects make up the majority of the people who currently identify with the label. (Stewart 2018).

In Germany there are about 950,000 allotment gardens with a total area of more than 45,000 ha and about 400 community gardens organized by local urban dwellers (Dietrich 2014). If one calculates at least two to three users for each allotment garden, one arrives at between two and three million urban allotment garden users with a certainly not insignificant "dark figure" of sporadic co-users from the family environment. The Federal Federation of German Allotment and Leisure Gardeners (Bundesverband Deutscher Gartenfreunde) assumes even higher figures, including home gardens. According to its calculations, 17 million hobby gardeners, that is, almost 21% of the total population and about 30% of the population over 18 years of age, cultivate 1.9% of the national area in home gardens and allotment gardens (Bundesverband Deutscher Gartenfreunde 2008, p. 11). **This means that almost every third German adult is a gardenerer!**

The number of allotment gardeners in Germany is slightly decreasing but the allotment gardeners are getting younger. Almost every second allotment plot is located in Eastern Germany. Here alone there are 5871 allotment garden associations with 259,159 gardens. The number of plots in the whole of Germany decreased by 50,000 between 2011 and 2015.

It is true that the management of an allotment garden is still mainly an activity of retired people who spend a lot of time and have a lot of time for this hobby of creating garden nature. However, in the case of new leases the proportion of young families with children is around 40%. This is certainly not yet a trend reversal, but it is a turn towards the family garden in the city as a "green island" with individual design options. The associations strengthen the contact with day-care centres and schools, because gardens awaken the interest of the youngest in nature, perhaps permanently. Allotment gardens are also attractive for migrants. Their share in the allotment garden community is 7.5% – with an increasing tendency and growing integration potential, especially as many migrants bring with them even closer ties to gardens and agriculture.

The hobby gardeners see themselves as "the true greens", without being perceived as such by politics and the public. Gardening close to nature according to the rules of sustainability is promoted in the associations and already practiced extensively (DPA 2015).

Although many aspects of allotment gardening have changed in the course of development, its core, the creative interaction with nature, has been preserved and is still relevant in modern urban life. From the point of view of ecologically oriented urban development, the maintenance of human health and nature-related leisure activities in urban space, especially in large cities, allotment gardening continues to be of great importance at the beginning of the twenty-first century (Breuste 2007, 2010; Breuste and Artmann 2015; Breuste et al. 2016).

Allotment gardening is today a European phenomenon with worldwide "outposts". Organised allotment garden associations, starting from ideas developed in Germany (Leipzig) (see ▶ Sects. 2.6 and 3.6), spread rapidly in Central Europe,

▣ Table 5.7   Federated allotments in Germany (2013/2015) and the UK (2008)

	Number of allotments	Number of allotment garden associations	Area in km^2
Germany	950.000	5871	466.40
United Kingdom	330.000	–	–
Austria	35.500	364	8.96
Saxony	220.000	4000	90.00
Berlin	67.961	808	31.37
Leipzig	40.000	290	9.63
Hamburg	33.000	316	–
Dresden	23.400	366	7.67
Hall	11.847	132	4.79
Cologne	13.000	115	–

Ab-in-den-Urlaub-de 2013; Breuste 2010; Breuste et al. 2016

Western Europe and Scandinavia between 1886 and 1910. In the interwar period, Eastern and Southern European countries adopted the idea. With the development of the environmental movement in the 1970s, urban gardening also spread to southern Europe (Bell et al. 2016).

In most cases the allotment gardeners are organised as associations in allotment garden sites (from a few dozen to several thousand allotment gardens) (▣ Table 5.7).

Wildlife Gardening (Nature-Oriented Gardening)

The near-natural garden is a model of the reintegration of nature into the process flows of gardening and the garden structures. This is becoming increasingly attractive as a way of actively counteracting denaturalisation in the private sphere. Thus, near-natural gardening is also part of a lifestyle and a value system that has become established in society.

A near-natural garden lets nature do some of the gardening. It provides a habitat for animals and wild plants. Maintenance is reduced in favour of natural processes. Natural materials are used. For the nature-oriented gardener, the good feeling arises of contributing something to a healthy nature and to the environment.

Natural gardening included elements such as:

- **Plant selection:** Planting with robust species and wild forms and varieties
- **Maintenance:** no perfect order, reduced maintenance, little frequent mowing of lawns, wild meadows, low sealing only with natural materials (joint greening), sand, chippings or gravel for paths, for example, permaculture, composting

- **Habitats:** for insects, bees, butterflies, birds and small mammals, "insect hotels"
- **Fertilization:** no artificial fertilizers, produce fertilizer yourself, no insecticides or pesticides
- **Elements:** perennials, flower beds, herb spiral, fruit trees, use old fruit varieties, shrubs, mainly native species,natural materials for fences and borders, water areas
- **Soil:** Soil improvement and maintenance by natural means only.

Urban gardening is not, as Müller even states in the subtitle of her 2011 book, "the retun of gardens to the city", because they have never disappeared from it.

» The biggest difference between the traditional institution of allotment gardens and the new urban gardens is not the sparse set of rules or the stronger focus on local food production of the "youngsters", nor is it the lack of fences. Rather, the new garden consciously places itself in relation to the city, enters into a dialogue with it and wants to be perceived as a genuine component of urbanity, not as an alternative to it – and **last but not least as a place where one wants to recover from the city.** (Müller 2011, p. 23, translated).

Community Gardens

A community garden is a piece of land used collectively as a garden, which is cultivated jointly by a group of people and is often open to the public. Often fallow land is converted into (garden) use. The legal status varies. The gardeners, acting as a community, are united by interest and pleasure in gardening, especially in producing their own healthy food. In addition to gardening, community gardeners are united by the need for collective, integrative action to achieve social, environmental or socio-political goals.

The idea of *community gardens* was developed in the 1970s in the USA and established in Europe in the 1990s, often in the context of integration goals (intercultural gardens) (Rosol 2006; Müller 2011; Endlicher 2012; Larson 2012).

Dietrich (2014) cites as new garden forms:
- Community Supported Agriculture (CSA)
- Regional subscription boxes
- Community gardens
- Intercultural gardens
- Neighborhood gardens
- Educational gardens
- Self-harvesting gardens
- Guerrilla gardening

Community garden associations also want to set political signals with their activities, such as actively and concretely contributing to the "energy and cultural turn-around" by jointly using and designing green spaces, growing food, fruit and vegetables. They thus also serve as a field of experimentation for new forms of society (Reimers 2010; Egloff et al. 2014; Transition Regensburg im Wandel 2018).

» To create a social meeting place in the city where we can exchange and spend time in the green. Sharing knowledge, learning about and applying permaculture (implementation in energy cycles, learning from nature). Start Guerilla Gardening actions, reclaim the green in the city, increase the self-sufficiency share of the city, On the way to a conscious, "grandchild-friendly" and regional lifestyle!. (Transition Regensburg in Transition 2018, translated).

» Whoever designs a garden creates a wishful image of the world. You take from nature what cannot run away, the soil and the plants, and imprint your will on it. One transforms the land for the sake of the people, from the most diverse intentions that complement or contradict each other, and already one is in the midst of the conflicts of politics. Horst Günther (Reimers 2010, p. 7, translated).

Urban Gardening has gained great and growing public attention in recent years and has been reflected in a large number of articles in the media. This attention has also been joined by local and regional politics. This has promoted the topic of "urban gardens"as a whole. However, it has also led to the fact that the
allotment garden system, which has existed for more than a hundred years in Germany, for example, has lost some of its importance. About one million German allotment gardeners are currently supported by a few hundred urban gardeners. Both together stand for many common and some different motives for the unbroken approval and importance of gardening in cities.

In countries where urban allotment gardening has no tradition, community gardening has become the most important approach to urban gardening. This movement is spreading the most worldwide (see ◘ Figs. 5.29 and 5.30).

◘ **Fig. 5.29** Guerrilla Gardening in Lodz, Poland, on an occupied unused piece of urban land in the city. (© Breuste 2010)

Fig. 5.30 Karl's Garden on Karlsplatz in the center of Vienna, Austria, "show and research garden for urban agriculture" (► www.karlsgarten.at). (© Breuste 2015)

The Community Gardens Began as Guerrillas

The community garden movement began in the 1970s, prompted by the structural decay and abandonment of land uses generated by the financial crisis. In New York's Lower East Side, a group of "Green Guerillas" dropped "seed bombs" (seeds, fertilizer, and water) over fences onto abandoned open spaces (public or private) to beautify them with plants. It didn't stop there, but the movement developed a program that found participants in neighborhoods. Properties were cleared of trash, fenced, and beautified by volunteers. The "Bowery Houston Community Farm and Garden" was legalized in 1974 by the City Office of Housing Preserva-

tion and Development through a land lease ($1 a month) and became the first community garden (in the world). The gardeners established 60 vegetable beds and planted fruit trees. The Community Garden got more and more attention, won awards. The Green Guerillas began offering workshops, experimenting with plant material adapted to urban conditions, and exchanging plant material with other gardeners in the city. In 1986, the Community Garden was named "Liz Christy's Bowery-Houston Garden" in honor of its founder, who died of cancer in 1985 (NYC Parks 2018) (see **Fig. 5.29**).

Urban Gardening Brings Home-Grown Natural Products into Cities as a New Experience in Dealing with Nature – The Example of Shanghai

Currently, there are already about 20 community gardens in residential areas, schools and parks in Shanghai. They are the result of superordinately promoted,

self-organized efforts to actively shape nature to produce food and learn about dealing with nature, in which children are also involved. This harmoniously

5

combines the traditional values of Chinese culture with the "Western" idea of the urban gardener.

The Knowledge and Innovation Community Garden (KICG) on Wujiaochang Street in Shanghai's Yangpu District is a typical community garden in a residential neighborhood, organized and maintained by local residents and their families.

The "Knowledge and Innovation Community Garden" (KICG) consists of planting beds that can be rented by the residents, jointly cultivated and used beds, play and green areas and a recreation and training room with washrooms and toilets. Great importance is attached to environmental education, especially for children, and training on the subject of "gardening". The garden was built on a "leftover" piece of land provided by the developer as part of a large construction project as a *community interaction space*, thus becoming Shanghai's first community garden as a neighbourhood facility

in 2016. The 220 m² facility is part of the Knowledge and Innovation Community (KIC) Park, a 1 million m² high-tech industrial cluster with information technology at its core. The district government has jointly implemented this with a Hong Kong company. This technology project was to be complemented by a public space with a nature education function. Meanwhile, the garden is also already a highly perceived social communication space. Since KICG is supervised in organization and administration by a professional *design team* (District, Tongji University, etc.), a real local self-administration by the residents is missing so far. The moderation of different interests among themselves and the assumption of their own tasks through their own coordination are still unfamiliar to most. These are aspects that are still likely to change with the transformation of Chinese society from *public* to *private* (Liu et al. 2017b) (see ◘ Fig. 5.31, ◘ Tables 5.8 and 5.9).

◘ **Fig. 5.31**   Knowledge and Innovation Community Garden in Shanghai, China. (© Breuste 2017)

■ **Table 5.8** Comparison of allotment gardens and community gardens in Germany based on selected characteristics

	Allotments	**Community gardens**
Number of clubs in Germany	5871	Over 650
Area in Germany in ha	Approx. 450,000	Approx. 60,000 (no official statistics)
Area per association in m^2	Approx. 10,000	100 to several 1000
Number of gardeners in Germany	950.000	10–15.000
Number of gardeners per association	50–600	10–50
Share of land used for food production (in %)	20–30	60–80
The gardeners	Predominantly pensioners, increasingly also families	Families with children, women, adults in working life, hardly any pensioners
Gardens existing since	Approx. 140 years	Approx. 20 years
Localization	In larger cities, often with an industrial tradition	Especially in big cities
Average share of areas for contemplative recreation (in %)	40–70	10–30
Arbours	Individual per plot (up to 21 m^2)	None or one per club
Structure	Individual plots	(mostly) no parceling
Organisational form	Non-profit association	Various legal forms, from non-profit association to GmbH
Ownership of the land	Mostly tenants	Mostly tenants
Toilets	Mostly WCs	Often composting toilets
Water connection	Always	Mostly
Useful bed structure	Various	Often raised beds or portable planters
Demarcation to the outside by fences or hedges	Yes	Yes

(continued)

5

◘ **Table 5.8**   (continued)

	Allotments	Community gardens
(Fruit) tree population	Mostly present	Barely available
User fees	Yes	Yes
Security	Generally given, partly uncertain, urban development as competitor	Low, residential construction as a competitor
Public relations work of the clubs	Low, internet presence hardly	Intensive, internet presence and social media are the rule
Lobbying	Seldom, hardly any political support	Frequent and aspirational, often political support, for example, from Green parties
External funding, donations and sponsoring	Barely	Frequent, important intake
Political perception (municipalities)	Low, hardly any public relations	Strong, often very good acceptance in city councils, actions in politics, advertising and campaigns
Community spirit	Pronounced	Very strong
Motivation for gardening	Recreation and relaxation, joy of gardening, contact with nature	Enjoyment of gardening, social contact, environmental improvement
Common lifestyles	Not necessarily	Frequently present
Networking of the associations	In the association, nationwide umbrella organization	In various ways, no nationwide umbrella organization Local garden networks
Nature conservation and education as goals of the association	Little	Mostly

Breuste 2007; Breuste and Artmann 2015; Anstiftung 2018

□ Table 5.9 Valuation of ecosystem services by gardeners in Barcelona

Ecosystem services	Importance rating (0–5)
Provisioning services	
Food production	3.75
Production of medicinal and aromatic plants	3.40
Regulating services	
Air purification	4.08
Local climate regulation	4.01
Global climate regulation	3.86
Increase of soil fertility	4.36
Pollination	4.27
Supporting and Habitat services	
Biodiversity	4.26
Cultural services	
Social community and integration	4.40
Identification	4.62
Political message	4.14
Nature promotion	4.65
Food quality	4.57
Aesthetics	4.46
Experience of nature and spirituality	4.51
Relaxation and stress relief	4.62
Entertainment and leisure	4.53
Activity and physical recreation	4.35
Learning and education	4.51
Support for cultural heritage	4.55

Modified according to Camps-Calvet et al. (2016) 0 = Unimportant, 5 = Great importance

5.3.2 Urban Gardens as Services Providers

Urban gardens are an important component of urban nature, a cultural factor, a place for learning, recreation and social encounters (Gilbert 1989a, b). They are green spaces that make the built-up areas habitable in the first place (Schiller-Bütow 1976). Allotment gardens have primarily a social function. They meet the human need for an appropriation of nature and self-designing life in nature, which is subordinate to human principles of order, risk-free and inspiring and conveys a feeling of well-being and health. No other natural component of cities is used as intensively as the garden, especially the urban allotment garden.

Gardening as a hobby is a tradition in Germany. It is ranked seventh among the ten favourite leisure activities in a 2003 survey. Creativity, nature and activity combine to create great acceptance and an awareness of being active in a way that promotes health (Menzel 2006; Rasper 2012).

The individual, self-determined and creative interaction with nature, which is seen as recreation and relaxation, as a counter-image to the rest of urban life, is at the forefront of urban gardening (Breuste and Artmann 2015; Hufnagl 2016).

As part of the green system of large cities, allotment gardens can, among other things, improve the urban climate and air hygiene, increase biodiversity by providing habitats and more contact with nature. The following services of the gardens are cited in particular:

Provisioning services:
- Provision of food

Regulating services:
- Reduction of the air temperature
- Reduction of air pollution
- Reduction of noise pollution
- Reduction of soil and groundwater pollution
- Reducing the contribution to climate change

Cultural services:
- Physical and mental recuperation
- Health through physical activity and mental well-being
- The emotional experience of nature and acquisition of knowledge about nature
- Spiritual or aesthetic appreciation (Madlener 2009; Endlicher 2012; Borysiak and Mizgajski 2016; Borysiak et al. 2017; Breuste and Artmann 2015; Speak et al. 2015; Bell et al. 2016; Schram-Bijkerk et al. 2018).

The services as habitat for plants and animals (biodiversity) are dealt with separately in ► Chap. 6. Social effects of the interactions of the individuals involved are undeniable, but cannot be recognized as ecosystem services.

The **provision of healthy, home-grown food** and the **privacy of the use of the services** are the main differences in performance and perception of performance in urban gardens compared to all other service providers of urban nature (park, forest, water bodies, etc.).

Dietrich 2014 lists a number of benefits of allotments, but not food production:

- **Social functions**
  - Garden culture, cultural knowledge and knowledge management
  - Improving the quality of life
  - Integration
  - Recreation
  - Nature experience and environmental education
  - Upgrading of the residential environment

- **Natural balance and biodiversity**
  - Agrobiodiversity
  - Biotope network
  - Biotope function
  - Ecological compensation

- **Scenery and recreational provision, ecosystem services, climate change**
  - Climate adaptation
  - Climate protection

- **Provisioning services**

Food is produced in urban gardens and the producers are consumers of their own products. These are mainly vegetables and fruit. To a small extent, farm animals (especially rabbits, chickens, pigeons) are kept in allotment gardens. Health regulations of the city prohibit this formerly widespread keeping of animals. In other countries, this is possible on a broader scale, for example, in France (Langemeyer et al. 2016).

For allotment gardeners, the largest group of gardeners in Germany and Austria (Breuste and Artmann 2015), the food supply from their own garden covers only a small part of the eastern and vegetable consumption of allotment gardener households. A survey of 156 allotment gardeners in Salzburg (Austria) showed that the majority of gardeners (52%) produce about 10% of their fruit needs and 44% of their vegetable needs themselves in the allotment garden. The reason given for food production is not the reduction of expenses but the healthy quality of the self-produced food (47%) and its better taste (41%).

The most widely produced fruits (apples, cherries, berries, etc.), lettuce, onions, cabbage, carrots, tomatoes and strawberries are mainly enjoyed fresh (81%) (41% also mainly preserve fruit).

Most gardeners have invested a lot of work and energy to increase (often first to achieve) fertility of their gardens (soil improvement, compost fertilization (85%), tree planting (54%), bush planting (82%)) and cultivate 76% fruits and vegetables.

The trend is towards a reduction in maintenance and the transformation of the kitchen garden into a predominantly (but not exclusively) recreational garden. This is in contrast to the results of studies in other, for example, Mediterranean countries, such as Barcelona, where an increase in the maintenance intensity of gardens

Fig. 5.32 Vegetable cultivation between roadways in Chiang Saen, Thailand. (© Breuste and Astner 2018)

5

consisting almost exclusively of vegetable beds can currently be observed (Hufnagl 2016).

Food provision is not the main motivating factor for gardening in any of the known studies (for example, Breuste 2007, 2010; Breuste and Artmann 2015; Langemeyer et al. 2016), even where gardens consist almost exclusively of vegetable plots (Hufnagl 2016).

However, supply services are becoming increasingly important as trust in industrially produced food from the trade has declined and sensitivity to healthy food has grown. The controlled self-production of food is an alternative for this (Hoffmann 2002).

It can be expected that the interest in food production in community gardens goes even beyond that in allotment gardens. The reason can be found in the higher motivation for food production and in the structure of the community gardens based on this.

In times of crisis and war and economically weak countries, urban gardens are an important source of general food security (Barthel and Isendahl 2013) (see
■ Figs. 5.32, 5.33, 5.34, 5.35, 5.36, 5.37, 5.38, and 5.39).

The British Royal Horticulture Society (2013) identifies eight reasons for urban *allotment gardening* in the UK. Five of these reasons, including reason number one, have to do with producing fresh food. In the UK, the focus of gardening is clearly on healthy, home-grown fruit and vegetables.

■ Rest and relaxation

Main motivations for urban gardening among allotment gardeners, based on a survey of 156 allotment gardeners in Salzburg (Breuste and Artmann 2015) and 93 respondents in Barcelona (Hufnagl 2016), are (Salzburg/Barcelona):
- Relaxation and recreation (80%/83%)
- Contact with nature (65%/61%)
- Gardening as a hobby (65%/65%)
- Rest and retreat from everyday life (57%/47%)

■ Fig. 5.33 Picture plate at the entrance to the Hort Can Masdeu community garden: "We are autonomous" in Barcelona, Spain. (© Hufnagl 2015)

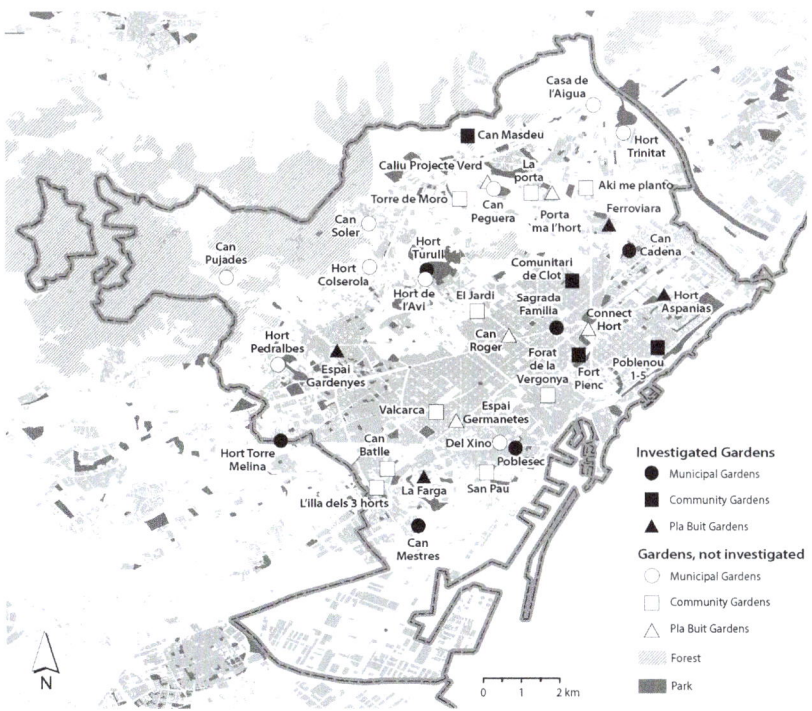

■ Fig. 5.34 Urban garden types in Barcelona, Spain, design: J. Breuste, drawing: W. Gruber 2017

5

◘ **Fig. 5.35** Urban garden (Xarxa d'horts municipal, Municipal Garden) Hort Torre Melina in Barcelona, Spain. (© Hufnagl 2015)

◘ **Fig. 5.36** Urban Garden (Xarxa d'horts comunitaris, Community Garden) Hort Polenou 4 in Barcelona, Spain. (© Hufnagl 2015)

- Establishing social contacts (15%/**52%**)
- Self-sufficiency in fruit and vegetables (**46%**/14%).

Only in the case of social contacts and self-sufficiency with fruit and vegetables do the figures differ significantly. In most respects, gardening is also comparably motivated in different European cultural circles and different garden forms.

For community gardeners in Berlin, Fritsche et al. 2011, based on individual interviews, give as reasons for motivation:

Fig. 5.37 Urban garden (Pla Buit Garden) Hort Espai Gardenyes in Barcelona, Spain. (© Hufnagl 2015)

Fig. 5.38 Tenants' garden complex Silbergrund 1988, Halle-Silberhöhe residential gardens, created in 1988. (© Breuste 2008)

Fig. 5.39 Urban gardening began as early as the 1990s in the USA. Vegetable garden in the Fifth Ward neighborhood of Houston, Texas, USA. (© Breuste 1997)

- Joy in gardening,
- Socialization with other like-minded people and neighborhood contacts,
- Improvement of the green situation in the neighbourhood, with which there is dissatisfaction,
- Safe play and adventure space for your own children.

The establishment of social contacts is not the main motive of allotment gardeners. Although they form associations and cooperate in them, they are more concerned with individuality in their own garden. Only 15.3% of allotment gardeners state the establishment of social contacts as a motivating factor in a Salzburg study (Breuste and Artmann 2015).

Considerable parts of the urban population spend their free time as tenants or their family members in allotment gardens. The majority of allotment gardeners are pensioners with a relatively large amount of free time. Co-users are their younger family members. No other public green space is visited and used only remotely as intensively as the allotment garden (Breuste 2007; Breuste et al. 2016).

In the allotment garden, preferably in the summer half of the year, but also in winter, most of the personal leisure time is spent outdoors, relaxing and gardening. The garden is the second centre of life. The 269 allotment gardeners in Salzburg interviewed in 2005 (Breuste 2007) were two-thirds pensioners who used gardens of 300–350 m² on average. The frequent use of allotment gardens is encouraged by a relatively close spatial relationship with the place of residence, which is nowadays more and more dissolved. Almost all allotment gardeners (95.5%) need less than 30 min to walk from their home (see also Breuste and Breuste 1994).

In the Salzburg study (Breuste and Artmann 2015), the majority spent three to 6 hours a day in the garden (42.8%). High is the number of those who spend six to 9 hours in their own garden (24.9%). 11.9% report spending over 9 hours a day in Salzburg's gardens. 13.8% use the garden on weekdays for 1–3 hours, less than 1.0% of the allotment gardeners spend a maximum of 1 hour a day in the garden. On average 66.0% of the respondents spend more than 6 hours a day in the garden at the weekend.

Allotment gardens thus play an important role in outdoor recreation. The allotment garden replaces other places of recreation in the city or the surrounding countryside. The vast majority of respondents (80.7%) feel particularly close to nature. Allotment gardeners have a particularly **high level of satisfaction** with the urban nature they use, which is not matched anywhere else in urban green spaces. 79.9% are fully satisfied with the allotment garden and express no wishes for improvement or change.

■ **Climate regulation and air pollution control**
These ecosystem services are based on the above-mentioned characteristics of vegetation areas (see ▶ Sects. 5.1 and 5.1.2). The tree population in particular is effective here as a service provider. The effectiveness of this performance can be expected from two factors: First, many larger areas are also located in the middle of the urban area and surrounded by built-up areas. They thus form cool and clean air islands in the building areas. However, their effect on the surrounding area should

not be overestimated. Depending on the vegetation structure, park or forest ecosystems surpass garden areas in their regulatory services (Szumacher 2005; Baró et al. 2014). Some studies indicate that for example, cooling capacity is also related to additional irrigation (Hof and Wolf 2014; Hof 2015). The second fact is the high residence time of gardeners in their gardens. They thus enjoy the benefits for a much longer time than, for example, most visitors to parks. Langemeyer et al. 2016 consider the service area "climate and air pollution control" to be significant, without giving further reasons for this. A third fact is the additional evaporative cooling caused by quite intensive irrigation during the growing season, which is also the main residence period. It can be assumed that this supplementary irrigation corresponds to supplementation of the annual precipitation by several hundred millimetres, which can be particularly effective in areas and years with low rainfall (for example, central German drylands, northern Bohemian basin, Upper Rhine Graben) (Tessin 2001; Freitag 2002; Breuste 2007; Neumann 2013; Breuste and Artmann 2015).

■ **Physical activity and health**

Outdoor physical activity is of particular health importance for older people, still the main group of gardeners in Europe. Browne (1992) and Park et al. (2011) show this importance for older people, who are motivated to outdoor physical activity by gardening more than by any other urban nature form. Gardening has thus been shown to contribute to healthy lifestyles for a particularly sensitive age group.

■ **Experiencing nature and urban gardens as urban places of learning**

Urban gardens offer a variety of opportunities to experience nature and are particularly well suited as places of learning about nature and natural processes for the younger generation. This is mainly due to the fact that they are delimited, manageable natural spaces where, under the guidance and supervision of the older generation(s), learning about natural processes can take place in a self-designing way. As school gardens and educational gardens, this has long been part of the teaching spectrum of schools in various countries. Urban gardens are diverse, which makes learning about different facts related to nature easy. Learning takes place through experimentation, emotionally, across generations and through the acquisition of knowledge. These are ideal conditions for sustainable environmental learning, which also lead to a change in values in relation to nature in general and to an overall appreciative behaviour towards nature, without the usual media-accompanying emotional transfiguration of nature, but supported by one's own practical experiences (Eisel 2012; Freitag 2002; Hoffmann 2002).

» "The allotment garden is a place where children and adults without prior knowledge can quickly and directly access nature" (SenStadt 2012, p. 21). "Allotment gardens are as places of "recollection of "elementary experiences of nature" (Hoffmann 2002, p. 97, translated).

In a study on gardens as places of learning and on nature awareness of allotment gardeners in Salzburg (Breuste and Artmann 2015) 156 gardeners were interviewed. Two-thirds stated that they had learned more about nature through gardening,

especially about natural processes (in relation to care) and had increased their contact with nature (one third each stated this). 78% of gardeners believe that these
experiences are especially important for the younger generation and should be
passed on, which in many cases also takes place through intergenerational knowledge transfer (children are especially taught by grandparents how to deal with
nature). Birds, but also insects and amphibians are perceived as dominant.
Observations of nature are also exchanged among gardeners. Three-quarters of
the gardeners state that they observe nature together with children. Those who are
gardeners observe nature especially in the garden (80%), far less in forests (34%)
and public urban parks (9%). 70% of gardeners consider themselves to be mostly
environmentally aware and to have a special affinity with nature.

This intensity of nature perception through intergenerational learning and joint
activities can hardly be observed in any other urban natural space. This makes
urban gardens particularly valuable as spaces for experiencing nature and as places
of learning (see ◘ Fig. 5.38).

Urban gardening is now taking place in many cities around the world, even
outside Europe. While in smaller cities it is often still initially an activity to supplement the food supply with fresh vegetables, in large cities, especially in emerging
markets, it is already an aspect of a health-conscious lifestyle among the middle
classes. For more and more people, learning more about growing healthy vegetables, herbs and fruits (workshops and teachings are offered by NGOs and committed people) and growing their own vegetables is now part of a modern,
health-conscious lifestyle. A health-conscious lifestyle is part of the trend and can
be afforded financially and in terms of time by more and more people, even outside
Europe. Examples from the metropolises of Bangkok and Shanghai confirm this
(see ◘ Figs. 5.40 and 5.41).

In parts of politics and parts of the public allotment gardeners are unjustifiably
considered as petit-bourgeois, complacent, unworldly fellow citizens and "anach-
ronistic relicts" (Müller 2011, p. 22), as the poet Erich Weinert caricatured them in
1930. This is to be contradicted (Tessin 2001; Neumann 2013) (see ◘ Figs. 5.42
and 5.43).

◘ **Fig. 5.40** Lad Phrao Urban
Experimental and Educational
Garden in a neighborhood in
Bangkok, Thailand. (© Breuste
2016)

◘ **Fig. 5.41** School garden on the roof of a school building in Huay Kwang, Bangkok, Thailand. (© Breuste 2016)

◘ **Fig. 5.42** Small garden as tree-screen planting in San Francisco, USA. (© Breuste 1997)

◘ **Fig. 5.43** Small gardens can also be created where space is limited, tree slab greening in Halle (Saale). (© Breuste and Astner 2018)

5

Eight Reasons for Allotment Gardening in England (RHS 2013)

1. *Get the **freshest produce:** the flavor and freshness of food straight from the plot is streets ahead of most supermarket produce.*

2. ***Save money:** A bag of salad costs as much as a packet of rocket seed, and sometimes a lot more! One packet of seed will give you dozens of bags-worth of tasty salads.*

3. ***Get some exercise** in your own 'green gym': Getting outside in the garden is a proven winner for health and stress relief. 'Allotments are the ultimate stress-buster'.*

4. ***Avoid additives:** If you care about what goes into and onto your food, growing your own organically is the best way of taking control. You can avoid chemical additives that are sometimes found in shop-bought food.*

5. ***Get to know neighbors:** Having an allotment is one of the best ways of getting to know people in your local area. 'Allotment communities are genuine communities, with people from all sorts of backgrounds and ages.'*

6. ***Save food miles:*** Think of the carbon saved by growing your own; a smaller distance from 'plot to plate'also means tastier, fresher food.

7. ***Grow the food you enjoy:*** The number of varieties of fruit and vegetable available to home gardeners is huge compared to the number available in shops.

8. ***A great escape:*** Sometimes it's just great to get away from the house, and normal day-to-day chores! For many, allotments are a perfect stress-buster!"

Barcelona – New Trend – Gardening in Different Garden Forms

Linked to the global financial crisis, since 2008 community gardens have become established as a new component of the green structure of cities in many southern European cities (Keshavarz and Bell 2016). In Spain, however, the first urban gardens were established as early as the 1980s, stimulated by neighbourhood organisations and supported by city councils. They were part of an urban regeneration process aimed at overcoming the deficiencies of public facilities and green spaces in often peripherally located social housing neighbourhoods. This process was also accompanied by the illegal seizure of derelict land and its cultivation through gardens. Mainly pensioners, unemployed people and immigrants from agricultural areas tried to overcome economic crises in this way. In 1997, the "Barcelona City Garden Network", a municipal programme aimed primarily at pensioners, was launched.

This gave rise to different forms of urban gardens in Barcelona:

1. Urban gardens for pensioners (Xarxa d'horts municipals)
2. Community gardens (Xarxa d'horts comunitaris), 11 established between 2002 and 2015
3. Pla Buits Gardens (Gardends established by the municipality by the plan for social inclusion and upgrading of brownfield sites in public ownership for the population), nine since 2013

Community gardens became a form of resistance to the privatization of public open space and offered opportunities for experimentation with new urban lifestyles (Camps-Calvet et al. 2015) (see ◘ Figs. 5.33 and 5.34).

In 2014, Camps-Calvet et al. 2016 surveyed 201 Barcelona urban gardeners about their assessments of ecosystem services provided by the urban gardens they managed (see ◘ Table 5.9).

In a study led by J. Breuste, Salzburg (Hufnagl 2016), 93 urban gardeners in all types of urban gardens in Barcelona were interviewed in 2016 about their food production.

East and vegetable gardening in each type of garden does not differ significantly. It is the main goal of gardening.

The gardens are predominantly vegetable gardens. Ornamental plants hardly play a role. Between 50 and 80% of the gardeners grow fruit and vegetables and consume them fresh themselves or distribute them among family and friends. Preservation plays only a minor role (less than 1/3 of the respondents). The main vegetables grown are tomatoes, lettuce, onions, broad beans (all over 50% of the mentions) and peppers, garlic, potatoes and beans (20–50% of the mentions) and many other vegetables. All forms of gardens have seen an increase in the amount of care required in recent years, often considerably, but the 'occupied gardens' have seen the least. The gardeners consider their lifestyle to be particularly environmentally friendly due to their organically produced vegetables, composting of organic waste (both approx. 80%) and the use of public transport.

However, the motive for gardening is not primarily self-sufficiency in fruit and vegetables. Only 10–23% give this reason. Much more important are relaxation and recreation, the hobby of "gardening", the activity in fresh air and contact with nature. Except in the Pla-Buits gardens, social contacts also play only a subordinate role (see ◘ Figs. 5.35, 5.36, and 5.37).

5

The White House Vegetable Garden in Washington

In the grounds of the White House, there are various vegetable gardens since 1800. Honey from a hive was also used to brew "White House Ale" in the past. Mrs. Roosevelt planted a "Victory Garden" during World War II. Hillary Clinton planted a garden on the roof. On March 20, 2009, Michelle Obama planted the largest vegetable garden to date. Foundations supported this with US$ 2.5 million.

Obama's vegetable garden is 100 m^2 in size and features 55 varieties of vegetables that are also used for White House meals. A portion is earmarked for the local soup kitchen and the Food Bank Organization for the needy. Children from Bancroft Elementary School in Washington were involved.

In 2012, the former First Lady published a book about the White House vegetable garden and healthy food: American Grown: The Story of the White House Kitchen Garden and Gardens Across America (Obama 2012). This encouraged further spread of the garden idea across the US. Since 2017, however, the campaign has not been continued under the former First Lady (Obama 2012; Wikipedia 2018) (□ Fig. 5.39).

5.4 Urban Waters

5.4.1 Natural Element Urban Waters

Wetlands are of great importance for biodiversity, climate and flood protection. Urban water bodies and wetlands are often also part of the urban nature. Many cities are located on rivers and coasts. Textbooks on urban ecosystems devote special attention to urban water bodies in their own sections (for example, rivers, canals, ponds, lakes, reservoirs and water mains, Gilbert 1989a, b, urban water bodies, Leser 2008; Forman 2014, hydrosphere: urban still and flowing waters, groundwater and surface water, wetlands and water in the urban environment, Niemelä et al. 2011; Endlicher 2012; urban water bodies, Breuste et al. 2016). Wetlands, however, are rarely addressed in textbooks. Often, the two, water bodies and wetlands, are considered separately (Forman 2014). This may also be because there is no universally accepted definition of the term "wetland" and the specific ecosystems are assigned differently depending on national traditions. Open water bodies such as lakes, rivers and streams are often not considered to belong to wetlands.

However, good guidance is provided by the generally accepted RAMSAR Convention on the Protection of Wetlands, which classifies water bodies as wetlands.

Wetlands (RAMSAR Definition, Art. 1)

For the RAMSAR Convention, wetlands are wet meadows, bogs and marshes, or bodies of water, natural or artificial, permanent or temporary, standing or flowing, of fresh, brackish or salt water origin, including those marine areas not exceeding a depth of six metres at low water (Ramsar 1971).

Urban Waters

Flowing and still waters that are subject to characteristic influences of urban use (commercial use, flood protection, aesthetic design, pollution, eutrophication, etc.) can be designated as urban waters (Schuhmacher 1998). The particular use leads to significant changes in ecologically relevant water body characteristics compared to water bodies of the same type outside cities (Breuste et al. 2016). Examples of urban waters include small water bodies, ponds, lakes, park waters, stormwater retention basins, streams, rivers, drainage ditches, canals and harbour basins. Urban waters thus do not represent a uniform type in the sense of the classical water body typology (Gunkel 1991).

Wetlands in the Narrower Sense

Wetlands in the narrower sense are transitional areas between terrestrial habitats and permanently wet ecosystems. This covers various habitat types with a year-round surplus of water, such as floodplain, swamp forest, wet meadow, bog, fen, marsh or marshland. Open waters such as lakes, rivers, and streams are not included in wetlands here, although interactions exist and open waters are often enclosed by wetlands (for example, Forman 2014; Naju 2018).

In cities, open water bodies are more common as wetlands in the narrower sense. Stillwaters are naturally formed small water bodies, ponds, lakes, but also artificial, designed water bodies and rainwater retention basins. The transition between "natural" and "artificial" is fluid, especially in cities, and "renaturation" is often a development goal (see ▶ Sect. 8.2) (Gunkel 1991; Gilbert 1989a, b).

Water bodies in the city have a high level of acceptance among the population. Prerequisites are the limitation or exclusion of risks associated with the medium of water. Particularly worth mentioning here are:

— Flood Hazard,
— Danger of drowning, especially for children,
— Health hazard due to contamination and
— Olfactory and visual impairments, for example, due to discharged sewage and waste.

The high attractiveness of the waters is mainly due to

- The peculiarity of the medium of water as a counter-image to the known and familiar land
- The visual aspects (light reflection, reflections, view over the waters, etc.)
- The possibility of observing the process character of the flow of water (natural processes that are effective for a short time and thus directly observable can hardly be observed elsewhere in a similarly impressive way)
- Special ability to observe life forms and processes at the water (animals – birds, fish, insects etc. – in the habitat "water and wetland", natural vegetation development in the transition area between water and land)
- Exclusive accessibility of the medium "water" by means of water vehicles (boat trips, water sports, etc.)
- The "expansion" and visual attractiveness of the public space.

Water is directly associated with a high quality of life in cities. Thus, urban water areas offer suitable offers of use for all age groups. Combined with green spaces in the city, they form an attractive "green and blue infrastructure". An advantage, especially in the case of flowing waters, is their linear structure, which, together with the accompanying vegetation, can form naturally preformed natural corridors through the cities. The prerequisite is that planning and management are aware of this advantage and have not used or do not use precisely these corridors primarily for traffic routes, as this is where the least conflicts of use occur.

Natural and/or designed water bodies are often elements of city parks, sometimes even characterising them (for example, Summer Garden Beijing, West Lake Hangzhou, English Garden Munich).

Coastal areas, partly beaches, designed for recreational use, make up particularly attractive urban natural areas (for example, Clifton Beach in Karachi, Copacabana Beach in Rio de Janeiro, Bondi Beach in Sydney, Blouberg Beach in Cape Town). Cities that have consciously exploited this advantage are among the most attractive cities in the world and derive a good part of their preference from this, also by tourists (for example, Rio de Janeiro, Cape Town, Sydney, Tel Aviv, Casablanca, Stockholm, Helsinki, etc.).

Wet meadows, bogs and marshes, the classic elements of wetlands, are often protected natural areas (often nature reserves) even in cities, with limited accessibility for this reason. Here there are particularly good conditions for nature observation. The main function, the preservation of flora and fauna, need not be significantly impaired by this if well managed. Cities with wetlands are not at all uncommon, but often these areas are little in the awareness of city dwellers and often little visited and used for nature observation. This is not always seen as problematic by nature conservationists, as disturbance by humans may well also affect habitat characteristics and their absence could support nature conservation. Important urban wetlands include parts of Chongming Island in Shanghai (RAMSAR site), Ljubljana Marsh in Ljubljana, the Venice Lagoon, wetlands of the Sabana de Bogotá in Bogota, bogs in Salzburg and many others.

Water bodies and wetlands are often not optimally integrated into the urban landscape in terms of function. In urban development, their development as settlement space or as agricultural supplementary space to settlements occurred relatively late. Unlike other "natural obstacles" such as forests, their drainage required considerable technical effort, which was either not feasible or involved too much expensive compared to the expected benefits. Thus, wetlands often remained unused for a long time as insular natural elements in the developing urban area (for example, moorland development in Salzburg only in the eighteenth/nineteenth century, city (re)founding at the end of the seventh century).

Rivers, as transport routes and sources of energy, were integrated into the urban use cycle at an early stage and were thus also reshaped and built upon. New bodies of water were created for the growing demands of use in the form of canals, reservoirs, ponds and, in some cases, lakes. Existing water bodies were drained in some towns and used for settlement purposes. Drainage of wetlands and water bodies was also undertaken to combat malaria, long a serious health hazard emanating from wetlands.

Pollution of urban waters as a result of unregulated or uncontrolled use is already in decline in many cities. This has been brought about by stronger public awareness and controlled legal regulations. Water pollution has largely lost its latent acceptance. Urban waters need not be sewers. Especially in many developing countries, any further use of water services is ruled out because of health hazards and low aesthetic quality. Efficient monitoring of water quality is necessary to detect water pollution at an early stage and to prevent it in the long term (Kasch 1991).

However, the construction of urban watercourses is still accepted. This is based on the justified concern about flood risks. To minimise or eliminate these risks, the main focus is on technical flood protection through the development of watercourses. This is usually in conflict with the objectives of near-natural watercourse design. Where these conflicting objectives have been identified, measures have been introduced which take both aspects into account in a balanced manner, in some cases also in spatially differentiated sections of watercourses. Technical flood protection is usually indispensable in cities, but this does not exclude near-natural watercourse design, including the dismantling of structures at suitable locations.

Biological impoverishment, lack of self-purification capacity, and lingering flood hazards are often negative characteristics of urban water bodies (Gunkel 1991; Kasch 1991; DVWK 1996, 2000; Schuhmacher and Thiesmeier 1991; Schuhmacher 1998; Leser 2008; Endlicher 2012).

Technical construction also leads to the isolation of watercourses and the reduction of habitat functions. One of the main problems of urban water bodies and wetlands is that their accessibility is often still limited or in some cases not possible. This has not only to do with the lack of attention to this type of urban nature, but often with the necessary effort of access with minimal risk for visitors and the fauna and flora to be protected. The isolated location and low accessibility are therefore often a reason for low use today. However, where these barriers do not exist and wetlands and waterbodies are accessible, their use by visitors is often very high. Sometimes it even needs to be regulated for this reason (Stiftung Die Grüne Stadt 2018).

5.4.2 Services of Urban Waters

Urban wetlands and water bodies in combination with terrestrial natural elements such as parks, forests or grassland are among the most attractive natural spaces in cities due to their diversity of natural components.

Their ecosystem services are of great importance. The use profile of these services has changed again and again in line with societal and individual requirements (see ▶ Chap. 2) (DVWK 1996).

Potential ecosystem services of urban waters include:

5

— Recreation
— Nature experience
— Habitat for flora and fauna
— Climate moderation
— Provision of drinking and process water
— Energy source
— Discharge of wastewater (this often exceeds the disposal potential of the watercourse)
— Cityscape beautification (Endlicher 2012; Breuste et al. 2016) (see ◘ Fig. 5.44)

The functions of urban water bodies have changed significantly over 250 years in many industrially developed cities (see ▶ Sect. 2.7). Today, these are mainly:

— Recreational and leisure use (dominant)
— Disposal (runoff absorption, not sewage absorption)
— Enhancement of the residential environment and the cityscape
— Habitat for plants and animals
— Service water supplier
— Drinking water supply
— Supply of energy (Kaiser 2005; Breuste et al. 2016)

◘ **Fig. 5.44**   Functions of urban waters. (© DVWK 1996 (Breuste et al. 2016, ▶ Fig. 4.13, p. 105))

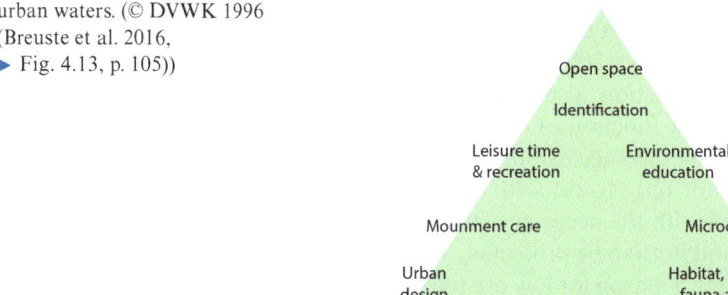

SOCIO-CULTURAL

Open space

Identification

Leisure time & recreation   Environmental education

Mounment care   Microclimate

Urban design   Habitat, retreat for fauna and flora

Hydro power

ECONOMIC   ECOLOGICAL

In many cities in Latin America, Africa and Asia, urban watercourses in particular are used for the following functions:

‐ Service water, especially for agricultural irrigation
‐ Disposal of waste water and solid and fluid waste
‐ Drinking water supply (to a small extent) (see ◘ Figs. 5.45, 5.46, 5.47, 5.48, 5.49 and 5.50)

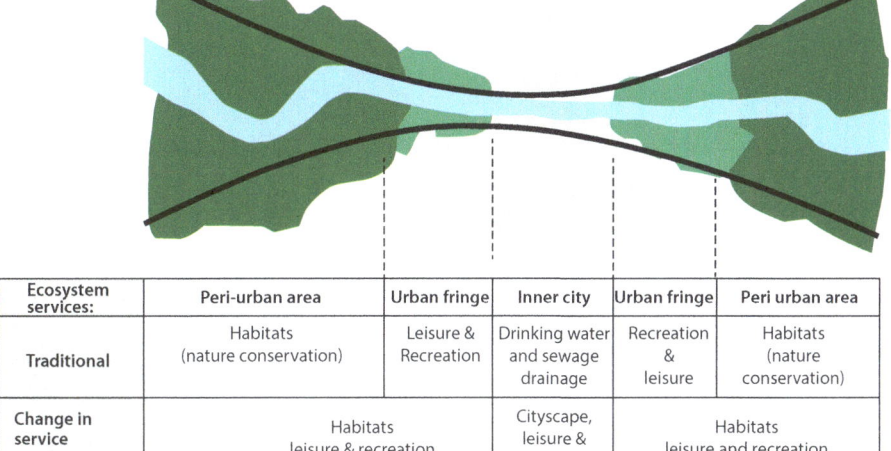

Ecosystem services:	Peri-urban area	Urban fringe	Inner city	Urban fringe	Peri urban area
Traditional	Habitats (nature conservation)	Leisure & Recreation	Drinking water and sewage drainage	Recreation & leisure	Habitats (nature conservation)
Change in service requirements	Habitats leisure & recreation		Cityscape, leisure & recreation	Habitats leisure and recreation	

◘ **Fig. 5.45** Ecosystem services along urban watercourse across an urban area, design: J. Breuste, drawing: W. Gruber

◘ **Fig. 5.46** Urban flowing water as a sewer in Mae Sai, Thailand. (© Breuste and Astner 2018)

🔲 **Fig. 5.47** Urban watercourse used for waste disposal, Karachi, Pakistan. (© Breuste 2004)

🔲 **Fig. 5.48** Rural life at fish ponds in Luang Prabang city, Laos (UNESCO World Heritage City). (© Breuste 2015)

🔲 **Fig. 5.49** User information on the water basin at the Stadthalle, Chemnitz, Germany ("Urban Water, Acess of own risk, Attention! No drinking water, Bathing ban for pets, Danger for slipping risk of injury, water is not disinfected, Parents are responsible for their children"). (© Breuste and Endlicher 2017)

◘ **Fig. 5.50** Part of the Riachuelo River eutrophied with organic polluted wast water from meat processing plants and from residential areas in the suburban area of Buenos Aires, Argentina. (© Breuste and Wiesinger 2013)

The WWT London Wetland Centre – A Wetland Oasis for Nature and People in the Middle of London

The Wildfowl and Wetlands Trust is a UK charity and the largest organisation internationally dedicated to the conservation and protection of wetlands and their biodiversity. Since its foundation in 1946, the WWT has grown to over 200,000 members. Its strategies include acquiring the land of endangered wetlands and having them placed under protection. The WWT owns ten protected areas in the UK with a total area of about 20 square kilometres, habitat for about 150,000 breeding birds. For widespread publicity about wetlands, each of the protected areas has a visitor centre (Wetland Centre), visited by over a million people a year. They offer well-organised, commercial nature experience stays for individuals, groups, families and children.

The WWT London Wetland Centre is located only 40 minutes drive or 10 km from Buckingham Palace and thus from the heart of London. Open to visitors since 2000, the nature reserve consists of several disused Victorian water reservoirs covering more than 40 hectares. In 2002, the Barn Elms Wetland Centre was opened as a "Site of Special Scientific Interest" on an area of 29.9 ha.

The designed wetland is the first of its kind in the UK – and it is an urban wetland, primarily for Londoners. It is habitat for a rich variety of bird life, which is also accessible to observing visitors in its habitat. Regular talks, tours and other events are held to communicate this nature. In 2005 it was featured in the BBC series Seven Natural Wonders to a worldwide audience as "the natural wonder of London".

In 2012, the WWT London Wetland Centre received the BBC Countryfile Magazine Award as the most popular nature reserve in the UK. This shows that nature conservation and nature education can work well together and that wetlands in urban areas are particularly suitable for this purpose (WWT 2018).

5.4.3   Restoration of Ecosystem Services of Urban Waters

Restoring or redeveloping ecosystem services is a management activity for urban waters in many cities. In Germany, for example, the Water Resources Act and state water laws require the renaturation of streams and rivers if no overriding interests are affected. This can be done simply and inexpensively as part of normal maintenance or by means of development measures in an official procedure (with planning approval and planning permission).

With the means of **watercourse restoration,** lost services are to be restored or new services of the urban waters are to be generated. For this purpose, performance targets for this urban nature must be determined before the measures begin. Normally, the original condition cannot be the reference target. Instead, a "**new, near-natural condition**" is defined and supported by initially technical measures. The priority is to improve water quality, followed by the removal of unnecessary obstructions and the (partial) restoration of the natural processes of the watercourse. Increasing low water discharge and combining flood protection with renaturation measures are current challenges in urban watercourse management (DVWK 2000; Breuste et al. 2016). Nature conservation and nature development, flood protection and recreational use must be brought together again in the management of urban waters.

> **Renaturation of Urban Waters**
>
> Renaturation refers to the return of an ecosystem to a condition comparable to that before anthropogenic disturbance or, according to the EU Water Framework Directive, to at least significant improvements towards good ecological status (Directive 2000/60/EC 2000).

The goals of urban stream restoration are:
- Improvement of the water retention capacity
- Maintenance and regeneration of the ecological capacity (biotope function, refugial function, self-purification capacity)
- Water-bound process dynamics in a corridor with the development of natural small biotopes (rushing areas, still water areas, shallow water zones)
- Improvement of usability of the potential for nature experience and tourism (NRC 1992; Breuste et al. 2016) (see ◘ Figs. 5.51 and 5.52).

In addition to restoring natural water conditions and processes to increase biodiversity (see ► Sects. 6.2 and ► 7.3), recreation in nature and experiencing and learning about nature are important ecosystem services that can be perceived especially in connection with water bodies and wetlands. This is due to water as a living medium and the process character, especially of flowing water (see ◘ Figs. 5.53 and 5.54).

Fig. 5.51 Technically managed Pleißemühlgraben water stream in Leipzig, Germany. (© Breuste and Mohr 2000)

Fig. 5.52 Vision for an ecological design of the technically developed Pleißemühlgraben water stream in Leipzig, Germany. (Breuste and Mohr 2000)

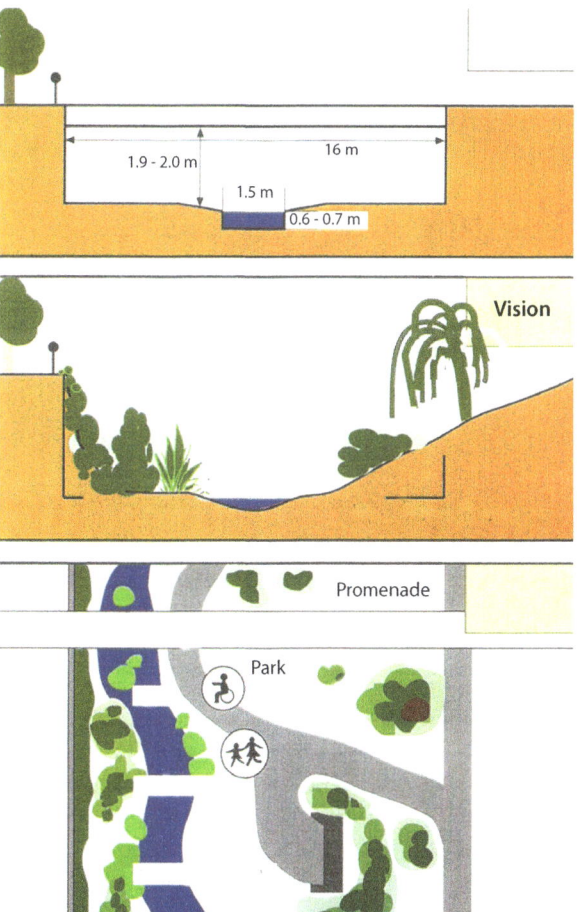

5

■ Fig. 5.53 Section of the renaturalised Isar in Munich. (© Breuste 2015)

■ Fig. 5.54 Children catching frog larvaes at a park water body in Chengde, China. (© Breuste and Faggi 2014)

Renaturation of the Isar in Munich (2000–2011)

The idea of renaturalizing a section of the Isar in Munich (Isar Plan) arose in 1988 with the participation of citizens, associations and political committees. From 1995 onwards, a plan was developed for this purpose. In 2000, the renaturation measures began with the following goals:

- Improved flood protection
- More space and nature-near conditions for the river landscape
- Improvement of recreational and leisure functions

They were successfully completed in 2011 along a length of eight kilometers. The Isar floodplain with its islands, gravel banks, meadows, floodplain forests and parks is considered the most attractive recreational area for cycling, walking, jogging, sunbathing, swimming, barbecuing, playing and experiencing nature for the whole of Munich.

The widening of the riverbed by moving back the flood dikes improved the flood flow and allowed flat, partly terraced, accessible banks, gravel areas and natural bank formations with recreational opportunities and new, interesting possibilities for experiencing nature to emerge. Today, the Isar is much-visited bathing water in Munich with high water quality. At the same time, a near-natural habitat for fauna and flora has been developed. The dynamics of the further development of the river bed are deliberately included. In 2007, the successful project was awarded the DWA (German Association for Water, Wastewater and Waste) Water Development Prize for exemplary measures for the preservation, near-natural design and development of water bodies in urban areas.

The realised project shows that flood protection, improved habitat diversity for typical river animal and plant species, nature experience and recreational use can also be realised together in cities.

The costs for the project (flood protection and renaturation measures) amounted to approximately €35 million They were covered 55% by the Free State of Bavaria and 45% by the City of Munich (Wasserwirtschaftsamt München 2011; Breuste et al. 2016) (see ◘ Fig. 5.53).

The Imperial Summer Residence Chendge, China – A Designed Water Landscape for Imperial Edification

In Chinese garden architecture, emotional impressions were often emphasized and deliberately evoked when combining water, vegetation and pavilions. This is the case, for example, in the 5.6 km^2 landscaped pond landscape of the imperial summer residence in Chengde, China, 250 km northeast of Beijing, the world's largest imperial park. Emperor Qianlong (1711–1799) designated 36 specific places, many of them associated with emotions at bodies of water, with poems on landscape sensation (see ◘ Fig. 5.55).

◘ **Fig. 5.55** Pond landscape in the park of the Chengde summer residence, China, by Emperor Qianlong with poem-stele on landscape perception. (© Breuste and Faggi 2014)

5

The Paraná Delta – Attractive Wetland in Greater Buenos Aires, Argentina

The 14,000 km² delta of the Paraná River into the Rio de la Plata, Argentina (Delta de Paraná), is directly adjacent to the megacity Buenos Aires and only 30 km away from its center. It is the most important habitat for plants and animals and at the same time the most important recreational area of the city of 16 million inhabitants.

The Predelta National Park, created in 1992, covers 24.58 km² of islands, river courses, lagoons, marshes and floodplains in the Paraná Delta. In 2000, the Paraná Delta Biosphere Reserve with a total of 886 km² was created, a unique natural area with extraordinary biodiversity and low human influence (2600 inhabitants). The biosphere reserve aims to combine sustainable development and nature conservation.

The city of Tigre (approx. 100,000 inhabitants) is the gateway to the delta nature for the inhabitants of Buenos Aires. Every weekend, thousands of recreationists arrive here by public transport and disperse to different natural areas, mostly with the aim of water-based recreation.

The city's upper middle and high class has built its gated communities with their own internal water nature amidst this nature. Approximately 25,000 residents live in the nine neighborhoods of the Nordelta residential area alone, which was built in 1999 (Nordelta 2018) (see ◘ Figs. 5.56 and 5.57).

◘ **Fig. 5.56 The** delta of the Paraná, in the agglomeration of Buenos Aires, Argentina, habitat for native species and space for nature experience and recreation. (© Breuste and Wiesinger 2013)

◘ **Fig. 5.57** Nordelta Gated Community – Living in an Uuban waterscape in the Delta of the Paraná River, in the Agglomeration of Buenos Aires, Argentina. (© Breuste 2015)

Río Matanza-Riachuelo in the Agglomeration of Buenos Aires, Argentina – The Dirtiest River in Latin America Is to Become Clean

The Río Matanza-Riachuelo is an only 80 km long river, crosses with shallow depth (0.3–0.5 m) and low flow rate (8 m³/s) the agglomeration of Gran Buenos Aires from the agricultural hinterland, through the suburbs and suburban area, to the industrial core of Buenos Aires, where it flows into the Rio de La Plata. It has a considerable urban catchment area of small water bodies of 2046.56 km². The mainly organic pollution of the river is caused by the meat factories in the surrounding area and the municipalities, which discharge their wastewater almost untreated into the river, which is only small, and the inhabitants, who dispose of household waste in it. This critical situation is now no longer accepted by society, and NGOs and communities are working to turn the river back into a valuable ecological asset. To this end, the Autoridad de Cuenca Matanza-Riachuelo, ACUMAR, a community organization responsible for the entire catchment area, was founded in 2006. Since then, there have been efforts, slow but steady, to improve water quality and make the riparian areas more natural – with some success. The majority of residents are dissatisfied with the situation and would like to see a clean and well-maintained river in their neighborhood. A comprehensive study by an interdisciplinary Argentine-Austrian research team is developing management perspectives on this (Breuste and Faggi 2014; Faggi and Breuste 2015) (see ◘ Table 5.10).

In a project with schoolchildren in the catchment area led by Prof. Faggi and Prof. Breuste, they drew a) the real river in their perception and b) an ideal river of their imagination (see ◘ Figs. 5.58, 5.59, 5.60 and 5.61), thus demonstrating that awareness of the problem of water pollution has already reached even the youngest children.

■ **Table 5.10**   Use of the Río Matanza- Riachuelo catchment, Buenos Aires, Argentina

Use	Area (ha)	Area (%)
Urban	45.305	22.14
Periurban	18.901	9.24
Suburban	14.476	7.07
Rural	111.631	54.55
Afforestation	5828	2.85
Water areas	8515	4.16
Total	**204.656**	**100.00**

■ **Fig. 5.58**   The most polluted river in Latin America, the Río Matanza-Riachuelo River in the agglomeration of Buenos Aires, Argentina, **a** surrounding area. (© Faggi and Breuste 2015)

■ **Fig. 5.59**   The most polluted river in Latin America, the Río Matanza-Riachuelo River in the agglomeration of Buenos Aires, Argentina, **b** Suburban area. (© Faggi and Breuste 2015)

◘ **Fig. 5.60** The most polluted river in Latin America, the Río Matanza-Riachuelo River in the agglomeration of Buenos Aires, Argentina, **c** industrial district at the mouth in La Boca. (© Faggi and Breuste 2015)

◘ **Fig. 5.61** The Río Matanza-Riachuelo river in the agglomeration of Buenos Aires, Argentina, **a** as perceived and **b** as wished for by Briza, a 9-year-old schoolchild from a schhol in the surrounding

Cheonggyecheon Restoration Project – An Almost Eleven-Kilometre-Long Artificial Waterway Transforms the Megacity of Seoul, South Korea

In the expanding and largely denatured centre of Seoul, the capital of South Korea with a population of 25.6 million, a unique renaturation project was implemented in 2003–2005 that set the trend for urban development. An almost 11 km long section of the Cheonggye River, which is covered by a busy road and flows underground in a canal, was opened up and landscaped with accompanying vegetation and footpaths. The waterway had fallen victim to traffic development during the city's modernization and industrialization, and was covered by a concrete structure and used as a road in 1958. In 1976 a highway was built over it. The small river was completely drained. This development represented the modernization and economic boom of the town and the country. Nature had no place, especially in cities. With the opening of the canal, the necessary change of the traffic routing and the (re-)naturalisation, a new way of urban development was taken by the city administration. Although actually no more "renaturation" could take place and even the river water was no longer available and had to be pumped from the Han River, its tributaries and the groundwater into the canal and fed back into the Han River (approx. 120,000 m^3 of water daily), new nature was nevertheless created around the medium "water" in the middle of the city.

The project is exceptionally well received by residents and tourists and is an example of reintegrating nature into cities. The Cheonggyecheon Restoration Project Headquater and the Cheonggyecheon Restoration Research Corps moderate conflicts with the Seoul Metropolitan Government, the transport system, the adjacent businesses and the residents, implement the restoration plan and constantly monitor its realization.

The fact that the canal is not only "renaturalized" by natural elements, but also closed off from vehicular traffic by its location below street level, makes it a pedestrian-only zone of retreat from the otherwise very busy city life and of devotion to nature and relaxation. As such, it very much meets the actual needs of city residents and forms part of an intended network of paths reserved for pedestrians. Overall, the project is part of a larger urban renewal and beautification project in which urban nature is to play a co-determining role.

Creating a natural oasis with clean water and habitat for plants and animals, as well as a corridor with reduced summer temperatures, was the goal sought and achieved.

The high project costs of US$ 281 million (6.5 times more expensive than the 8 km long section of the Isar river in Munich, which was renaturalized in 2011 at much greater expense), the annually increasing maintenance costs and the lack of historical and ecological authenticity and the merely symbolic character of the measure in the face of extensive de-naturalization in Seoul were criticized. A much more extensive ecological revitalization of nature in the entire Cheonggye Basin (approx. 50 km^2) was demanded by environmental organizations.

Despite the small size of the project compared to the massive denaturation that has characterized the city of Seoul in recent decades, it has become a symbol for more nature in the city. The broad approval and intensive use show that urban development must continue in this way and not as before. This has also contributed to a rethinking in the city administration. After the realisation of this project, it will not be possible to return to the guiding principles of previous urban development (Park 2010) (see ◘ Fig. 5.62).

◘ **Fig. 5.62** Section of the renaturalized Cheonggye River in Seoul, South Korea. (© Breuste and Faggi 2014)

References

Ab-in-den-Urlaub-de (2013) Studie: Kleingärten in Deutschland. www.ab-in-den-urlaub.de/.../studie-532-557-kleingarten-in-den-131-grosten-stadten-deutschlands-spitzenreiter-ist-berlin--leipzig-hamburg-und. Accessed 16 March 2018

Aevermann T, Schmude J (2015) Quantification and monetary valuation of urban ecosystem services in Munich, Germany. Z Wirtschaftsgeographie 59(3):188–200

Alleyne A (2017) China unveils plans for world's first pollution-eating 'Forest City'. www.cnn.com/style/article/china-liuzhou-forest-city/index.html. Accessed 9 March 2018

Annerstedt M, Ostergren P-O, Bjork J, Grahn P, Skarback E, Wahrborg P (2012) Green qualities in the neighbourhood and mental health – results from a longitudinal cohort study in Southern Sweden. BMC Public Health 12:337

Anstiftung (Hrsg) (2018) Offene Werkstätten, Reparatur-Initiativen, Interkulturelle und Urbane Gemeinschaftsgärten. https://anstiftung.de/.../1417-erste-schritte-wie-baue-ich-einen-interkulturellen-gemeinschaftsgarten-auf. Accessed 17 March 2018

Appenzeller-Winterberger C, Kaufmann-Hayoz R (2005) Wald und Gesundheit. Schweiz Z Forstwes 7:234–238. https://www.gantrisch.ch/.../Wald-und-Gesundheit_Schweizerischer-Forstverein_2005. Accessed 4 Aug 2019

Artmann M, Sartison K (2018) The role of urban agriculture as a nature-based solution: a review for developing a systemic assessment framework. Sustainability 10:1–32. https://doi.org/10.3390/su10061937. https://www.mdpi.com/2071-1050/10/6/1937/pdf. Accessed 2 Jan 2019

Barbosa O, Tratalos JA, Armsworth PR, Davies RG, Fuller RA, Johnson P, Gaston KJ (2007) Who benefits from access to green space? A case study from Sheffield, UK. Landsc Urban Plan 83:187–195

Baró F, Chaparro L, Gómez-Baggethun E, Langemeyer J, Nowak DJ, Terradas J (2014) Contribution of ecosystem services to air quality and climate change mitigation policies: the case of urban forest in Barcelona, Spain. Ambio 43(4):466–479

Barthel S, Isendahl C (2013) Urban gardens, agriculture, and water management: sources of resilience for long-term food security in cities. Ecol Econ 86:224–234

Bell S, Fox-Kämpfer R, Keshavarz N, Benson M, Caputo S, Noori S, Voigt A (eds) (2016) Urban allotment gardens in Europe. Routledge, London

Bernath K, Roschewitz A, Studhalter S (2006) Die Wälder der Stadt Zürich als Erholungsraum. Besucherverhalten der Stadtbevölkerung und Bewertung der Walderholung. Eidgenössische Forschungsanstalt für Wald, Schnee und Landschaft (Hrsg) Birmensdorf. http://www.waldwissen. net/themen/wald_gesellschaft/unentgekldliche_waldleistungen/wsl_wald_erholung_zuerich_DE. Accessed 11 March 2018

Bernatzky A (1958) Die Beeinflussung des Kleinklimas (Temperatur und relative Luftfeuchtigkeit) durch Grünanlagen. Stadtehygiene 10:191–194

BfN (Bundesamt für Naturschutz) (2010) "Stadtgärtnerei-Holz" – eine neue Waldfläche für Leipzig. Presse Pressearchiv. www.bfn.de. Accessed 21 Dec 2013

Bonauioto M, Aiello A, Perugini M, Bonnes M, Ercolani AP (1999) Multidimensional perception of residential environment quality and neighborhood attachment in the urban environment. J Environ Psychol 19:331–352

Bongardt B (2006) Stadtklimatologische Bedeutung kleiner Parkanlagen, dargestellt am Beispiel des Dortmunder Westparks. (= Essener Ökologische Schriften 24) Hohenwarsleben, p 228

Borysiak J, Mizgajski A (2016) Cultural services provides by urban allotment garden ecosystems. Ekonomia Srodowisko 4(59):292–305

Borysiak J, Mizgajski A, Speak A (2017) Floral biodiversity of allotment gardens and its contribution to urban green infrastructure. Urban Ecosyst 20(2):323–335. https://doi.org/10.1007/s11252-016-0595-4

Brecht B (2018a) Geschichten vom Herrn Keuner. www.cvd-gs.de/uploads/media/AP_Brecht_Herr_K_01.pdf. Accessed 9 March 2018

Brecht B (2018b) Die Pappel vom Karlsplatz. Genius. https://genius.com/Bertolt-brecht-die-pappel-vom-karlsplatz-annotated. Accessed 11 January 2018

Breuste J (2004) Decision making, planning and design for the conservation of indigenous vegetation within urban development. Landsc Urban Plan 68:439–452

Breuste J (2007) Stadtnatur der "dritten Art" – Der Schrebergarten und seine Nutzung. Das Beispiel Salzburg. In: Dettmar J, Werner P (Hrsg) Perspektiven und Bedeutung von Stadtnatur für die Stadtentwicklung. Schriftenreihe des Kompetenznetzwerkes Stadtökologie, Darmstadt, pp 163–171

Breuste J (2010) Allotment gardens as a part of urban green infrastructure: actual trends and perspectives in Central Europe. In: Müller N, Werner P, Kelcey J (Hrsg) Urban biodiversity and design – implementing the convention on biological diversity in towns and cities. Wiley-Blackwell, Oxfort, pp 463–475

Breuste J, Artmann M (2015) Allotment gardens contribute to urban ecosystem service: case study Salzburg, Austria. J Urban Plan Dev 141(3):A5014005

Breuste J, Astner A (2018) Which kind of nature is liked in urban context? A case study of solarCity Linz, Austria. Mitt Österr Geogr Ges 158:105–129

Breuste I, Breuste J (1994) Ausgewählte Aspekte sozialgeographischer Untersuchungen zur Kleingartennutzung in Halle/Saale. Greifswalder Beiträge zur Rekreationsgeografie/Freizeit- und Tourismusforschung 5:171–177

Breuste J, Endlicher W (2017) Stadtökologie. In: Gebhardt H, Glaser R, Radtke U, Reuber P (eds) Geographie – Physische Geographie und Humangeographie, 3rd edn. Elsevier, München, pp 628–638

Breuste J, Faggi A (2014) Urban waters – ecosystem services, restoration and people's perception. In: Proceedings of the 4th International conference urban biodiversity and design, 9–12 October 2014, Incheon, Korea, pp 47–56

Breuste J, Mohr B (2000) Stadtökologisches Gutachten. In: Hochschule für Technik, Wirtschaft und Kultur (eds) Pädagogisch-didaktisches Konzept zur Umweltbildung und – kommunikation am Beispiel der Pleiße-Öffnung, vol 2, Hochschule für Technik, Wirtschaft und Kultur, Leipzig

Breuste J, Rahimi A (2015) Many public urban parks but who profits from them? – the example of Tabriz. Iran Ecol Process 4(6):1–15. https://doi.org/10.1186/s13717-014-0027-4

Breuste J, Wiesinger F (2013) Qualität von Grünzuwachs durch Stadtschrumpfung- Analyse von Vegetationsstruktur, Nutzung und Management von durch Rückbau entstandenen neuen Grünflächen in der Großwohnsiedlung Halle-Silberhöhe. Hallesches Jahrb Geowiss 35:1–26

Breuste J, Pauleit S, Haase D, Sauerwein M (2016) Stadtökosysteme. Funktion, Management, Entwicklung. Springer Spektrum, Berlin

Browne CA (1992) The role of nature for the promotion of well-being of elderly. In: Relf D (ed) The role of horticulture in human well-being and social development. Timber Press, Portland, pp 75–79

Bundesinstitut für Bau-, Stadt- und Raumforschung (BBSR) (Hrsg) (2018) Handlungsziele für Stadtgrün und deren empirische Evidenz. Indikatoren, Kenn- und Orientierungswerte. Bonn, Bundesinstitut für Bau-, Stadt- und Raumforschung

Bundesverband Deutscher Gartenfreunde (eds) (2008) Artenvielfalt: Biodiversität der Kulturpflanzen in Kleingärten. Studie, Berlin

Burkhardt I, Dietrich R, Hoffmann H, Leschner J, Lohmann K, Schoder F, Schultz A (2008) Urbane Wälder. Abschlussbericht zur Voruntersuchung für das Erprobungs- und Entwicklungsvorhaben "Ökologische Stadterneuerung durch Anlage urbaner Waldflächen auf innerstädtischen Flächen im Nutzungswandel – ein Beitrag zur Stadtentwicklung". Naturschutz und Biologische Vielfalt 63:3–214

BUWAL (1999) Gesellschaftliche Ansprüche an den Schweizer Wald – Meinungsumfrage, Bd 309. BUWAL, Bern

Camps-Calvet M, Langemeyer J, Calvet-Mir L, Gómez-Baggethun E, March H (2015) Sowing resilience and contestation in times of crises: the case of urban gardening movements in Barcelona. Partecipazione e Conflicto (PACO) 8(2):417–442

Camps-Calvet M, Langemeyera J, Calvet-Mir L, Gómez-Baggethun E (2016) Ecosystem services provided by urban gardens in Barcelona, Spain: insights for policy and planning. Environ Sci Pol 62:14–23

Carfan AC, Galvani E, Nery JT (2014) Land use and thermal comfort in the county of Ourinhos, SP. Curr Urban Stud 2:140–151

Carp F, Carp A (1982) Perceived quality of neighborhoods: development of assessment scales and their relation to age and gender. J Environ Psychol 2:295–312

Chaudhry P, Tewari VP (2010) Role of public parks/gardens in attracting domestic tourists: an example from city beautiful of India. Tourismos 5(1):101–110

Chiesura A (2004) The role of urban parks for the sustainable city. Landsc Urban Plan 68:129–138

Comber A, Brunsdon C, Green E (2008) Using a GIS-based network analysis to determine urban green space accessibility for different ethnic and religious groups. Landsc Urban Plan 86:103–114

Coombes E, Jones AP, Hillsdon M (2010) The relationship of physical activity and overweight to objectively measured green space accessibility and use. Soc Sci Med 70:816–822

Copestake M, City of Ottawa (2017) Urban forest management plan. https://ottawa.ca/en/urban--forest-management-plan. Accessed 14 February 2018

Crow T, Brown T, De Young R (2006) The riverside and Berwyn experience: contrasts in landscape structure, perceptions of the urban landscape, and their effects on people. Landsc Urban Plan 75:282–299

Deutsche Vereinigung für Wasserwirtschaft, Abwasser und Abfall (ed) (2000) Gestaltung und Pflege von Wasserläufen in urbanen Gebieten. GFA – Gesellschaft zur Förderung der Abwassertechnik, Hennef

Deutscher Verband für Wasserwirtschaft und Kulturbau e. V. (DVWK) (1996) Urbane Fließgewässer – I. Bisherige Entwicklung und künftige städtebauliche Chancen in der Stadt. – DVWK-Materialien 2, Hennef

Die Welt (2001) Auf jeden Münchner kommen 33,8 Quadratmeter Grün. Interview mit Münchens Oberbürgermeister Uhde. Die Welt. https://www.welt.de. Accessed 6 February 2018

Dietrich R (2013) 2. Fachsymposium "Stadtgrün" 11–12. Dezember 2013 in Berlin-Dahlem. Urbaner Wald Leipzig. Stadtplanungsamt Leipzig, Leipzig. https://www.julius-kuehn.de/.../FS-2--Stadtgruen_2.4_Dietrich_Urbaner_Wald_Leipzi. Accessed 13 February 2018

5

Dietrich K (2014) Urbane Gärten für Mensch und Natur. Eine Übersicht und Bibliographie. BfN-Skripten, Bonn-Bad Godesberg, p 386

Directive 2000/60/EC (2000) The European parliament and of the council of 23 october 2000 establishing a framework for community action in the field of water policy. Off J Eur Comm. L327, 1–71, https://eur-lex.europa.eu/resource.html?uri=cellar:5c835afb-2ec6-4577-bdf8-756d3d694eeb.0004.02/DOC_1&format=PDF. Accessed 7 Dec 2021

Dohlen M (2006) Stoffbilanzierung in urbanen Waldökosystemen der Stadt Bochum (= Bochumer Geographische Arbeiten 73) Paderborn, pp 1–125

DPA (2015) Weniger Kleingärtner in Deutschland – der Migrantenanteil wächst. https://www.derwesten.de/.../weniger-kleingaertner-in-deutschland-der-migrantenanteil-waechst-id11058316.html. Accessed 15 March 2018

Dunnett N, Swanwick C, Woolley H (2002) Improving urban parks, play areas and open spaces. Department for Transport, Local Government and the Regions, London

Dwyer FF, Nowak DJ, Noble MH, Sisinni SM (2000) Connecting people with ecosystems in 21st. Century: an assessment of our nation's urban forest. General Technical Report PNW-GTR-490. US Departmant of Agriculture, Forest Service, Pacific Northwest Research Station, Portland

Eastern Urban Center (EUC) Sectional Planning Area (SPA) (ed) (2009) Urban parks, recreation, open space, and trails plan, San Diego

EEA (ed) (2012) The contribution of transport to air quality. TERM 2012: transport indicators tracking progress towards environmental targets in Europe. Report No 10. ISSN 1725–9177. Office for official publications of the European Union, Kopenhagen. https://www.eea.europa.eu/.../transport-and-air-quality-term-2012. Accessed 2 January 2019

Egloff L, Eichenberger U, Siegenthaler T (2014) LOCONOMIE. Die Gemüsekooperative ortoloco Widerspruch – Beiträge zu sozialistischer Politik 64:120–127

Eisel U (2012) Gespenstische Diskussion über Naturerfahrung. In: Kirchhoff T, Vicenzotti V, Voigt A (Hrsg) Sehnsucht nach Natur. transcript, Bielefeld, S 263–285

Eliasson I, Upmanis H (2000) Nocturnal airflow from urban parks – implications for city ventilation. Theor Appl Climatol 66(1–2):95–107

Endlicher W (2012) Einführung in die Stadtökologie. Grundzüge des urbanen Mensch-Umwelt-Systems. Ulmer, Stuttgart

Faggi A, Breuste J (eds) (2015) La Cuenca Matanza-Riachuelo – una mirada ambiental para recuperar sus riberas. Buenos Aires, Universidad de Flores (UFLO), 68. http://img.uflo.edu.ar/a/cuencamatanza.pdf. Accessed 13 April 2018

Fanan U, Dlama KI, Oluseyi IO (2011) Urban expansion and vegetal cover loss in and around Nigeria's Federal Capital City. J Ecol Nat Environ 3:1–10

FAO (1998) Forest resource assessment (RA) 2000: terms and definitions. FRA Working paper No. 1. FAO, Rome. http://www.fao.org/forestry/fo/fra/index.jsp. Accessed 8 March 2018

FAO (Food and Agriculture Organization of the United Nation) (Hrsg) (2018) Food for the cities. www.fao.org/. Accessed 18 March 2018

Fisch I (1966) Zur hygienischen Bedeutung von Grünanlagen für das Klima in Großstadtgebieten. Dissertation, Medizinische Fakultät Universität, Berlin

Forman R (2014) Urban ecology. Science of cities. Cambridge University Press, Cambridge

Freitag G (2002) Kleingärten in der Stadt – ein Beitrag zum ökologischen Ausgleich für den Naturhaushalt. Grüne Schriftenreihe des Bundesverbandes Deutscher Gar-tenfreunde 158:49–65

Fritsche M, Klamt M, Rosol M, Schulz M (2011) Social dimensions of urban restructuring: urban gardening, residents' participation, gardening exhibitions. In: Endlicher W, Hostert P, Kowarik I, Kulke E, Lossau J, Marzluff J, von der Meer E, Mieg H, Nützmann G, Schulz M, Wessolek G (eds) Perspectives in urban ecology. Studies of ecosystems and interactions between humans and nature in the metropolis of Berlin. Springer, Berlin, pp 261–295

Gälzer R (2001) Grünplanung für Städte. Planung, Entwurf, Bau und Erhaltung. Ulmer, Stuttgart

Gasser K, Kaufmann-Hayoz R (2005) Wald und Volksgesundheit. Literatur und Projekte aus der Schweiz. Umwelt Materialien No 195, Bundesamt für Umwelt, Wald und Landschaft, Bern, p 34

Gaston K (Hrsg) (2010) Urban ecology. Cambridge University Press, Cambridge

Gatti L (2017) Bosco Verticale, Nanjing Vertical Forest Towers. https://www.stefanoboeriarchitetti.net/en/portfolios/vertical-forest/. Accessed 6 Oct. 2017

Gebhardt H, Glaser R, Radtke U, Reuber P (eds) (2007) Geographie – Physische Geographie und Humangeographie, 1st edn. München, Elsevier

Geiss S, Kemper J, Krings-Heckemeier MT (2002) Halle Silberhöhe. In: Deutsches Institut für Urbanistik (Hrsg) Die Soziale Stadt. Eine Erste Bilanz des Bund-Länder-Programms "Stadtteile mit besonderem Entwicklungsbedarf- die soziale Stadt". Deutsches Institut für Urbanistik, Berlin, pp 126–137

Gilbert OL (1989a) The ecology of urban habitats. Chapmann & Hall, London

Gilbert OL (1989b) Städtische Ökosysteme. Neumann, Radebeul

GreenSurge (2015) A typology of urban green spaces, ecosystem provisioning services and demands. no place rep

Greiner J, Gelbrich H (1975) Grünflächen der Stadt, 2nd edn. VEB Verlag für Bauwesen, Berlin

Greiner J, Karn H (1960) Freiflächen in Städten. Schriftenreihe Gebiets-, Stadt- und Dorfplanung. Deutsche Bauakademie, Berlin

Grunewald K, Richter B, Meinel G, Herold H, Syrbe R-U (2016) Vorschlag bundesweiter Indikatoren zur Erreichbarkeit öffentlicher Grünflächen Bewertung der Ökosystemleistung "Erholung in der Stadt". Naturschutz und Landschaftsplanung 48(7):218–226

Grunewald K, Richter B, Meinel G, Herold H, Syrbe R-U (2017) Proposal of indicators regarding the provision and accessibility of green spaces for assessing the ecosystem service "recreation in the city" in Germany. Int J Biodivers Sci Ecosyst Serv (Manage Special Issue: Ecosystem Services Nexus Thinking) 13(2):26–39

Gunkel G (1991) Die gewässerökologische Situation in einer urbanen Großsiedlung (Märkisches Viertel, Berlin). In: Schumacher H, Thiesmeier B (eds) Urbane Gewässer. Westarpp Wissenschaften, Essen, pp 122–174

Haase D, Larondelle N, Andersson E, Artmann M, Borgström S, Breuste J, Gomez-Baggethun E, Gren A, Hamstead Z, Hansen R, Kabisch N, Kremer P, Langemeyer J, Rall EL, McPhearson T, Pauleit S, Qureshi S, Schwarz N, Voigt A, Wurster D, Elmqvist T (2014) A quantitative review of urban ecosystem service assessments: concepts, models, and implementation. Ambio 43:413–433

Hannig M (2006) Wieviel Wildnis ist erwünscht? Stadt+Grün 55(1):36–42

Haq SMA (2011) Urban green spaces and an integrative approach to sustainable environment. J Environ Protect 2:601–608

Helms JA (1998) The dictionary of forestry. Society of American Foresters, Bethesda

Hof A (2015) Modelling gardens as social-ecological systems using geodata – the example of watering and landscaping of urban ecosystems. In: Jekel T, Car A, Strobl J, Griesebner G (eds) GI_ Forum 2015. Geospatial minds for society. Wichmann, Berlin, pp 614–624

Hof A, Wolf N (2014) Estimating potential outdoor water consumption in private urban landscapes by coupling high-resolution image analysis, irrigation water needs and evaporation estimation in Spain. Landsc Urban Plan 123:61–72

Hoffmann H (2002) Urbaner Gartenbau im Schatten der Betonriesen. Grüne Schriftenreihe des Bundesverbandes Deutscher Gartenfreunde 158:84–98

Horbert M (1983) Die bioklimatische Bedeutung von Grün- und Freiflächen; VDI-Berichte Nr. 477, Berlin

Horbert M, Kirchgeorg A (1980) Stadtklima und innerstädtische Freiräume am Beispiel des Großen Tiergartens in Berlin. Bauwelt 36:270–276 (bzw. Stadtbauwelt HJ. 67)

Huber I (2016) Indikator-basierte Untersuchungen zur strukturellen Biodiversität & Erholungsnutzung des innerstädtischen Grünraums Mönchsberg in Salzburg. Master Thesis, Faculty for Natural Sciences, Department Geography and Geology, Univ. Salzburg, Salzburg Stadtgärten in Barcelona: Untersuchungen zur aktuellen Situation und Nutzung. University Salzburg, Österreich, Master Thesis Natural Science. Fac

Jauregui E (1990) Influence of a large urban park on temperature and convective precipitation in a tropical city. Energ Buildings 15–16:457–463

Javed MA, Ahmad SR, Ahmad A, Taj AA, Khan A (2013) Assessment of neigbourhoud parks using GIS techniques in Sheikhupura city. Pak J Sci 65(2):296–302

Jim CY (2011) Urban woodlands as distinctive and threatened nature-in-city patches. In: Douglas I, Goode D, Houck M, Wang R (Hrsg) The Routledge handbook of urban ecology. Routledge, London, pp 323–337

5

Johnson CW, Baker FS, Johnson WS (1990) Urban and community forestry. USDA Forest Service, Ogden

Jorgensen E (1970) Urban forestry in Canada. In: Proceedings of the 46th International shade tree conference. University of Toronto, Faculty of Forestry, Shade Tree Research Laboratory, Toronto, pp 43–51

Jorgensen E (1986) Urban forestry in the rearview mirror. Arboric J 10(3):177–190

Kabisch N, Qureshi S, Haase D (2015) Human–environment interactions in urban green spaces – a systematic review of contemporary issues and prospects for future research. Environ Impact Assess Rev 50:25–34

Kabisch N, Strohbach M, Haase D, Kronenberg J (2016) Urban green spaces availability in European cities. Ecol Indic 70:586–596

Kaiser O (2005) Bewertung und Entwicklung von urbanen Fließgewässern. Ph.D Thesis, Faculity for Fortestry and Environmental Sciences, Albert-Ludwigs-University Freiburg i. Br., Freiburg i. Br

Kaplan R, Austin ME, Kaplan S (2004) Open space communities: resident perceptions, nature benefits, and problems with terminology. J Am Plan Assoc 70(3):300–312

Kasch H (1991) Ökologische Grundlagen der Sanierung stehender Gewässer. In: Schuhmacher H, Thiesmeier R (eds) Urbane Gewässer, 1st edn. Westarp, Essen, pp 72–87

Katz B (2017) White-house-magnolia. www.smithsonianmag.com/.../white-house-magnolia-tree-planted-andrew-jackson-be-cut-down-180967657. Accessed 31 Dec 2017

Keshavarz N, Bell S (2016) History of urban gardens in Europe. In: Bell S, Fox-Kämper R, Keshavarz N, Benson M, Caputo S, Boori S, Voigt A (eds) Urban allotment gardens in Europe. Routledge, London, pp 8–32

Kessel A, Green J, Pinder P, Wilkinson P, Grundy C, Lachowycz K (2009) Multidisciplinary research in public health: a case study of research on access to green space. Public Health 123(1):32–38

Konijnendijk CC (2008) The forest and the city. The cultural landscape of urban woodland. Springer, Heidelberg

Konijnendijk CC, Nilsson K, Randrup TB, Schipperijn J (eds) (2005) Urban forests and trees. A reference book. Springer, Berlin

Konijnendijk CC, Richard RM, Kenney A, Randrup TB (2006) Defining urban forestry – a comparative perspective of North America and Europe. Urban For Urban Green 4(3–4):93–103

Konijnendijk CC, Annerstedt M, Nielsen AB, Maruthaveera S (2013) Benefits of urban parks. A systematic review. A report for IFPRA (International Federation of Parks and Recreation Administration). Copenhagen/Alnarp. www.ifpra.org. Accessed 5 Feb 2018

Kowarik I (1995) Zur Gliederung anthropogener Gehölzbestände unter Beachtung urban-industrieller Standorte. Verh Ges Ökol 24:411–421

Kowarik I (2005) Wild urban woodlands: towards a conceptual framework. In: Kowarik I, Körner S (eds) Wild urban woodlands. New perspectives for urban forestry. Springer, Heidelberg, pp 1–32

Kowarik I, Körner S (eds) (2005) Wild urban woodlands. New perspectives for urban forestry. Springer, Heidelberg

Krause H-J, Bos W, Wiedenroth-Rösler H, Wittern J (1995) Parks in Hamburg. Ergebnisse einer Besucherbefragung zur Planung freizeitpädagogisch relevanter städtischer Grünflächen. Münster, Waxmann (= Politikberatung vol 1)

Kuttler W (1998) Stadtklima. In: Sukopp H, Wittg R (eds) Stadtökologie. Gustav Fischer, Stuttgart, pp 125–167

LAF (ed) (1995) Wald und Klima. Schriftenreihe der Sächsischen Landesanstalt für Forsten. part 2, Graupa

Langemeyer J, Latkowska M, Gómez-Baggethun EN (2016) Ecosystem services from urban gardens. In: Bell S, Fox-Kämpfer R, Keshavarz N, Benson M, Caputo S, Noori S, Voigt A (eds) Urban allotment gardens in Europe. Routledge, London, pp 115–141

Larson JT (2012) A comparative study of community garden system in Germany and the United States and their role in creating sustainable communities. Arboric J Int J Urban For 35:121–141

Leser H (2007) Umweltproblemforschung: Wissenschaft und Anwendung aus Sicht von Geographie und Landschaftsökologie. Gaia 16(3):200–207

Leser H (2008) Stadtökologie in Stichworten, 2nd edn. Gebrüder Borntraeger, Berlin

Library of the European Parliament (Hrsg) (2013) Geräuschpegel von Kraftfahrzeugen. Library Briefing 31/01/2013. www.europarl.europa.eu/.../Sound-level-of-motor-vehicles-DE.pdf. Accessed 11 March 2018

Liu S (2015) Urban park planning on spatial disparity between demand and supply of park service. In: International Conference on Advances in Energy and Environmental Science (ICAEES 2015), 1167–1171. www.SLiu–2015–researchgate.net. Accessed 11 February 2018

Liu Y, Li J, Li S (2017a) An evaluation on urban green space system planning based on thermal environmental impact. Curr Urban Stud 5:68–81

Liu Y, Yin K, Wei M, Wang Y (2017b) New approaches to community garden practices in high-density high-rise urban areas: a case study of Shanghai KIC garden. Shanghai Urban Plan Rev 2:29–33

Liu H, Hu Y, Li F, Yuan L (2018) Associations of multiple ecosystem services and disservices of urban park ecological infrastructure and the linkages with socioeconomic factors. J Clean Prod 174:868–879

Maas J, Verheij RA, Groenewegen PP, De Vries S, Spreeuwenberg P (2006) Green space, urbanity, and health: how strong is the relation? J Epidemiol Commun Health 60:587–592

Maas J, Van Dillen SME, Verheij RA, Groenewegen PP (2009) Social contacts as a possible mechanism behind the relation between green space and health. Health Place 15:586–595

Madlener N (2009) Grüne Lernorte. Gemeinschaftsgärten in Berlin, 1st ed. Ergon, Würzburg (= Erziehung – Schule – Gesellschaft; vol. 51)

Magistrat der Stadt Halle (ed) (n.d.) Die Dölauer Heide – Waldidylle in Großstadtnähe. Eigen, Halle

Matsuoka R, Kaplan R (2008) People needs in the urban landscape: analysis of landscape and urban planning contributions. Landsc Urban Plan 84(2):7–19

Menzel P (2006) Der Garten als Lebensraum – nicht nur für den Menschen. Der Fachberater: Verbandszeitschrift des Bundesverbandes Deutscher Gartenfreunde 56(1):13

Meyer FH (ed) (1982) Bäume in der Stadt. Ulmer, Stuttgart

Milcu AI, Hanspach J, Abson D, Fischer J (2013) Cultural ecosystem services: a literature review and prospects for future research. Ecol Soc 18:44

Mitchell R, Popham F (2008) Effect of exposure to natural environment on health inequalities: an observational population study. Lancet 372(9650):1655–1660. http://eprints.gla.ac.uk/4767/. Accessed 2 Jan 2019

Mitscherlich G (1981) Wald, Wachstum und Umwelt: Eine Einführung in die ökologischen Grundlagen d. Waldwachstums. Waldklima und Wasserhaushalt, vol 2, J.D. Sauerländers, Frankfurt a. M.

Mougeot LJA (2006) Growing better cities: urban agriculture for sustainable development. International Development Research Centre, Ottawa

Müller C (eds) (2011) Urban Gardening – Über die Rückkehr der Gärten in die Stadt. oekom, München

Nagendra H, Gopal D (2010) Street trees in Bangalore: density, diversity, composition and distribution. Urban For Urban Green 9(2):129–137

Naju (2018) Feuchtgebiete. Naturschutz-Wiki. www.naju-wiki.de/index.php/Feuchtgebiet. Accessed April 1, 2018

Nath S (2012) Urban forests. A comparative perspective on urban forestry terminology in India, Europe and the United States of America. https://sanchayanwrites.wordpress.com/.../a--comparative-perspect. Accessed 10 March 2018

National Research Council (NRC) (1992) Restoration of aquatic ecosystems: science, technology, and public policy. The National Academies Press, Washington, DC. https://doi.org/10.17226/1807, https://www.nap.edu/.../restoration-of-aquatic-ecosystems-science-technology-and-public--policy. Accessed 11 April 2018

Neumann K (2013) Hat das Kleingartenwesen eine Zukunft? Grüne Schriftenreihe des Bundesverbandes Deutscher Gartenfreunde 227(2):9–27

Newton J (2007) Wellbeing and the natural environment: a brief overview of the evidence. DEFRA & ESRC, London. resolve.sustainablelifestyles.ac.uk/sites/.../JulieNewtonPaper.pdf. Accessed12 Febr 2018

Niemelä J, Breuste J, Elmqvist T, Guntenspergen G, James P, McIntyre N (eds) (2011) Urban ecology, patterns, processes, and applications. Oxford University Press, Oxford

Nordelta (2018) Nordelta. http://www.nordelta.com/. Accessed 12 April 2018

Nowak DJ (2002a) The effects of trees on the physical environment. In: COST Action E12 "Urban forests and trees". In: Proceedings Nr. 1

Nowak DJ (2002b) The effects of urban trees on air quality. https://www.nrs.fs.fed.us/units/urban/local.../Tree_Air_Qual.pdf. Accessed 11 March 2018

Nowak DJ, Heisler GM (2010) Air quality effects of urban trees and parks. National Recreation and Park Association. Research series 2010. www.nrpa.org/.../nrpa.org/.../nowak-heisler-research-paper.pdf. Zugegriffen: 11. März 2018

Nowak DJ, Greenfield EJ, Hoehn RE, Lapoint E (2013) Carbon storage and sequestration by trees in urban and community areas of the United States. Environ Pollut 178:229–236

NYC Parks (2018) History of the community garden movement. https://www.nycgovparks.org/. Accessed 16 March 2018

Obama M (2012) American grown. How the white house kitchen garden inspires families, schools, and communities. Crown, New York

Park JK (2010) Fluss als städtebauliches und architektonisches Element der Stadterneuerung. Dissertation, TU Berlin, Fakultät Planen, Bauen und Umwelt, Berlin

Park SA, Lee KS, Son KC (2011) Determining exercise intensities of gardening tasks as a physical activity using metabolic equivalents in older adults. Hort Sci 46(2):1706–1710

Parsanik Z, Maroofnezhad A (2017) Assessing urban parks of district 13 of Mashhad municipality. Open J Geol 7:457–464

Priego C, Breuste JH, Rojas J (2008) Perception and value of nature in urban landscapes: a comparative analysis of cities in Germany, Chile and Spain. Landsc Online 7:1–22

Puliafito SA, Bochaca FR, Allende DG, Fernandez R (2013) Green areas and microscale thermal comfort in arid environments: a case study in Mendoza, Argentina. Atmos Clim Sci 3:372–384

Pütz M, Bernasconi A (2017) Urban Forestry in der Schweiz: fünf Herausforderungen für Wissenschaft und Praxis (Essay). Schweiz Z Forstwes 168(5):246–251

Pütz M, Schmid S, Bernasconi A, Wolf B (2015) Urban forestry. Definition, Trends und Folgerungen für die Waldakteure in der Schweiz. Schweizerische Z Forstwesen 166(4):230–237

Qureshi S, Kazmi SJH, Breuste JH (2010) Ecological disturbances due to high cutback in the green infrastructure of Karachi: analyses of public perception about associated health problems. Urban For Urban Green 9:187–198

Qureshi S, Breuste JH, Jim CY (2013) Differential community and the perception of urban green spaces and their contents in the megacity of Karachi. Pakistan Urban Ecosyst 16:853–870

Ramsar (1971) Übereinkommen über Feuchtgebiete, insbesondere als Lebensraum für Wasser- und Watvögel, von internationaler Bedeutung. Ramsar 2(2):1971. www.ramsar.org. Accessed 30 March 2018

Randrup TB, Konijnendijk CC, Kaennel Dobbertin M, Prüller R (2005) The concept of urban forestry in Europe. In: Konijnendijk CC, Nilsson K, Randrup TB, Schipperijn J (eds) Urban forests and trees. Springer, Berlin, pp 9–21

Rasper M (2012) Vom Gärtnern in der Stadt: die neue Landlust zwischen Beton und Asphalt. oekom, München

Reimers B (ed) (2010) Gärten und Politik. Vom Kultivieren der Erde. oekom, München

Rink D (2004) Ist wild schön? Garten+Landschaft 2:16–18

Rink D, Arndt T (2011) Urbane Wälder: Ökologische Stadterneuerung durch Anlage urbaner Waldflächen auf innerstädtischen Flächen im Nutzungswandel. UFZ-ed, Leipzig

Roloff A (2013) Bäume in der Stadt. Ulmer, Stuttgart

Roschewitz A, Holthausen N (2007) Wald in Wert setzen für Freizeit und Erholung. Situationsanalyse. Umwelt-Wissen Nr. 0716. Bundesamt für Umwelt, Bern, 39. http://www.waldwissen.net/themen/wald_gesellschaft/unentgekldliche_waldleistungen/wsl_wald_erholung_zuerich_DE. Accessed 11 March 2018

Rosol M (2006) Gemeinschaftsgärten in Berlin: Eine qualitative Untersuchung zu Potenzialen und Risiken bürgerschaftlichen Engagements im Grünflächenbereich vor dem Hintergrund des Wandels von Staat und Planung. Mensch & Buch, Berlin

Rowntree RA, McPherson, EG, Nowak DJ (1994) The role of vegetation in urban ecosystems. In: United States Department Agriculture, Forest Service (eds) Chicago's urban forest ecosystem: results of the Chicago urban forest climate project. General Technical Report NE-186, pp 1–2

Royal Horticultural Society (2013) Eight reasons to get an allotment. RHS Flower Show Tatton Park. www.rhs.org.uk. Accessed 25 July 2013

Saito I, Ishihara O, Katayama T (1990) Study of the effect of green areas on the thermal environment in an urban area. Energ Buildings 15(3–4):493–498

Sawidis T, Breuste J, Tsigaridas K, Mitrovic M, Pavlovic P (2011) A comparative study of heavy metal pollution in Salzburg, Belgrade and Thessaloniki city using trees as bioindicators. Environ Pollut 159(12):3560–3370. https://doi.org/10.1016/j.envpol.2011.08.008

Schetke S, Qureshi S, Lautenbach S, Kabisch N (2016) What determines the use of urban green spaces in highly urbanized areas? – examples from two fast growing Asian cities. Urban For Urban Green 16:150–159

Schiller-Bütow H (1976) Kleingärten in Städten. Patzer, Hannover-Berlin

Schipperijn J, Stigsdotter UK, Randrup TB, Troelsen J (2010) Influences on the use of urban green space – a case study in Odense, Denmark. Urban For Urban Green 9:25–32

Scholz T, Ronchi S, Hof A (2016) Ökosystemdienstleistungen von Stadtbäumen in urbanindustriellen Stadtlandschaften – analyse, Bewertung und Kartierung mit Baumkatastern. In: Strobl J, Zagel B, Griesebner G, Blaschke T (eds) AGIT 2-2016 Journal für Angewandte Geoinformatik. Wichmann, Berlin, pp 462–471

Schram-Bijkerk D, Otte P, Dirven L, Breure AM (2018) Indicators to support healthy urban gardening in urban management. Sci Total Environ 621:863–887

Schuhmacher H (1998) Stadtgewässer. In: Sukopp H, Wittig R (eds) Stadtökologie, 2. Aufl. Gustav Fischer, Stuttgart, S 201–218

Schuhmacher H, Thiesmeier R (eds) (1991) Urbane Gewässer, 1st edn. Essen, Westarp

Schwarz A (2005) Der Park in der Metropole. Urbanes Wachstum und städtische Parks im 19. Jahrhundert. transcript, Bielefeld

Senatsverwaltung für Stadtentwicklung und Umwelt Berlin/SenStadt (2012) Das bunte Grün. Kleingärten in Berlin, Berlin

Sermann H (2006) Kleingärten als Beitrag für ökologische Stadtentwicklung. Grüne Schriftenreihe des Bundesverbandes Deutscher Gartenfreunde 2006:49–53

Speak AF, Mizgajski A, Borysiak J (2015) Allotment gardens and parks: provision of ecosystem services with an emphasis on biodiversity. Urban For Urban Green 14:772–781

Spronken-Smith RA (1994) Energetics and cooling in urban parks. University of British Columbia, Vancouver, p 204

Spronken-Smith RA, Oke TR (1998) The thermal regime of urban parks in two cities with different summer climates. Int J Remote Sens 19(11):2085–2104

Stadt Halle (2007) ISEK – Integriertes Stadtentwicklungskonzept. Stadtumbaugebiete. Halle, pp 75–89. http://www.halle.de/VeroeffentlichungenBinaries/266/199/br_isek_stadtumbaugebiete_2008.pdf. Accessed 19 March 2018

Stanners D, Bourdeau P (eds) (1995) The urban environment. Europe's environment: The dobris assessment. Copenhagen, European Environment Agency, pp 261–296

Stefano Boeri Architetti (ed) (2018) https://www.stefanoboeriarchitetti.net/en/.../liuzhou-forest-city/. Accessed 9 March 2018

Stewart N (2018) Urban gardening in Germany. www.young-germany.de/topic/live/.../urban--gardening-in-germany. Accessed 16 March 2018

Stiftung 7000 Eichen (ed) (2018) 7000 Eichen. www.7000eichen.de/. Accessed 9 March 2018

Stiftung Die Grüne Stadt (2018) Nachhaltige Infrastruktur. Wasser in der Stadt, Schwerpunkt. www.die-gruene-stadt.de/. Accessed 1 April 2018

Stigsdotter UK, Ekholm O, Schipperijn J, Toftager M, Kamper-Jorgensen F, Randrup TB (2010) Health promoting outdoor environments – associations between green space, and health, health-related quality of life and stress based on a Danish national representative survey. Scand J Public Health 38:411–417

Swanwick C, Dunnett N, Woolley H (2003) Nature, role and value of green space in towns and cities: an overview. Built Environ 29(2):94–106

5

Swintion S, Lupi MF, Proberstson GP, Hamilton SK (2007) Ecosystem services and agriculture: cultivating agricultural ecosystems for diverse benefits. Ecol Econ 64:245–252

Szumacher I (2005) Funkcije ekologiczne parków miejskich (Ökologische Funktionen von Stadtparks). Prace i Studia Geograficzne 36. Wydawnictwa Uniwersitetu Warszawskiego (in Polnisch), Universität Warschau, Warschau

Tajima K (2003) New estimates of the demand for urban green space: implications for valuing the environmental benefits of Boston's big dig project. J Urban Aff 25:641–655

Tessin W (2001) Nachhaltige Entwicklung in urbanen Räumen unter besonderer Berücksichtigung des Kleingartenwesens. Grüne Schriftenreihe des Bundesverbandes Deutscher Gartenfreunde 151:7–22

TGL 113-0373 (1964) Freiflächen – Grundsätze und Richtlinien für die generelle Stadtplanung. Fachbereichsstandard Bauwesen. Berlin

Toftager M, Ekholm O, Schipperijn J, Stigsdotter U, Bentsen P, Gronbaek M, Randrup TB, Kamper-Jorgensen F (2011) Distance to green space and physical activity: a Danish national representative survey. J Phys Act Health 8:741–749

Transition Regensburg im Wandel (2018) Gardening. https://www.transition-regensburg.de/gruppen/gardening. Accessed 14 March 2018

Travel + Leisure (eds) (2014) The world's most-visited tourist attractions (2014). Landmarks + Monuments. www.travelandleisure.com. Accessed 6 Feb 2018

Tyrväinen L, Silvennoinen H, Kolehmainen O (2003) Ecological and aesthetic values in urban forest management. Urban For Urban Green 1:135–149

Tyrväinen L, Pauleit S, Seeland K, de Vries S (2005) Benefits and uses of urban forests and trees. In: Konijnendijk CC, Nilsson K, Randrup TB, Schipperijn J (eds) Urban forests and trees. A reference book. Springer, Heidelberg, pp 81–114

Unal M, Uslu C, Cilek A (2016) GIS-based accessibility analysis for neighbourhood parks: the case of Cukurova district. J Digital Landsc Archit 1:45–56

Urban Forest Innovations Inc., Beacon Environmental Ltd., Kenney WA (2016) Putting down roots for the future: city of Ottawa – urban forest management plan 2018–2037. Ottawa

Van Herzele A, Wiedemann T (2003) A monitoring tool for the provision of accessible and attractive urban green spaces. Landsc Urban Plan 63:109–126

Voigt A, Kabisch N, Wurster D, Haase D, Breuste J (2014) Structural diversity: a multi-dimensional approach to assess recreational services in urban parks. Ambio 43:480–491

Vollrodt S, Frühauf M, Haase D, Strohbach M (2012) Das CO2-Senkenpotential urbaner Gehölze im Kontext postwendezeitlicher Schrumpfungsprozesse. Die Waldstadt-Silberhöhe (Halle/Saale) und deren Beitrag zu einer klimawandelgerechten Stadtentwicklung. Hallesches Jb Geowiss 34:71–96

von Stülpnagel A (1987) Klimatische Veränderungen in Ballungsgebieten unter Berücksichtigung der Ausgleichswirkungen von Grünflächen. Ph.D Thesis, Department 14, Technical University of Berlin

Wang H, Qureshi S, Qureshi BA, Qiu J, Freidman CR, Breuste JH, Wang X (2016) A multivariate analysis integrating ecological, socioeconomic and physical characteristics to investigate urban forest cover and plant diversity in Beijing, China. Ecol Indic 60:921–929

Ward Thompson C, Roe J, Aspinall P, Mitchell R, Clow A, Miller D (2012) More green space is linked to less stress in deprived communities: evidence from salivary cortisol patterns. Landsc Urban Plan 105(3):221–229

Wasserwirtschaftsamt München (ed) (2011) Neues Leben für die Isar. Faltblatt. http://www.muenchen.de/rathaus/Stadtverwaltung/baureferat/projekte/isar-plan.html. Accessed 18 Nov 2015

Weber D, Anderson D (2010) Contact with nature: recreation experience preferences in Australian parks. Ann Leis Res 13:46–69

Wikipedia (2018) White House vegetable garden. https://en.wikipedia.org/wiki/White_House_Vegetable_Garden. Accessed March 19, 2018

Wildfowl and Wetlands Trust (WWT) (2018) https://www.wwt.org.uk/wetland-centres/london/. Accessed 15 April 2018

Wittig R (2002) Siedlungsvegetation. Ulmer, Stuttgart

Yang J, McBride J, Zhou J, Sun Z (2005) The urban forest in Beijing and its role in air pollution reduction. Urban For Urban Green 3(2):65–78

ZALF (Leibniz-Zentrum für Agrarlandschaftsforschung e. V.) (2013) Urbane Landwirtschaft und "Green Production" als Teil eines nachhaltigen Landmanagements. Diskussionspapier Nr. 6: Müncheberg

Zippel S (2016) Urban parks in Shanghai study of visitors' demands and present supply of recreational services. Master Thesis TU Dresden, Faculty Envirnomental Sciences, Dresden

Zipperer WC, Sisinni SM, Pouyat R (1997) Urban tree cover: an ecological perspective. Urban Ecosyst 1(4):229–246

What Constitutes Urban Biodiversity?

Contents

6.1   Urban Biodiversity:
        A Paradigm Shift? – 230

6.2   How Can Urban Biodiversity
        Be Measured? – 234
6.2.1   Integration Levels of
        Biological Diversity – 234
6.2.2   Structure of Urban Habitats
        as Indicator and Basis of Spatial
        Assessment – 235
6.2.3   Species Diversity as Indicator – 238

6.3   How Is Urban Biodiversity
        Perceived? – 244

6.4   Urban Biodiversity and
        Ecosystem Services – 249

        References – 251

© Springer-Verlag GmbH Germany, part of Springer Nature 2022
J. Breuste, *The Green City*,
https://doi.org/10.1007/978-3-662-63976-4_6

Biodiversity is often perceived only as species diversity. However, the structure of habitats is part of it and makes biodiversity comprehensible. This also applies to urban biodiversity. Like urban nature as a whole, designed biodiversity is based on urban structures and their maintenance and management. Biodiversity is shaped in the city and can therefore also be increased or reduced.

There is often a broad view of urban biodiversity that always sees it (only) in the context of ecosystem services and thus subsumes all the services of urban nature under the name "biodiversity". However, this devalues the concept of biodiversity and detaches it from its content.

Biodiversity in cities, like biodiversity in general, is diversity of species and habitats and genetic diversity of species. That this has a positive effect on people as city dwellers is often claimed, but needs concretisation, also to improve and expand this effect. People usually perceive biodiversity only fragmentarily and "in passing", so to speak, but often do not know what lies behind the term. This is confirmed by many studies. However, they perceive and value the diversity of natural structures and (observable) species very accurately. If diversity is valued positively, this also applies to biological diversity. Contact with nature is generally seen as positive and is in demand, even in cities.

Can biological diversity be measured? If so, how can structural and species diversity be quantified? Do only native species belong to urban biodiversity? Conservation is oriented towards the protection of native species outside cities. The City Biodiversity Index (CBI), a measurement and comparison tool used by planners and scientists, also reduces urban *biodiversity* to *native biodiversity* User's Manual 2014. The CBI thus deliberately excludes a not inconsiderable part of urban nature and reduces it to exclusively native species. Justified doubts may be raised as to whether this approach meets the demands for a broad, integrative design and appreciation of urban nature. Some ecologists are already calling for a necessary paradigm shift in the way biodiversity is viewed in relation to cities (e.g., Kowarik 2011).

6.1   Urban Biodiversity: A Paradigm Shift?

The Convention on Biological Diversity (CBD) was adopted in 1992 at the UN Conference on Environment and Development to ensure the conservation and sustainable use of biological diversity. This goal, known as "biodiversity" for short, has since been incorporated into the policies of many countries and accepted and recognised by the general public as a desirable and even necessary undertaking (United Nations 1992; COP 2009, 2010; TEEB 2011; Naturkapital Deutschland TEEB DE 2016).

The concept of diversity was divided by Whittaker (1972, 1977) into alpha (number of species/area, ecotope, plant population), beta (dimensionless via similarity values), along a gradient or for temporal comparisons, and gamma diversity (number of species/area, larger area, landscape, country) in order to characterize species diversity at different scale levels. In 1992, the above-mentioned International Convention on Biological Diversity CBD (United Nations 1992) extended the term to the level of ecosystems. This also includes, for example, the diversity of cultivated plants. However, the transfer to urban areas was not yet intended and only became an issue later on. In simplified terms, biodiversity is often equated with species diversity or species richness, especially in the media, and thus reduced.

Biodiversity

The term "biodiversity" is synonymous with "biotic diversity".

"... "biological diversity" means the variability among living organisms from all sources, including, inter alia, terrestrial, marine and other aquatic ecosystems and the ecological complexes of which they are part; this includes diversity within species and between species and diversity of ecosystems" (United Nations 1992, Article 2, p. 3).

This includes species diversity, the diversity of ecosystems and genetic diversity within the various species. The variability of spatial, temporal and functional properties of natural elements of different hierarchical assignments is one aspect of biodiversity (Beierkuhnlein 1998).

At the same time, biodiversity is a measurable and operationalizable ecological concept with concrete goals for nature and species conservation (United Nations 1992; Hobohm 2000; Wittig and Niekisch 2014) (see ► Sect. 7.3 and ◘ Fig. 6.1).

The spectrum of conservation and development activities for biodiversity in cities ranges from biodiversity research to the promotion of contact with nature, nature conservation and the maintenance of urban nature. Urban biodiversity and efforts to conserve it cannot focus on the few relict habitats and species of native flora still present in urban areas. Kowarik (2011, p. 1) rightly calls for a **"paradigm shift"** to include all urban ecosystems and also their non-native species, referring to their ecosystem services, social benefits and contribution to biodiversity conservation. In Germany, the National Biodiversity Strategy Germany includes this: "Native species find a substitute habitat here and warmth-loving immigrant species settle" (BmU 2007, p. 42).

Urban biodiversity

Urban biodiversity is based on the special nature of urban ecosystems. This concerns all species and all habitats, that is, different levels of integration of biological diversity (Beierkuhnlein 1998). Urban biodiversity does not refer exclusively or primarily to native species and their urban relict habitats, but also includes the diversity of cultivated plants and non-native species. Urban biodiversity is thus not only a result of natural processes, but also of conscious or unconscious design by humans, especially their ways of using urban ecosystems. Urban biodiversity is not "found", but designed. This means a paradigm shift of traditional concepts of nature conservation, with the protection goal of habitats that are little touched and exclusively native species.

6

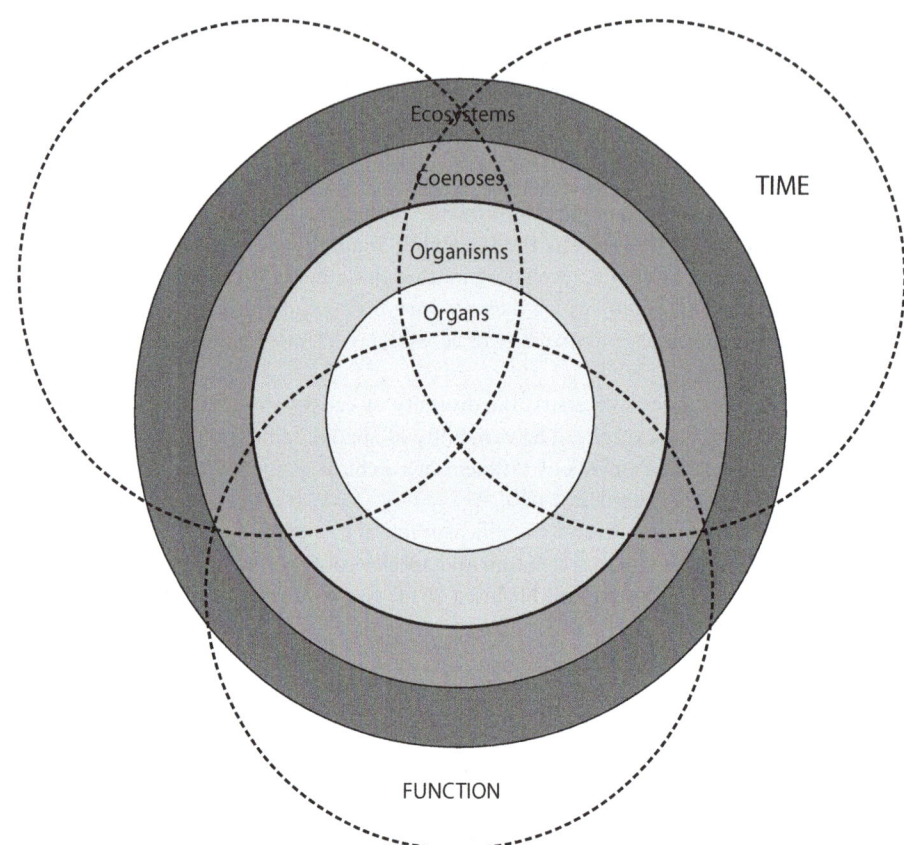

□ Fig. 6.1   Scheme of biotic diversity, circles denote biotic integration levels. (© After Beierkuhnlein 1998, p. 84, modified, design: J. Breuste, drawing: W. Gruber)

Cities: Urban Hotspots of Biodiversity

With their high species diversity and species density, cities are often designated as regional biodiversity hotspots (Werner and Zahner 2009). Kühn et al. (2004) state for Central Europe that for urban areas larger than 100 km² and with more than 200,000 inhabitants, approximately 1000 plant species in total and 30 to 600 plant species per km² can be expected. This far exceeds the species diversity of the intensively used cultivated landscape. The high number of species in cities can be explained by site diversity, often extreme and special ecological site conditions. In most cases, only spontaneous vegetation is recorded statistically, without differentiating between indigenous and hemerochore species. The comparison of plant diversity with near-natural ecosystems, in which mostly only indigenous species occur, shows that urban biodiversity is often characterized to a not insignificant extent by immigrant species (neophytes) (Breuste et al. 2016).

In the Frankfurt area (BioFrankfurt network), 1675 ferns and flowering plants are distributed. With an area share of only 0.06% of the German territory, this is about half of all known species in Germany. In the Taunus Mountains, which are eleven times larger, there are only 1250 plant species (Lehmhöfer 2010).

Since 1985, the Senckenberg Research Institute and Nature Museum has been systematically recording Frankfurt's urban nature through scientific mapping in the Biotope Mapping Working Group on behalf of the City of Frankfurt's Environment Department. Urban planning and development make their decisions on the basis of these biodiversity surveys and analyse their effects on biodiversity. The results

on biodiversity are also made accessible to a broad public through publication (see ■ Fig. 6.2). Despite these successes in integrating urban biodiversity into the decision-making processes of urban development, the assessment of the scientists of the Senckenberg Research Institute regarding its protection is not very optimistic:

» "Nevertheless, urban nature, with its effect and significance usually perceived only indirectly and in the long term, is often on the losing end in the face of economic constraints that have a very immediate impact" (Ottich et al. 2009, p. 158, translated).

■ **Fig. 6.2**   BioFrankfurt – concept of the network for biodiversity. (© BioFrankfurt)

6.2 How Can Urban Biodiversity Be Measured?

6.2.1 Integration Levels of Biological Diversity

The analysis of biodiversity is concerned with combining and delimiting natural elements, that is, biotic compartments of different levels of organisation, into abstract units. With increasing integration (organization), the complexity of the natural elements grows. In addition to species, superordinate units (biocenoses, ecosystems, formations) and the diversity of their various properties play a special role here. In the case of ecosystems (systemic level), it is possible to analyse and compare the diversity of units in an area in order to derive statements about (bio) diversity at this level of integration. According to Beierkuhnlein (1998), species (second integration level), biocenoses as communities of organisms of different species (third integration level) and together with their habitat "biotope" as ecosystems (fourth integration level) are practically usable integration levels for quantifying biodiversity. Assessments of biodiversity usually refer to a specific area or scale of observation (Beierkuhnlein 1998). These ideas are also transferable to urban zones and ecosystems (see ☐ Figs. 6.3 and 6.4).

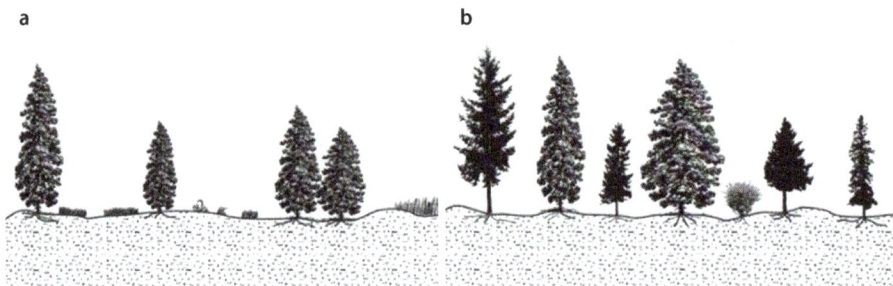

a   b

☐ **Fig. 6.3** Schematic representation of low **a** and high **b** diversity at the species level using the example of plant communities with different species diversity, for example, in an urban park. (© Drawing: W. Gruber, using Beierkuhnlein 1998, p. 86)

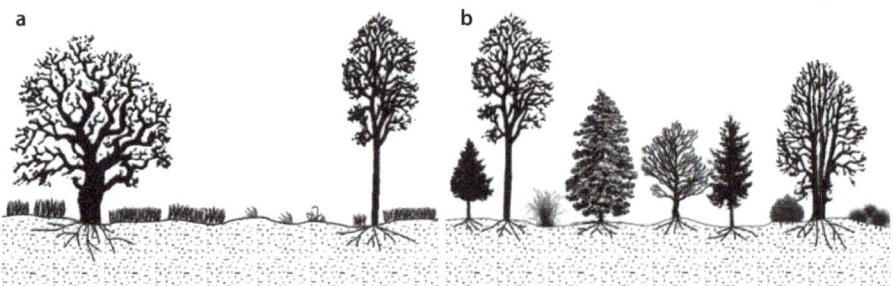

a   b

☐ **Fig. 6.4** Schematic representation of low **a** and high **b** structural diversity using the example of differently structured vegetation units, for example, in an urban park. (© Drawing: W. Gruber, using Beierkuhnlein 1998, p. 88)

6.2.2 Structure of Urban Habitats as Indicator and Basis of Spatial Assessment

Biodiversity can be determined in terms of spatial properties in the area, for example, the formation of spatial patterns of plant communities, and according to stand stratification. These spatial properties can be further complemented in diversity by functional and temporal properties. However, this requires further investigation, so that the relatively easy-to-record spatial properties by similarity of species composition and formation of vegetation patterns are more commonly used as indicators at different scales. For entire cities, the survey scale of 1:5000–10,000 determines the structural content that can be mapped. Here, biotopes are recorded as type habitats with similar internal vegetation structures, primarily characterised by use (biotope mapping). They form the balance sheet basis for descriptions of their inventory (species, structures, uses, etc.) and assessments of their "biotope value" as, for example, "species-rich", "habitat of rare and protected species" or "worthy of protection".

» "In the settled area, it is primarily the uses that shape the distribution pattern of organism species. The basis of nature conservation work in the city is therefore to systematically record the most important types of use and to describe their species populations and their ecological conditions of existence. The final result shows the extent to which individual uses of certain types contribute to the conservation of species in the populated area. It is also possible to identify which uses are characterised by a marked lack of species and may require measures for "renaturalisation"" (Sukopp et al. 1980, p. 565, translated).

Biotope mapping

"Biotope mapping" refers to a process of recording habitats (biotopes) and their actual or potential species inventory through maps and descriptions. Because of the close relationship to the use of the areas, biotope mapping is often based on use type mapping underpinned by further characteristics. They are usually commissioned by authorities, cities and municipalities, but also by nature conservation associations, and are used for decision-making processes in urban planning and development and in nature conservation management. They also serve as a basis for research work. The survey is carried out either on an area-wide, representative basis (representative areas) or selectively ("valuable" biotopes). The practical scale for the recording of land use types in urban areas is 1:5000–10,000 (Arbeitsgruppe 1986, 1993).

Since 1979, the use-related area-wide biotope survey by means of biotope mapping in urban areas has been in use in Germany to record urban biodiversity. A total of 21 meetings of a supra-regional and interdisciplinary working group "Biotope Mapping in Settled Areas" for the supra-regional coordination of the concept took place until 2004, followed by another five meetings of a Competence Network

6

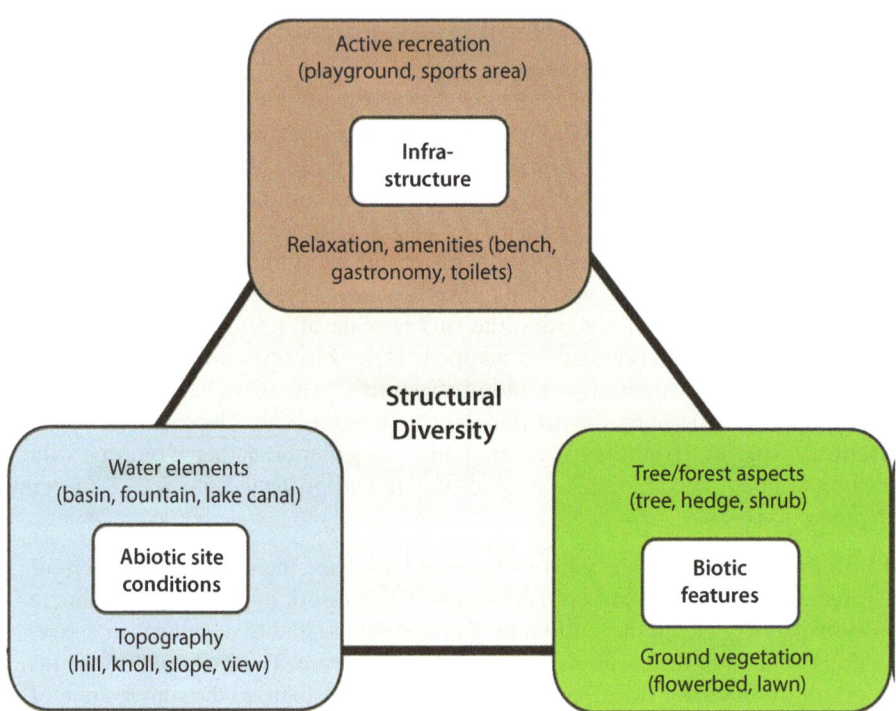

□ **Fig. 6.5**   Conceptual model of structural diversity using the example of an urban park. (© Design: J. Breuste, Drawing: W. Gruber after Voigt et al. 2014, p. 482)

Urban Ecology (CONTUREC) in 2006–2011. In 1986, a standard procedure was developed for this purpose in Germany (Basic Programme for Area-wide Biotope Mapping in Settled Areas (Working Group 1986), extended in 1993 (Working Group 1993) for determining species and habitat diversity, their localisation and temporal development, often combined with practical instructions for action. In this way, biodiversity is "measurable" over space and time, mapped, balanced and made available for practical action (Breuste 1994a, b). The 25 independent cities in Bavaria, for example, were already completely mapped by 1989. The procedure is now also being adapted and successfully applied in other countries (for example, Bedé et al. 1997 in Brazil). It is recommended that mapping be repeated regularly at short intervals for biotopes and longer intervals for species.

In the scale of observation, parks, for example, are at best distinguished by the intensity of maintenance "intensive" or "extensive", but not in their internal structure. This can be considered as structural diversity consisting of biotic, abiotic and infrastructure (see □ Figs. 6.5 and 6.6).

Only a part of this structural diversity is the diversity of areal vegetation patterns, which can be typified as indicators of biodiversity.

The **structural diversity of vegetation** of urban habitats is suitable as an indicator of biodiversity (Whitford et al. 2001). This is particularly true for birds (Sandström et al. 2006). Small structures and especially trees play a special role as a habitat factor (Chace and Walsh 2006).

■ **Fig. 6.6** Diversity of natural structures in Lednice Landscape Park in Moravia, Czech Republic. (© Breuste 2017)

The positive relationship between species diversity and size of vegetation patches is discussed contradictorily, especially when different animal groups are analysed. The factor "size" usually stands for a combination of other factors (for example, Raino and Niemelä 2003; Deichsel 2007). Tscharntke et al. 2002 argue that area size is not a simple parameter to predict species richness. This is probably also true for the often-named positive relationship between patch age and species richness. Sattler et al. (2011) refer to the age of green spaces in Swiss cities as the most important indicator of biodiversity. However, the factor "age" probably often stands for the factor "habitat quality" (Kowarik 1998).

Connectivity is achieved either by spatial proximity or by direct connection. In cities, connecting corridors can be watercourses, transport routes (especially railway lines) and park-like green corridors (Werner and Zahner 2009). Marzluff and Ewing (2001) already pointed out that generalists and invasive species benefit most from connectivity structures. Angold et al. (2006) emphasize that specialists (wetlands) are not promoted by connectivity structures. This is also true for waterbirds (Werner 1996).

For their biodiversity assessment, the diversity of biotic structure types, area share and value weights of habitat structures (according to type-related quantity of species inventory) can be used. Since the horizontal and vertical structuring of vegetation stands must be recorded for this assessment, remote sensing surveys (aerial and satellite data) must be supplemented by on-site mapping (see ■ Fig. 6.7).

Which ecosystems have high or low biodiversity in the assessment can be determined through an assessment key. This then also allows a comparison of ecosystems with each other, which can contribute to decision-making, for example, on biodiversity protection, increasing biodiversity and generally defining nature conservation measures.

Biodiversity is not uniformly distributed in cities, but depends on structural features of urban ecosystems, disturbance and management, and other characteristics. However, it can be assigned to typical urban structures (urban structure

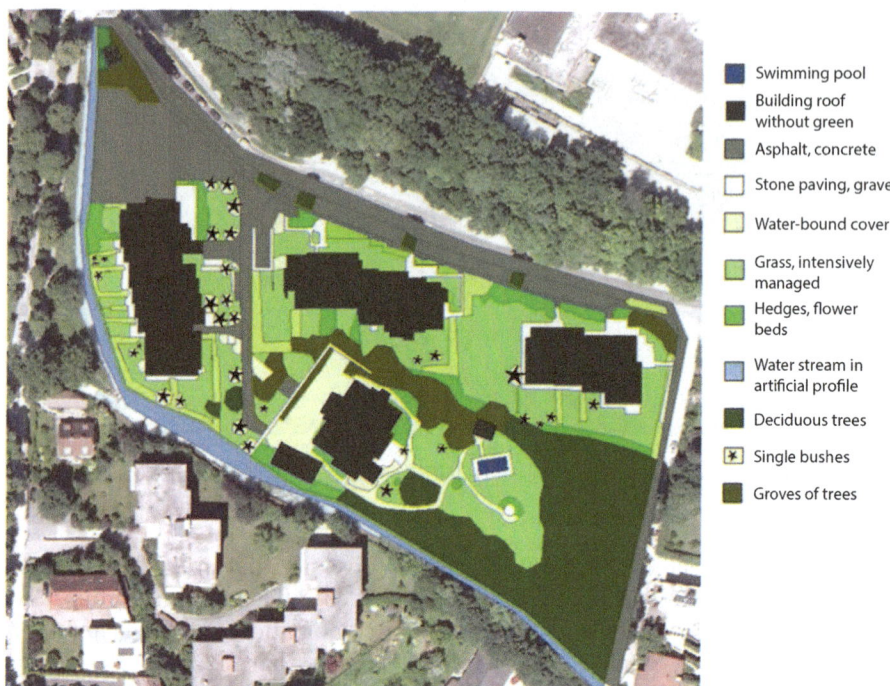

Swimming pool

Building roof without green

Asphalt, concrete

Stone paving, gravel

Water-bound cover

Grass, intensively managed

Hedges, flower beds

Water stream in artificial profile

Deciduous trees

Single bushes

Groves of trees

◘ Fig. 6.7   Area-specific recording of vegetation and build- up structure types as landscape elements and carriers of biodiversity and ecosystem services at a large scale (greater than 1:5000). (© Wurster et al. 2014)

types) in order to gain at least a rough overview of the spatial distribution of bio-diversity in cities. Plant or animal species can be used as indicators for this purpose.

Studies in Linz (Breuste et al. 2013a, b; Weissmair et al. 2000–2001) show that certain urban structure types have significantly higher biodiversity, measured by the number of breeding bird species and number of individuals typically found in them (see ◘ Table 6.1).

The results show that not only urban woodlands and parks (see ◘ Fig. 6.8), but also residential areas of low built-up density and with much green, trees and gardens have comparably high biodiversity values due to their structural features. Tree and woody stands in particular act as biodiversity-enhancing elements, at least for breeding birds, the majority of which are tree and cavity breeders in the city (see ◘ Table 6.2). As expected, the lowest biodiversity was found in the areas of intensive agriculture.

6.2.3   Species Diversity as Indicator

The recording of species in reference areas, their habitats or the entire city is a frequently used practice to map urban biodiversity. In this process, population density values are determined for species groups or individual species.

● **Table 6.1** Biodiversity of urban structure types in Linz (indicator breeding birds)

Urban structure type	Typical number of species Breeding birds	Typical number of breeding bird individuals
Urban woodland	More than 60	More than 10,000
Park	More than 60	More than 10,000
Residential areas with low built-up density (single houses with gardens)	More than 60	More than 10,000
Densely built-up residential areas (multi-storey buildings with little green space)	50–60	5000–10,000
Commercial areas	50–60	5000–10,000
Agricultural grassland	Under 50	Under 5000
Arable land	Under 50	Under 5000

Czermak (2008) and Weissmair et al. (2000–2001, supplemented)

● **Fig. 6.8** Urban forest-like Freinberg Park in Linz, Austria. (© Breuste 2005)

Differentiations according to functional services and the lifespan of individuals (Beierkuhnlein 1998) are more rarely recorded in addition. For the general public, biodiversity is often described by the number of species (species diversity). However, this only provides a fragmentary representation of biodiversity (see ▶ Sect. 6.1). The species considered are usually those that can be recorded and identified with a reasonable amount of effort and available species knowledge. These species or groups of species serve as indicators of biodiversity as a whole, without this actually being certain.

Table 6.2 Biodiversity of Linz city parks, breeding birds and structural indicators

Parks	Size in ha	BS1	BS2	BS3	Number of breeding bird species	Structure diversity of vegetaton	Dominant structures of vegetation	Disturbance intensity
Farmer's Mountain	9,54	15	10	12	37	3	B	2
Freinberg West-East	6,5	12	12	8	32	3	B	2
Hummelhof Forest	7,81	10	8	13	31	3	B	3
Freinberg Aroboretum	9,0	9	8	8	24	2	R/B	2
Bergschlössl	2,79	1	14	9	25	1	R	2
Panuli Meadow	1,5	11	6	7	24	2	R	3
Water forest	70,86	10	10	3	22	3	B	2
Schlossberg	1,1	3	10	9	23	2	B	2
Danube Park	7,57	6	13	2	21	2	R/B	3
People's Garden	2,6	6	7	6	19	2	B	3
University Park	4,1	6	6	4	16	3	R	1
Pöstlingsberg	2,61	6	6	3	15	3	R	2
J. W. Kleinstrasse	1,35	4	5	1	10	2	R	3
Wag Park	1,67	4	5	1	10	1	R	3
Ecopark	1,12	7	1	2	10	1	B	1
Ing.star.street	1,02	2	2	5	9	1	R	3
Recreation park Urfahr	7,73	3	1	3	7	1	R	2
Peuerbachstraße	1,07	3	1	0	4	1	R	2
Harbach Park	1,39	0	0	1	1	2	B	3

BS 1 = breeding possible, BS 2 = breeding probable, BS 3 = breeding confirmed
Structural diversity of vegetation: 3 = high, 2 = medium, 1 = low
Disturbance intensity by visitors: 3 = high, 2 = medium, 1 = low
Dominant structures of vegetation: B = trees, R = lawns
Disturbance intensity (noise, users, maintenance): 3 = high, 2 = medium, 1 = low
Dark shading in the table: high biodiversity
Medium shading: medium biodiversity
No shading: low biodiversity

Weissmair et al. (2000–2001), Czermak (2008), and Breuste et al. (2013a, b, supplemented)

Vascular plants are usually recorded in their entirety as part of geobotanical plant surveys. The fauna is often only represented by individual species groups. Here, it is particularly the breeding birds that are used as indicators of biodiversity. Species diversity of breeding birds is used as an indicator of "landscape quality" (nationwide indicator "Species diversity and landscape quality"). Between 2004 and 2014, unlike in agricultural land, a significant increase in the number of species could be demonstrated in settlements as the only habitat in Germany (Wahl et al. 2015).

Other animal groups as indicators are studied much less frequently, for example, mammals, fish or amphibians, or herpetofauna (Werner and Zahner 2009). One of the best-studied species is the red fox *(Vulpes vulpes)* (for example, Gloor et al. 2001). Especially butterflies and carabids are used internationally as typical indicators for urban biodiversity studies (for example, GLOBENET, Niemelä et al. 2000).

Birds are considered the best-studied and most widely used bioindicators of urban biodiversity worldwide (for example, Biadun 1994; Müller et al. 2010).

The nationwide voluntary bird monitoring programmes are coordinated by the Federal Association of German Avifaunists (DDA) and evaluated in cooperation with the Agency for Nature Conservation and the bird protection agencies of the German federal states (Flade et al. 2008). They show that the most important factors influencing species diversity,

- Increase in intensity of use, change of use
- Drainage
- Abandonment of use (succession) and
- Sports and recreational uses are.

For breeding birds in Germany, a "target value 100" of species diversity has been determined, of which only 69% has been reached (Wahl et al. 2015).

In urban areas, parks and gardens in particular are the most species-rich habitats for breeding birds, in addition to forests (Jokimäki 1999; Peris and Montelongo 2014). "Wilderness" (for example, forest) and garden thus do not represent a biodiversity contradiction with regard to species diversity in cities.

Complete species lists (vascular plants and/or birds) are currently available for more than 180 cities as a basis for comparison (Werner 2011). The comparison of biodiversity between cities and urbanization processes and their effects on biodiversity is becoming increasingly important. It is becoming clear that the area size of a city has a positive relationship with the species diversity of vascular plants due to an increase in habitat offerings (Pysek 1998).

The quantity and quality of the vegetation components of the urban matrix and their distribution and connectivity promote both native species and "specialists". Urban biodiversity can therefore be "made" (see ◘ Table 6.3 and ◘ Fig. 6.9).

The diversity of plant species and that of breeding birds for the same urban area do not always coincide, which at least calls into question the choice of indicators for general statements on urban biodiversity.

Hand et al. (2016) developed a methodology to assess urban biodiversity multiscale with indicators, combining multiple species groups (birds, invertebrates, plants), plant life form types, vegetation cover and stratification, and degree of human impact in one operation.

6

◘ Table 6.3   Relationship between urban area and species numbers in breeding birds

City	Area km²	Number of bird species	% of the regional. Species pool
St. Petersburg (Russia)	1431	242	80
Warsaw (Poland)	517	146	65
Valencia (Spain)	135	232	62
Rome (Italy)	1272/300 (city)	120	50
Munich (Germany)	310	122	50

Werner (2011)

◘ **Fig. 6.9**   Long-tailed Shrike *(Lanius schach)* in the Knowledge and Innovation Community Garden in Shanghai, China. (© Breuste 2017)

Measuring Urban Biodiversity with Citizen Science: Hour of Garden Birds 2018

In citizen science, projects are carried out with the assistance or entirely by interested laypersons. They report, for example, observations of the species population. In May 2018, the German Nature and Biodiversity Conservation Union (NABU) called on birdwatchers for the 14th time to count birds observed in gardens and parks and to send them to NABU via the Internet or by post. 55,976 birdwatchers reported observations of 1,244,151 birds from more than 36,841 gardens and parks as part of the 2018 "Hour of Garden Birds" campaign.

However, the results show the lowest bird numbers since the start of the cam-

◼ **Table 6.4** Results of the 2018 hour of garden birds in Germany

Rank	Bird species	Number	Occurrence in % of gardens	Birds per garden	Comparison with previous year (trend in %)
1.	House Sparrow	180.597	65.28	4.90	+2
2.	Blackbird	121.039	94.85	3.29	−5
3.	Great Tit	94.788	82.04	2.57	−7
4.	Star	81.336	49.31	2.21	−10
5.	Tree Sparrow	78.832	35.12	2.14	−3
6.	Blue Tit	75.520	71.56	2.05	−8
7.	Magpie	55.596	65.27	1.51	−6
8.	Wood Pigeon	43.610	45.53	1.18	+4
9.	House Martin	43.377	20.54	1.18	−8
10.	Swift	40.017	21.08	1.09	−17

NABU (2018)

paign in 2004. 33.7 birds were recorded per garden or park on average. Overall, this is a drop of over 5% compared to the previous year and the long-term average. Among the 15 most common garden birds, seven species had the lowest numbers ever, including blackbird, great tit, blue tit, magpie, greenfinch, chaffinch and redstart. Only house sparrow, tree sparrow, wood pigeon and raven were observed with similar frequency as in previous years.

Species that feed their young with insects show particularly low numbers. The aerial insect hunters mealy swallow and swift only reach a value of 60% in relation to 2006. The two seed-eating finch species goldfinch and grosbeak continued their population increases.

The starling as the bird of the year 2018 comes to −16% compared to the previous year.

The results show that more needs to be done to protect birds and that birds are good indicators of biodiversity as a whole. The call to all urban dwellers is: "Everyone can start making their garden a mini nature reserve" (NABU 2018) (see ◼ Table 6.4).

Since 1989, the population development of all common breeding bird species throughout Germany has also been monitored using standardised methods. Since 2004, the surveys have been carried out on 1×1 km sample areas that are representative for the whole of Germany. The results are updated annually in the report "Birds in Germany" and are included in the indicator "Species Diversity and Landscape Quality" of the Federal Government (Federal Agency for Nature Conservation) as well as indicators at European level (including the "European Farmland Bird" indicator, EBCC) (NABU 2018).

6.3   How Is Urban Biodiversity Perceived?

The already complex relationships between people and biodiversity are often referred to as the **"people-biodiversity paradox"** (Fuller et al. 2007; Shwartz et al. 2014; Pett et al. 2016). This refers to a mismatch of

- People's biodiversity preferences and how these inclinations relate to personal subjective well-being;
- This refers to the limited ability of individuals to accurately perceive the biodiversity that surrounds them.

A clear distinction must be made between biodiversity and the perception of this biodiversity ("subjective biodiversity"), as shown by Hof and Keul 2017 using urban example studies. People can benefit from biodiversity, but they do not have to perceive (be able to perceive) or understand biodiversity in its complexity.

» "Much more research is needed to discern the links between exposure to biodiversity and how this might, ultimately, lead to shifts in underlying attitudes and behavior. Beyond education, understanding what individuals perceive as constituting a preferable biodiverse environment will allow for human-modified landscapes to be designed in a manner that delivers benefits to both people and biodiversity" (Pett et al. 2016, p. 580–581).

The study "Naturbewusstsein 2015" (Nature Awareness 2015) (BmU, BfN 2016) commissioned by the German Federal Government shows through 2000 representative respondents the social awareness of nature and biodiversity, a goal of the National Strategy on Biological Diversity (BmU 2007). This includes the following on urban nature (BmU, BfN 2016):

- 94% of respondents think that nature is part of a good life
- They associate nature with health, recreation, happiness and attachment to the region
- Urban nature is primarily perceived as well-kept greenery and above all vegetation, wildlife much less so
- Urban dwellers are increasingly dependent on services provided by urban nature (ecosystem services)

- Satisfaction with urban nature is high (80% of respondents are very or rather satisfied)
- Cities are increasingly becoming primary places for experiencing nature
- The diversity of urban nature is appreciated (92% of respondents agree)
- Public green spaces, trees, roadside green spaces, water bodies, woods and gardens are used less frequently (only 39% of respondents use them several times a week)
- Wilderness and agriculture in the city (Nature of the fourth kind, Kowarik 1992, see ► Sect. 1.3) is not an established concept (only 20% of respondents consider such nature very important (for public green spaces, Nature of the third kind, Kowarik 1992, see ► Sect. 1.3, the figure is 80%)
- A higher level of education does not influence a better relationship with urban nature
- Women and older people have a closer relationship with urban nature than men and younger, people
- In larger cities, urban nature is perceived as less important (BmU, BfN 2016).

With regard to biodiversity (in general), the following statements can be summarised as results of the nature awareness study (BmU, BfN 2016):
- 85% of respondents believe that biodiversity enhances their well-being and quality of life.
- They associate it first of all with the diversity of species (plants and animals), an ethical mandate of "heritage" to be protected, land consumption, but also "well-being and quality of life".
- For most, biodiversity takes place in protected areas, but it is not associated with urban nature. 92% of respondents are willing to make their personal contribution to the protection of biodiversity by best staying away from protected areas!
- Out of 2000 respondents, only 868 gave any information at all about what biodiversity is. More than half of the respondents (58%) do not even know what biodiversity means! Only for half of those who answered, it is more than diversity of species.
- Among those who are positively inclined towards biodiversity, upper classes and upper-middle classes are clearly overrepresented and this across all lifestyles (BmU, BfN 2016) (see ▫ Fig. 6.10)

These results of a representative survey in Germany (BmU, BfN 2016) are likely to apply similarly to other European countries. Thus, although there is a great devotion to urban nature in its cultivated form, there is little understanding of biodiversity outside the educational elites, despite broad efforts to educate the media about the environment. The message that biodiversity is important has been received by society and is reproduced on demand. However, its rationale is not fully understood and is reduced primarily to biodiversity. Biodiversity is almost not referred to urban nature at all, but is localized elsewhere in underutilized spaces left to

6

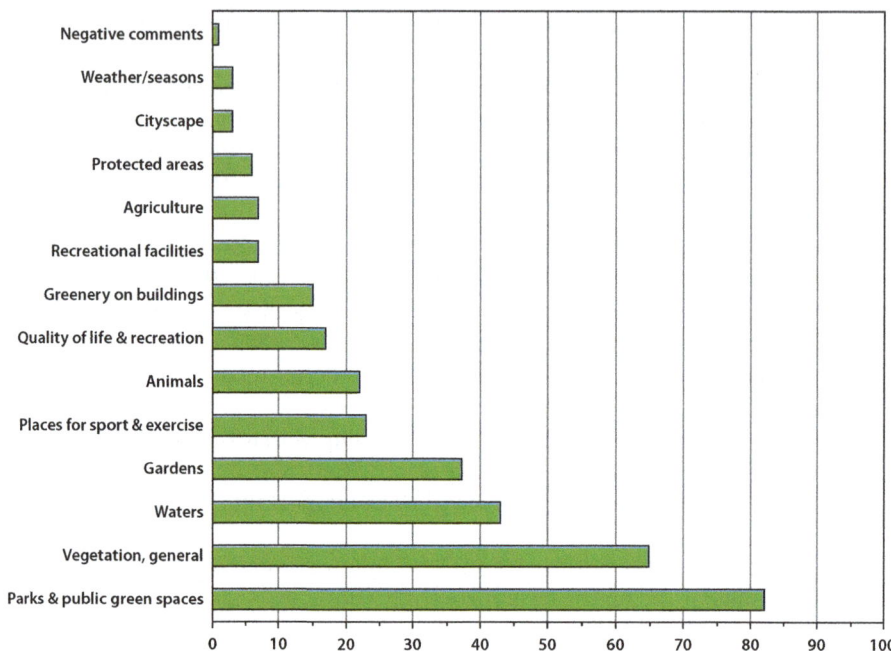

● **Fig. 6.10**   Nature awareness 2015, associations with urban nature among 2000 German respondents in the nature awareness survey, categorized. (According to BmU, BfN 2016)

natural processes. The fact that the best biodiversity-promoting behaviour of almost all respondents (92%!) is to avoid such spaces so as not to trigger any disturbances is evidence of an understanding of fragile nature that humans are best advised to stay away from. However, this is not an alternative overall and especially for cities, and is by no means the goal of biodiversity efforts (BmU, BfN 2016). This would further reduce contact with nature and real understanding of biodiversity individually. Since a similar survey in 2009, nothing significant has changed. This is alarming!

Broad public relations work in the media at national, regional and local level has been trying to counteract this for some time with moderate success. The city of Frankfurt am Main, for example, has been publishing the results of its biotope mapping by the Senckenberg Research Institute and Nature Museum for years, promoting "nature on your doorstep" and explaining how and where you can "experience biodiversity" (Ottich et al. 2009) (see ● Fig. 6.11).

Empirical studies in parks in Linz (Breuste et al. 2013a, b) show that the integration level of species is little perceived by visitors to urban nature. This is hardly surprising, since many species are not observable by visitors who are not professionally trained and prepared. In the flora it is mainly woody plants, especially trees, and in the fauna it is mainly birds that attract attention. Many inconspicuous species, even if they are special or even common, are hardly or not at all noticed. The primary motives for visiting parks are the diversity of their visual vegetation and their infrastructure. The integration level of biotopes and ecosystems

◘ Fig. 6.11 Postcard as part of the public relations activities of the BioFrankfurt biodiversity region. (© ► www.biofrankfurt.de) (Text: The beer tulip. Only one of 1400 plant species in the Rhine-Main Region)

(Beierkuhnlein 1998) is perceived more clearly as structural diversity (Voigt et al. 2014) than the integration level of species (species level).

Palliwoda et al. (2017) identified a variety of "interactions" between park visitors in Berlin and plants in park grasslands, many of which were not conscious and purposeful interactions. They were able to show that biodiversity is also perceived at the species level alongside other targeted activities. The majority of interactors were women (78%). This included 26 cultivated or spontaneous plant species (native and non-native) for eating, decoration and gaining experience (biodiversity experience). Of the activities observed and surveyed, 12% were biodiversity interactions following walking, resting, and walking dogs, still even more than were obtained from jogging (6%). Native as well as non-native species were included at about the same rate concerning their occurrence. This also suggests that hitherto little-noticed activities when visiting urban nature do indeed play a role in the perception of biodiversity. Consumption and decoration aspects, scents and visual impressions are likely to play a role and should be investigated further.

6

How Is Urban Biodiversity Perceived?: Results of Studies in Shanghai Parks

In a study on park use led by the author, 421 Shanghai park visitors over 15 years of age in roughly equal age groups (15–29, 30–49, over 50 years old) were surveyed in four parks (older parks: Fuxing Park, Changfeng Park, young parks: Lujiazui Park, Mengqing Park).

By far the most dominant motives for visiting the parks are enjoying fresh air (74.0%) and being in nature (58.2%). Learning about nature is the reason given by 19.3% of respondents, while 24.9% want to enjoy the diversity of the landscape.

Although the reason for visiting the park away from city life in "fresh air" and "to be surrounded by nature" dominates, about one in five people visit the park for emotional and cognitive nature sensations. About a quarter of the respondents (27.8%) state that they have no alternatives at all for the perception of nature in the otherwise highly condensed city of Shanghai.

More than half of the visitors come from less than 20 min walking distance and visit the park nature mainly for more than 1 hour (42.9%) in the morning (37.8%) and in the afternoon (22.1%).

Species-rich, multi-layered wooded areas with good infrastructure were most frequently visited in all four parks, the open lawns only in the morning hours for Tai Chi exercises.

Although the preferred areas are also those with high biodiversity, however, the motivation to visit is not focused on biodiversity, but on structural and impression diversity (Voigt et al. 2014). Enjoyment of life and a satisfying feeling of health are particularly felt during a visit to the park.

On a scale of 1 (rejection) to 5 (special appreciation), the following facilities achieve values above 4 (appreciation) (see ◙ Table 6.5).

Biodiversity is not perceived as species diversity, but rather as structural diversity, especially where this is also well accessible to users through good infrastructure quality (Zippel 2016) (see ◙ Fig. 6.12).

◙ **Table 6.5** Appreciation of structural elements of parks in Shanghai

Value judgement	Structural elements	Species diversity of vegetation structures
4.5	Tree groups and woods with undergrowth	High
4.4	Varied, diverse green spaces,	High
4.4	Open lawns	Low
4.2	Wide paths with resting places	No
4.1	Planted flower borders	Low
4.0	Areas around water surfaces	Medium

■ **Fig. 6.12**   Diversely structured green area of Freinberg Park in Linz, Austria. (© Breuste 2005)

6.4   Urban Biodiversity and Ecosystem Services

The importance of biodiversity for the "functioning" of ecosystems is still little studied (Beierkuhnlein 1998). This is still true now (Schwarz et al. 2017). It is not the recording of species numbers, which are often high for urban ecosystems compared to their agrarian surroundings, but the importance of this biotic diversity for ecological functions and ecosystem services that are of practical importance. Furthermore, much more attention needs to be paid to the spatio-temporal variability of biotic diversity. Because of the high dynamics of anthropogenically induced changes in urban ecosystems and the still limited knowledge of how long biotic compartments take to adapt to new conditions, significantly more attention must be paid to this issue when considering urban biodiversity.

In both scientific and environmental policy discussions, it is widely assumed that urban biodiversity is a prerequisite of ecosystem services in cities, or that its increase leads to an increase in ecosystem services (for example, Hand et al. 2016; Kabisch et al. 2016; Ziter 2016).

» "...biodiversity has been linked to providing multiple benefits ranging from supporting city sustainability to enhancing the health and well-being of individual residents" (Hand et al. 2016, p. 33).

However, there are hardly any empirical findings on this. Ecosystem services are often equated with biodiversity. The BioFrankfurt association, which is supported by 12 institutions from research, education and nature conservation, lists, for example, under teaching materials:

» "What is biodiversity and why is it so important? Biodiversity provides us with many of nature's "services" such as food, medicine and raw materials, without which we could not survive". ... "For humans, biodiversity is one of the most important foundations of life and guarantor of quality of life, on which we depend or benefit in many ways" (BioFrankfurt 2018, translated).

Positive relationships between biodiversity and ecosystem services have only been demonstrated in experiments and non-urban ecosystems (woodlands, grassland, wetlands) in a few studies (Schwarz et al. 2017). There is currently also insufficient empirical evidence as to whether, as assumed, the concepts of "Green Infrastructure" (European Commission 2012) and "Nature-Based Solutions" (European Commission 2015) actually promote urban biodiversity and ecosystem services (Schwarz et al. 2017).

Schwarz et al. (2017) examined the relationship between urban biodiversity and ecosystem services addressed in individual scientific studies between 1990 and 2017. A total of 317 publications were found, covering this subject 944 times. Only 228 of these (24%) were empirical studies. 119 treatments (52%) substantiated a positive relationship. Taxonomic groups such as plants, birds, or insects were used in 43% of the 228 urban biodiversity-ecosystem service relationships examined. Functional biodiversity relationships and the role of individual species, including non-natives, and specific functional traits were rarely examined. Thus, there is a demonstrable lack of empirical data to support the relationship between urban biodiversity and ecosystem services.

The results of the study show that the relationship between urban biodiversity and urban ecosystem services has so far been studied primarily on the basis of taxonomic data. More quantitative and empirically-based research, especially on functional relationships, is needed. Schwarz et al. (2017) are convinced that such traits, which are known to be sensitive to urbanization processes and at the same time important for ecosystem services, have received little attention in research so far. Instead, on a weak empirical basis, an application that is also widely supported by the European Commission (2015) is being promoted in planning and design of urban ecosystems. As planners are already increasingly incorporating ecosystem services into their designs and decisions, these significant knowledge gaps should be adequately addressed to find trade-offs and synergies between biodiversity on the one hand and the conservation and promotion of ecosystem services on the other (Schwarz et al. 2017) (see ◘ Fig. 6.13).

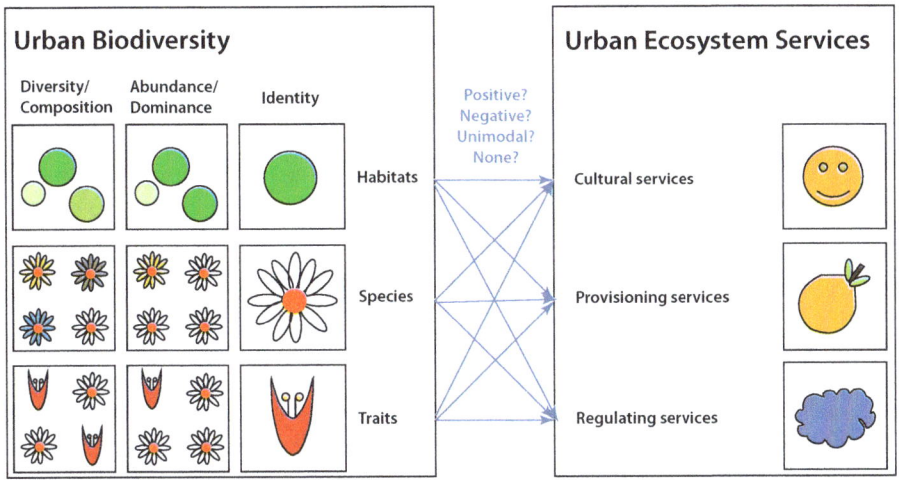

Fig. 6.13 Urban biodiversity – Urban ecosystem services study context. (Schwarz et al. 2017, p. 163)

References

Angold PG, Sadler JP, Hill MO, Pullin A, Rushton S, Austin K, Small E, Wood B, Wadsworth R, Sanderson R, Thompson K (2006) Biodiversity in urban habitat patches. Sci Total Environ 360:196–204

Arbeitsgruppe Methodik der Biotopkartierung im besiedelten Bereich (1986) Flächendeckende Biotopkartierung im besiedelten Bereich als Grundlage einer ökologisch bzw. am Naturschutz orientierten Planung: Grundprogramm für die Bestandsaufnahme und Gliederung des besiedelten Bereichs und dessen Randzonen. Natur Landsch 61(10):371–389

Arbeitsgruppe Methodik der Biotopkartierung im besiedelten Bereich (1993) Flächendeckende Biotopkartierung im besiedelten Bereich als Grundlage einer am Naturschutz orientierten Planung: Programm für die Bestandsaufnahme, Gliederung und Bewertung des besiedelten Bereichs und dessen Randzonen: Überarbeitete Fassung 1993. Natur Landsch 68(10):491–526

Bedé L, Weber W, Resende S, Piper W, Schulte W (1997) Manual para mapeamento de Biótopos no Brazil. Funacao Alexander Brandt, Belo Horizonte

Beierkuhnlein C (1998) Biodiversität und Raum. Erde 128:81–101

Biadun W (1994) The breeding avifauna of the parks and cemeteries of Lublin (SE Poland). Acta Ornithol 29:1–13

BioFrankfurt (2018) https://www.biofrankfurt.de/. Accessed 7 May 2018

Breuste J (1994a) "Urbanisierung" des Naturschutzgedankens: Diskussion von gegenwärtigen Problemen des Stadtnaturschutzes. Naturschutz und Landschaftsplanung 26(6):214–220

Breuste J (1994b) Flächennutzung als stadtökologische Steuergröße und Indikator. Geobotan Kolloquium 11:67–81

Breuste J, Schnellinger J, Qureshi S, Faggi A (2013a) Investigations on habitat provision and recreation as ecosystem services in urban parks – two case studies in Linz and Buenos Aires. In: Breuste J, Pauleit S, Pain J (eds) Stadtlandschaft – vielfältige Natur und ungleiche Entwicklung. Schriftenreihe des Kompetenznetzwerkes Stadtökologie. CONTUREC 5, Darmstadt, pp 5–22

Breuste J, Schnellinger J, Qureshi S, Faggi A (2013b) Urban ecosystem services on the local level: urban green spaces as providers. Ekologia 32(3):290–304

Breuste J, Pauleit S, Haase D, Sauerwein M (2016) Stadtökosysteme. Funktion, Management, Entwicklung. Springer Spektrum, Berlin

Bundesministerium für Umwelt, Naturschutz, Bau und Reaktorsicherheit (BmU) (2007) Nationale Strategie zur biologischen Vielfalt. Vom Bundeskabinett am 7. November 2007 beschlossen. Reihe Umweltpolitik, Berlin

Bundesministerium für Umwelt, Naturschutz, Bau und Reaktorsicherheit, Bundesamt für Naturschutz (BmU, BfN) (2016) Naturbewusstsein 2015. Bevölkerungsumfrage zu Natur und biologischer Vielfalt, Berlin

Chace JF, Walsh JJ (2006) Urban effects on native avifauna: a review. Landsc Urban Plan 74:46–69

Convention on Biological Diversity (COP) (2009) Report on the first expert workshop on the development of the City Biodiversity Index in the first expert workshop on the development of the City Biodiversity Index, Singapore City, February 10–12. http://www.cbd.int/doc/meetings/city/ewdcbi-01/official/ewdcbi-01-03-en.pdf. Accessed 22 Sept 2010

Convention on Biological Diversity (COP) (2010) Report on the second workshop on the development of the City Biodiversity Index in the second expert workshop on the development of the City Biodiversity Index, Singapore City, July 1–3. http://www.cbd.int/doc/meetings/city/ewdcbi-02/official/ewdcbi-02-03-en.pdf. Accessed 22 Sept 2010

Czermak P (2008) Ökologische Bewertung von Parkanlagen der Stadt Linz auf der Basis des Datenbestandes der Brutvogelkartierung. Master Thesis, Facutly Natural Sciences, University of Salzburg, Salzburg

Deichsel R (2007) Habitatfragmentierung in der urbanen Landschaft – Konsequenzen für die Biodiversität und Mobilität epigäischer Käfer (Coleoptera: Carabidae und Staphylinidae) am Beospiel Berliner Waldfragmente. Ph.D Thesis, Freie Universität Berlin, Berlin

European Commission, Directorate-General for Research and Innovation (2015) Towards an EU research and innovation policy agenda for nature-based solutions and re-naturing cities: final report of the horizon 2020 expert group on "Nature-Based Solutions and Re-Naturing Cities" (full version). Brussels, S 1–70. https://ec.europa.eu/programmes/horizon2020/en/news/towards-eu-research-and-innovation-policy-agenda-nature-based-solutions-re-naturing-cities. Accessed 7 Dec 2021

European Commission, European Commission's Directorate-General Environment (2012) The multifunctionality of green infrastructure, science for environment policy | In-depth reports | DG Environment News Alert Service. Brussels, pp. 1–37. www.ec.europa.eu/environment/nature/.../Green_Infrastructure.pdf. Accessed 7 May 2018

Flade M, Grüneberg C, Sudfeldt C, Wahl J (2008) Birds and biodiversity in Germany – 2010 target. Dachverband Deutscher Avifaunisten, Steckby

Fuller RA, Irvine KN, Devine-Wright P, Warren PH, Gaston KJ (2007) Psychological benefits of greenspace increase with biodiversity. Biol Lett 3(4):390–394

Gloor S, Bontadina F, Heggelin D, Deplazes P, Breitenmoser U (2001) The rise of urban fox populations in Switzerland. Mamm Biol 66:155–164

Hand KL, Freeman C, Seddon PJ, Stein A, van Heezik Y (2016) A novel method for fine-scale biodiversity assessment and prediction across diverse urban landscapes reveals social deprivation-related inequalities in private, not public spaces. Landsc Urban Plan 151:33–44

Hobohm C (2000) Biodiversität. UTB 2162. Quelle & Meyer, Wiebelsheim

Hof A, Keul A (2017) Objektive und subjektive Biodiversität städtischer Parks. AGIT J Angewandte Geoinform 3:364–373

Jokimäki J (1999) Occurence of breeding bird species in urban parks: effects of park structure and broad-scales variables. Urban Ecosyst 3(1):21–34

Kabisch N, Frantzeskaki N, Pauleit S, Artmann M, Davis M, Haase D, Knapp S, Korn H, Stadler J, Zaunberger K, Bonn A (2016) Nature-based solutions to climate change mitigation and adaptation in urban areas – perspectives on indicators, knowledge gaps, opportunities and barriers for action. Ecol Soc 21:39. https://doi.org/10.5751/ES-08373-210239. Accessed 13 Mar 2018

Kowarik I (1992) Das Besondere der städtischen Flora und Vegetation. In: Natur in der Stadt – der Beitrag der Landespflege zur Stadtentwicklung. Schriftenreihe des Deutschen Rates für Landespflege 61:33–47

Kowarik I (1998) Auswirkungen der Urbanisierung auf Arten und Lebensgemeinschaften – Risiken, Chancen und Handlungsansätze. Bundesamt für Naturschutz, Schriftenreihe für Vegetationskunde 29:173–190

Kowarik I (2011) Novel urban ecosystems, biodiversity and conservation. Environ Pollut 159(8–9):1974–1983

Kühn I, Brandl R, Klotz S (2004) The flora of German cities is naturally species rich. Evol Ecol Res 6:749–764

Lehmhöfer A (2010) Die üppig Blühende. Frankfurter Rundschau. 7 October, S R2

Marzluff JM, Ewing K (2001) Restoration of fragmented landscapes for the conservation of birds: a general framework and specific recommendations for urbanizing landscapes. Restor Ecol 9(3):280–292

Müller N, Werner P, Kelcey JG (2010) Urban biodiversity and design. Blackwell, Singapore

NABU (2018) Fast nur Verlierer unter den Gartenvögeln. News. https://www.nabu.de. Accessed 28 May 2018

Naturkapital Deutschland – TEEB DE (2016) Ökosystemleistungen in der Stadt – Gesundheit schützen und Lebensqualität erhöhen. Edited by Ingo Kowarik, Robert Bartz und Miriam Brenck. Technische Universität Berlin, Helmholtz-Zentrum für Umweltforschung – UFZ. Berlin, Leipzig

Niemelä J, Kotze J, Ashworth A, Brandmayr P, Desender K, New T, Penev L, Samways M, Spence J (2000) The search for common anthropogenic impacts on biodiversity: a global network. J Insect Conserv 4:3–9

Ottich I, Bönsel D, Gregor T, Malten A, Zizka G (2009) Natur vor der Haustür – Stadtnatur in Frankfurt am Main. E. Schweizerbart'sche Verlagsbuchhandlung, Stuttgart

Palliwoda J, Kowarik I, Moritz von der Lippe (2017) Human-biodiversity interactions in urban parks. Landscape and Urban Planning 157:394–406

Peris S, Montelongo T (2014) Birds and small urban parks: a study in a high plateau city. Turk J Zool 38(3):316–325

Pett TJ, Shwartz A, Irvine KN, Dallimer M, Davies ZG (2016) Unpacking the people – biodiversity paradox. A conceptual framework. Bioscience 66(7):576–583

Pysek P (1998) Is there a taxonomic pattern to plant invasions? Oikos 82:282–294

Raino J, Niemelä J (2003) Ground beetles (Colooptera: Carabidae) as bioindicators. Biodivers Conserv 12:487–506

Sandström UG, Angelstam P, Mikusinski G (2006) Ecological diversity of birds in relation to the structure of urban green space. Landsc Urban Plan 77:39–53

Sattler T, Obrist MK, Duelli P, Moretti M (2011) Urban arthropod communities: added value or just a blend of surrounding biodiversity? Landsc Urban Plan 103:347–361

Schwarz N, Moretti M, Bugalho MN, Davies ZG, Haase D, Hack J, Hof A, Melero Y, Pett TJ, Knapp S (2017) Understanding biodiversity-ecosystem service relationships in urban areas: a comprehensive literature review. Ecosyst Serv 27:161–171

Shwartz A, Turbé A, Simon L, Julliard R (2014) Enhancing urban biodiversity and its influence on city-dwellers: an experiment. Biol Conserv 171:82–90

Sukopp H, Kunick W, Schneider C (1980) Biotopkartierung im besiedelten Bereich von Berlin (West): Part II: Zur Methodik von Geländearbeit. Gart Landsch 7:565–569

TEEB (2011) TEEB Manual for cities: ecosystem services in urban management. http://www.naturkapitalteeb.de/aktuelles.html. Accessed 26 Aug 2014

Tscharntke T, Steffan-Dewenter I, Kruess A, Thiess C (2002) Characteristics of insect populations in habitat fragments: a mini review. Ecol Res 17:229–239

United Nations (1992) Convention on biological diversity, Rio de Janeiro. www.dgvn.de/fileadmin/user…/UEbereinkommen_ueber_biologische_Vielfalt.pdf. Accessed 17 Apr 2018

User's Manual on the Singapore Index on Cities' Biodiversity (also known as the City Biodiversity Index) (2014) https://www.cbd.int/…/city/…/subws-2014-01-singapore-index-m. Accessed 2 June 2018

Voigt A, Kabisch N, Wurster D, Haase J, Breuste J (2014) Structural diversity – a multi-dimensional approach to assess recreational services in urban parks. Ambio 43(3):480–491

Wahl J, Dröschmeister R, Gerlach B, Grüneberg C, Langgemach T, Trautmann S, Sudfeldt C (2015) Vögel in Deutschland 2014. Dachverband Deutscher Avifaunisten, Münster

Weissmair W, Rubenser H, Brander M, Schauberger R (2000–2001) Linzer Brutvogelatlas. Naturkundliches Jahrbuch der Stadt Linz 46(47):1–318

Werner P (1996) Welche Bedeutung haben räumliche Dimensionen und Beziehungen für die Verbreitung von Pflanzen und Tieren im besiedelten Bereich. Gleditschia 24(1–2):303–314

Werner P (2011) Stadtbiotopkartierung, City Biodiversity Index und Co.: Biodiversität der Städte im Spiegelbild von Indikatoren und Naturschutzzielen. 5. Conturec-Tagung: Stadtlandschaft – vielfältige Natur und ungleiche Entwicklung, Laufen und Salzburg 22.–24.09.2011, Presentation 22 Sept 2011 – Laufen

Werner P, Zahner R (2009) Biologische Vielfalt und Städte. In: Bundesamt für Naturschutz (ed) BfN-Skripten 245. Bundesamt für Naturschutz, Bonn

Whitford V, Ennos AR, Handley JF (2001) City form and natural processes – indicators for the ecological performance of urban areas and their application to Merseyside, UK. Landsc Urban Plan 57(2):91–103

Whittaker RH (1972) Evolution and measurement of species diversity. Taxon 12:213–251

Whittaker RH (1977) Evolution of species diversity in land communities. Evol Biol 10:1–67

Wittig R, Niekisch M (2014) Biodiversität. Grundlagen, Gefährdung, Schutz. Springer Spektrum, Berlin

Wurster D, Artmann M, Breuste J (2014) Meeting supply and demand of ecosystem services in built-up areas: an illusion? Korea, Unveröffentlichtes Vortragsmaterial URBIO Konferenz

Zippel S (2016) Urban parks in Shanghai. Study of visitors' demands and present supply of recreational services. Master Thesis, Faculty Environmental Sciences, Techn. University Dresden, Dresden

Ziter C (2016) The biodiversity-ecosystem service relationship in urban areas: a quantitative review. Oikos 125:761–768

6

What Constitutes Urban Nature in the Green City Concept?

Contents

7.1   The Green City, a
        Conceptual Mosaic of Action
        Objectives – 256

7.2   Green Infrastructure:
        The Local Basic Concept – 263

7.3   The Concept of Urban
        Biodiversity – 266

7.4   Wild Goes Urban: Wilderness
        as Part of Urban Nature – 283

7.5   The Green City Concept
        in European and Global
        Perspective – 302

        References – 309

© Springer-Verlag GmbH Germany, part of Springer Nature 2022
J. Breuste, *The Green City*,
https://doi.org/10.1007/978-3-662-63976-4_7

The Green City is a guiding principle and a vision. The broad concept of the Green City contains urban nature as its core. In the Green City concept, urban nature is "conceptualised". It is linked in parts or as a whole with development goals and prepared for planning; all this in order to achieve better living conditions for the city's inhabitants. In this conceptual mosaic of action goals for the Green City, a national policy in Germany has only been discernible since the 2017 White Paper on *Urban Greening* by the Federal Ministry for the Environment. Already the garden city movement gives urban nature a first face at the beginning of the twentieth century with the concepts of "light, air and sun", which should be granted to all. At the beginning of the twenty-first century, new fields of action are being developed for this purpose, because the problem situations (for example, population density, climate change, lack of open space) have become more acute. Systemic and networked thinking is giving rise to the concept of "green infrastructure" and making it the subject of planning and design action. High expectations are placed on the now recognised capacity of urban nature to benefit city dwellers, often too high (for example, reduction of air pollution). Biodiversity is becoming the focus of many considerations and objectives, and is even taking on a conceptual character in cities. Urban biodiversity, contrary to popular perception, is not just the diversity of species and can be consciously designed to benefit people in the city. The exclusively ethical perspective of protecting rare native species from humans is no longer the general way of looking at urban biodiversity. This also requires new paradigms.

The treatment of wildernesses in cities is being put to the test, as they also provide positive benefits for city dwellers and are often special hotspots of urban biodiversity. However, since the "new wildernesses" as wastelands and unused land are often associated with decay and disorder, they are not particularly popular in the minds of city dwellers, and are even largely rejected. Giving this neglected and hitherto little-noticed urban nature a secure place among all other urban nature species is a phenomenon currently being observed. It is justified by ecology as a science and flanked by sociology, which seeks to understand how people interact with this nature. This promises to become an even more exciting field of research, experience and design.

Urban greening strategies are currently still modest but perceptible at European and global level. Other priorities dominate this level of design. At the local level, on the other hand, urban nature has already arrived as a design goal and is being transformed into policy, oriented towards proactive role models. The impetus often comes from concepts of innovative planners and masterminds, for example, the vertical forest or the eco-city, where a new "balance with nature" is sought in the city and placed at the centre of urban development as a guiding principle.

7.1   The Green City, a Conceptual Mosaic of Action Objectives

National organisations and the European Landscape Contractors Association (ELCA), essentially carry the idea of a city built on urban nature as a "Green City". They and scientists and planners have been arguing for decades, without always using the term "Green City", for an urban development policy that includes urban nature as an essential component (see ▶ Sect. 1.1).

The Green City is a vision and concept

The Green City is a visionary objective to which individuals, organisations and decision-makers are committed and incorporate its conceptual ideas into their decisions. It is not a ready-made concept that simply needs to be applied. The idea of the Green City requires the ever-new adapted development of visions under concrete natural, spatial and social conditions. To this end, all actors must share the basic goals of this development model, understand urban nature as an essential component of a liveable, healthy and (biologically) diverse city, and define concrete goals for action. The Green City thus develops "from below" and "from above" in equal measure.

The term "ecologically oriented urban development" was already emerging in the 1980s especially in Germany, but remained diffuse and often merely placative. A national German policy on urban nature has been missing for a long time.

Principles for the promotion of urban nature in urban planning were developed in Germany by scientists as early as the 1980s as "Guidelines for the implementation of nature conservation in urban planning" (Sukopp and Sukopp 1987). They are still fully valid and applicable today as a guide for conserving and promoting urban nature.

In 2003, the Forum DIE GRÜNE STADT (▶ www.die-gruene-stadt.de) was founded in Germany, since 2009 it has been a foundation. The foundation provides a platform for organisations, companies, individuals, including health experts, building managers and architects, homeowners' associations, industrial companies, auditing firms, associations, Agenda 21 working groups, municipalities and universities who want to work together for more urban green. This is based on the conviction that greenery and urban nature have so far been too little in the political spotlight, that bundling of knowledge and the exchange of experience, the creation of public and private green spaces in the city and the sharpening of the nature awareness of citizens and decision-makers are necessary. The spectrum of activities ranges from indoor greening to private gardens, green spaces, parks, botanical gardens and street greenery. In the meantime, such "Green City Organizations" are also active in the Netherlands, Great Britain, France, Italy and Hungary in cooperation called "Green City Europe". The idea of the Green City is spreading across Europe (see for example, De Groene Stad (NL) (2018) ▶ www.degroenestad.nl and The Green City (UK) (▶ www.thegreencity.co.uk) and gaining a broad social base (DIE GRÜNE STADT 2018).

In the "National Strategy on Biological Diversity" (2007), the city and its nature are first carefully integrated. Here, the field of action C 9 "Settlement and transport" is created and land consumption and land fragmentation are named as priorities to be overcome. At the same time, however, the demand for further densification and the increase in sealing often counteract this objective in concrete terms. The first objective concerning urban nature is stated here (BMU 2007, p. 79):

» "Spaces for nature experience are also created in green areas that are as close to walking distance as possible in order to promote children's understanding of nature" (BmU 2007, translated).

The "concrete visions" either describe the current state or are still content with modest greening visions. Diverse urban nature is described with the term "green", which is easier to convey to the public:

» **"Our vision for the future is:** Our cities have a high quality of life for people and provide a habitat for many animal and plant species, including rare and endangered ones. Diverse greenery improves air quality and the urban climate. It offers extensive recreation opportunities, play and experiencing nature for young and old.

   Our goals are: By 2020, the greening of settlements, including green spaces close to residential areas (for example, courtyard green spaces, small green spaces, roof and façade green spaces) is significantly increased. Publicly accessible green spaces with a variety of qualities and functions are generally available within walking distance" (BmU 2007, p. 42, translated).

On the initiative of the former German Federal Ministry of Transport, Building and Urban Affairs (Department of Urban Development), interdepartmental cooperation on the topic of "Green in the City" was initiated in 2013 with the aim of putting the topic on the political agenda and initiating discussion processes. After the 1st Federal Congress "Green in the City" in 2015 in Berlin, a *Green Book Stadtgrün. Green in the City.- For a livable future* (BmU 2015) developed the topic of urban green (= urban nature) and its development as a forward-looking policy field and multi-faceted political design task. The Green Book expresses the interministerial view of the federal government and considers not only the potentials but also the areas of tension. It is thus a broad stocktaking of the subject. Building on this, the BMUB has initiated a broad dialogue on urban green space.

   The term **Urban Green** is now concretely defined and operationalized in political and planning terms. It includes urban nature of all kinds and tenures and largely corresponds to the term "urban nature" used in this book. It is only surprising that, unlike the urban nature term used here, even indoor greening is included, but water surfaces in cities are not named as part of this urban greening term. However, water bodies are specifically mentioned in the term "green infrastructure" (BBSR 2017, p. 8) (see ◘ Fig. 7.1).

Fig. 7.1 Urban green as an essential structural element. Germany's first garden city Dresden-Hellerau (1909). (© Breuste 2011)

Urban Green

"Urban Green includes all forms of green open spaces and green buildings. Green spaces include parks, cemeteries, allotments, fallow land, play areas and playgrounds, sports areas, street greenery and street trees, settlement greenery, green spaces on public buildings, nature conservation areas, woodland and other open spaces that need to be developed, maintained and cared for to structure and shape the city. Private gardens and agricultural land are also an essential part of urban green. Furthermore, building greenery with facade and roof greenery, interior greenery as well as plants at and on infrastructural facilities are included. Urban greenery also includes the network of paved paths, promenades, squares, service roads of water, forestry and agriculture in an urban context, as well as indirectly traffic-calmed streets and wide footpaths, which are a prerequisite to achieve urban greenery..." (BBSR 2017, p. 8).

Following the 2nd German Federal Congress "Green in the City" in Essen in 2017, urban greening and its development were developed as a forward-looking policy field. The 2nd Federal Congress is followed by concrete steps with a white paper and recommendations for action and possibilities for implementation.

The White Paper "Urban Greening" of the German Federal Ministry for the Environment, Nature Conservation and Nuclear Safety (BMUB 2017) and the recommendations for action of the Federal Institute for Research on Building, Urban Affairs and Spatial Development (BBSR 2017) based on it now open up a **National Green City Strategy in** Germany since 2017. Action targets are defined for ten fields of action, which are made measurable using indicators, characteristics and orientation values - a major step forward on the way to a Green City in Germany.

The White Paper on Urban Greening is intended to support municipalities and other actors in their efforts to create, develop and maintain urban greening as a natural aspect of integrated urban development and planning.

7

Towards the Green City: "Guidelines for the Implementation of Nature Conservation in Urban Planning"

Principle of:
1. Priority areas for environmental protection and nature conservation
2. Zonally differentiated focal points of nature conservation and landscape management
3. Consideration of nature development in the downtown area
4. Historical continuity
5. Preservation of large contiguous open spaces
6. Networking of open spaces
7. Preservation of site differences
8. Differentiated intensities of use
9. Preservation of the diversity of typical elements of the urban landscape
10. Elimination of all avoidable interventions in nature and landscape
11. Functional integration of structures into ecosystems
12. Creation of numerous ventilation corridors
13. Protection of all life media (Sukopp and Sukopp 1987, pp. 351–354).

Charter "Future City and Green"

A broad alliance of associations, foundations and companies in Germany, including NABU (Nature and Biodiversity Conservation Union.) and Association of German Landscape Architects (bdla, ▶ www.bdla.de), is campaigning for "more quality of life through urban green spaces", in a joint charter initiated by the Federal Landscape Contractors Association) (BGL) and the foundation DIE GRÜNE STADT.

Urban green should make a much greater contribution to sustainable urban development. To this end, eight fields of impact and action were identified and demands made:
- Mitigating the effects of climate change,
- Promotion of health,
- Securing social functions,
- Increasing the quality of the location,
- Protection of soil, water and air,
- Preservation of species richness,
- Promotion of research in the field of construction and vegetation,
- Create legal and fiscal incentives.

The charter "Future City and Green" is intended to call on those responsible, above all in politics and administration, but also in business, science and civil society, to significantly increase their efforts to develop urban nature and to cooperate better (DIE GRÜNE STADT 2018, ▶ www.die-gruene-stadt.de).

Fields of Action of the White Paper "Urban Greening" of the Federal Ministry for the Environment (BMUB 2017)

1. Integrated planning for the urban green
   - Strengthening the importance of urban greenery in planning
   - Further development of regional, landscape and green space plans
   - Strengthening urban green in planning practice
   - Making parking space regulations and statutes more flexible
   - Support integrated strategies for green spaces
   - Strengthening urban-rural relations
   - Integrating federal real estate into urban development concepts
   - Cities on the way to more open space quality

2. Qualifying green spaces and making them multifunctional
   - Strengthening urban greenery as a compensatory measure
   - Securing cemeteries as part of the urban green space
   - Develop orientation and characteristic values for green
   - Strengthening urban green spaces as part of urban development funding
   - Expanding the scope of funding for urban green spaces
   - Urban green is a piece of building culture
   - Promote multi coded green and open spaces
   - Implementing "green" urban development with garden exhibition

3. Strengthening climate protection and mitigating climate impacts with urban greenery
   - Taking climate-friendly urban greening into account in planning practice
   - Using climate protection programmes for urban green spaces
   - Limiting climate risks with vital urban greenery
   - Developing cities in a water-sensitive way
   - Orient rainwater management towards retention and evaporation - reduce sealing, promote unsealing
   - Expand retention areas for flood prevention
   - Use planning instruments for fresh and cold air supply
   - Integration of future-oriented mobility

4. Developing urban green spaces in a socially acceptable and health-promoting way
   - Strengthening social cohesion and the importance of the approach of environmental justice through urban greening in urban development funding
   - Ensure equitable distribution of green space in social areas
   - Making public green spaces safer
   - Creating accessibility in outdoor spaces
   - Exploiting the potential of urban gardens
   - Better linking urban greening and health

7

5. Green buildings
   - Strengthening the greening of buildings
   - Include building greening in certification systems
   - Upgrading the street as a green and living space
6. Plan, lay out and maintain diverse green spaces professionally
   - Develop label for urban green
   - Focusing more on location characteristics
   - Ensure the maintenance of the urban green
   - Strengthen historical urban greenery as cultural heritage with social, touristic and ecological functions
   - Support knowledge transfer
7. Attracting stakeholders, involving society
   - Making room for the private sector and civil society involvement
   - Strengthen citizen science approaches
   - Activate private actors through legal instruments and financial incentives
   - Create legal certainty for the opening of private land
   - Strengthening and networking public actors
   - Demonstrating the value of ecosystem services
8. Strengthening and networking research
   - Establish a research cluster "Green in the City" as part of the "Future City" innovation platform
   - Integrated research into different facets of green in the city
   - Testing new forms of use and types of open space

9. Expanding the federal government's exemplary function
   - Handling a scarce resource in an exemplary manner
   - Take biodiversity concerns into account in climate adaptation and green space management
   - Qualified design through "green architecture"
   - Developing exemplary green transport and waterways
   - Establishing river continuity
   - Qualifying conversion areas and areas adjacent to railways
   - Developing, securing and maintaining greenery on federal properties
   - Making federal green spaces more accessible to the public
   - Planning, implementing and certifying sustainable green spaces
10. Public relations and education
   - Hold competitions
   - Strengthen and expand public relations work
   - Launching the "Green Space Development in Integrated Urban Development" initiative
   - Raising ecological awareness in allotment gardening
   - Improve environmental and awareness-raising for the urban green space
   - Developing education, training and further education offers

Thematic areas for action targets (BBSR 2017)
1. Climate and health
2. Environment and natural space
3. Society and social space
4. Organization and financing
5. Urban space

7.2 Green Infrastructure: The Local Basic Concept

"Urban green infrastructure" is a concept that has its origins in planning. It was introduced to understand the urban "green space system" as a unified planning object (Sandström 2002; Tzoulas et al. 2007).

Urban Green Infrastructure

Urban green infrastructure is a network of all natural and designed elements (green and open spaces as well as water areas) in cities. This also includes nature in built-up and sealed areas. This network of different natural structures of varying size, location and ownership is to be planned, maintained and developed as a joint task of various state, economic and civil society actors. The aim is to ensure that all urban natural elements are integrated in the sense of socially, economically and ecologically sustainable urban development.

- Are available to all citizens
- Promote the health and well-being of citizens
- Together enable a high level of biodiversity and an experience of nature
- Together contribute to an attractive cityscape and a high quality of life
- Provide locally targeted ecosystem services to urban citizens.

Urban green infrastructure contributes significantly to the quality of life and public services in cities. Sealed and built-up areas can become part of the green infrastructure through unsealing, greening and planting trees.

At EU level, "green infrastructure" (not specifically urban) defines a strategically planned European network on a trans regional scale. It is composed of valuable natural, semi-natural and designed areas as well as other environmental elements that ensure important ecosystem services and contribute to biodiversity conservation (see also Dover 2015; Naumann et al. 2011; BfN 2017; BBSR 2017).

» "It (green infrastructure – *the author*) can be considered to comprise of all-natural, semi-natural and artificial networks of multifunctional ecological systems within, around and between urban areas, at all spatial scales" (Tzoulas et al. 2007, 169).

The Green City bases urban nature on a "green" (and "blue") infrastructure. This includes water bodies as part of urban nature, but is also more often underlined again with the complementary term "blue". This is not entirely unjustified, since water bodies are often "forgotten" to be included if they are not located within official green spaces. The term "green infrastructure" has been used more intensively for the last 10 years or so, even though its contents were already in view earlier but were designated differently. By combining "green" and "infrastructure", an attempt is made to give urban nature a significance similar to that of technical infrastructure and thus to make it more assertive. Infrastructure is understood as a necessary substructure without which the functioning of the whole is not guaranteed. This is precisely what is to be expressed, the necessity of a nature-based city.

Green infrastructure, also commonly referred to as "green and blue" infrastructure, describes a strategic planning network to promote nature at different scales.

The Green Infrastructure Network aims at the conservation of biodiversity, the strengthening and regenerative capacity of ecosystem functions and ecosystem services in the sense of a sustainable use of nature. Intensive land use and strong landscape fragmentation threaten biodiversity worldwide, especially in Europe. The concept of green infrastructure aims to counteract this (European Commission 2009; Neßhöfer et al. 2012). Green infrastructure also stands for an integrative approach to bring stakeholders together (BfN 2017).

The term "infrastructure", originally from the military, is used in the economic and social spheres. It includes all long-lasting facilities of material or institutional nature that promote the purposeful functioning of the division of labour. This is also applied to the city and its urban nature.

7

» "The term "green infrastructure" offers the opportunity to clarify the social value of urban greenery, because "infrastructure" is associated with being indispensable for the functioning of the economy and society" (Torsten Wilke, Stadt Leipzig, Amt für Stadtgrün und Gewässer), (quoted from BfN 2017, p. 5, translated).

At the supra-regional, for example, European scale, green infrastructure is only about natural and semi-natural areas. With the transition to the local urban scale, all urban natural areas are included. A scale-based paradigm shift occurs. Instead of natural and near-natural areas, the focus is on "near-natural and designed areas and elements" (BfN 2017, p. 3). This justifies the new term "urban green infrastructure".

Green infrastructure plays a special role in urban areas. Here, the fragmentation of green spaces due to sealing, for example, by transport and building infrastructure, and the associated loss of biodiversity is particularly pronounced. However, a wide range of ecosystem services can be provided, particularly in cities, if the concept of green infrastructure is pursued. For example, air quality can be significantly improved by parks and green spaces. Overgrown house walls can also make a major contribution by absorbing the heat generated by (summer) solar radiation on the houses. These green walls contribute, among other things, to reducing the effect of urban "heat islands" (Neßhöfer et al. 2012).

While in the economy a distinction is made between infrastructure created by the private sector and infrastructure designed by the state, this distinction is usually not made for urban green infrastructure. Thus, highly diverse actors, public authorities, private landowners and stakeholders are involved. This makes network development a complicated municipal coordination task.

The concept of urban green infrastructure thus stands for strategic and integrated planning, protection, development and management of urban nature. This requires city-wide, neighbourhood and object-related spatial concepts, as different scales. This clearly goes beyond traditional open space planning.

Safeguarding, management and development of urban green infrastructure will be carried out taking into account the following principles:

– Adapt usability and performance of nature to the demands
– Develop strategic plans for this purpose
– Connecting nature
– Promote multiple-use and functional diversity
– Allow uninfluenced natural development and reduce maintenance and management where possible
– Develop green infrastructures also in the building and sealing area
– Enter into cooperation and alliances of actors

BBSR (2017) focused its research on the following questions regarding urban green infrastructure:

» 1. "What systematic surveys of green amenities and qualities exist?
2. How has green space and accessibility developed, what trends are foreseeable? Which functions can be derived empirically?
3. Which action goals/standards for urban greening in urban development already exist today? Which green standards can be systematically derived for municipalities?
4. What cities/town types are working with green space goals?
5. Is it about (even) more urban green, if so where, and/or about better green qualities and if so, in which form?
6. How should Conference of Garden Directors (Gartenamtsleiterkonferenz [GALK]), list of 1973 (meaning the GALK's guideline values for green provision in cities of 1973 – the author) be assessed from today's perspective? Does it make sense to update them? What recommendations for green standards can be developed?
7. What central, politically communicable core messages/action goals can be derived?" (BBSR 2017, p. 9, translated)

The design of urban green infrastructure is primarily the responsibility of municipalities, which can develop appropriate strategies and partnerships for this purpose.

The understanding of urban nature as a system of internally interacting natural elements in exchange with its environment is now well established. If planned, developed and maintained with foresight as urban green infrastructure, this system has the potential to guide urban development while integrating economic growth, nature conservation and public health care (Walmsley 2006; Schrijnen 2000; van der Ryn and Cowan 1996; Breuste et al. 2013). Thus, urban green infrastructure can be key to the Green City. With biotope mapping in urban areas in the 1970s and 1980s, good foundations were laid for this in Germany and subsequently in several other European and non-European countries (for example, Japan and Brazil), which are comparable between cities (see ▶ Sect. 6.2.2, see 🔲 Fig. 7.2).

Fig. 7.2 In 1993, the journal *Natur und Landschaft (Nature and landscape)* devoted an entire issue to the instructions for urban biotope mapping. (Arbeitsgruppe Methodik der Biotopkartierung im besiedelten Bereich 1993)

7.3 The Concept of Urban Biodiversity

Urban biodiversity strategies are measurable and operationalizable ecological concepts for urban areas with concrete goals for nature and species conservation. They have a political and implementation dimension. To this end, they are drawn up at the national, regional and local level and applied in a policy-shaping manner. The conservation and development of biological diversity in cities is increasingly becoming a design goal that is being pursued as a holistic task and vision for cities, far beyond traditional nature conservation and environmental protection, and with quite different understandings of biodiversity and different justifications.

Biodiversity conservation is now often replacing classical nature conservation as species and biodiversity protection internationally. **Biodiversity conservation and climate protection** are increasingly at the forefront of arguments for improving and shaping our environment.

Biodiversity Conservation – Protection and Conservation of Biodiversity

Biodiversity Conservation aims to shape conservation policy and practice based on the available scientific research. This includes cooperation with a wide range of research and planning partners as well as with political decision-makers and the economy as partners. Topics include process conservation for wilderness development *(rewilding)*, conservation of protected species and their habitats, management of invasive species, and EU and national frameworks (Jedicke 2016; British Ecological Society 2018).

Applied to urban areas, biodiversity conservation must always be seen and practiced in relation to urban dwellers. This has been a fundamental position for a long time.

» "Nature conservation in the city does not primarily serve to protect endangered plant and animal species; rather, its task is to specifically preserve living organisms and biotic communities as a basis for the direct contact of city dwellers with natural elements of their environment" (Sukopp and Weiler 1986, p. 25, translated).

» Today "city parks"are usually either managed overwhelmingly for the enjoyment of people, or may also have an important nature protection value. Semi-natural vegetation in a park supports both nature and nature-based recreation (Forman 2014, S. 349).

Urban biodiversity conservation – Protection and conservation of urban biodiversity

Urban biodiversity protection as species and biotope protection cannot primarily be about relict protection or the protection of rare and/or indigenous species, even if these are found in cities. Instead, it is about a holistic approach that focuses on people with their needs for nature and the benefits that nature brings them in the city (Sukopp and Weiler 1986; Breuste 1994a, b).

The protection of urban biodiversity became an internationally communicated and negotiated goal with some delay in the wake of the Rio Summit in 1992, starting around the year 2000. International biodiversity strategies in the wake of the Rio Biodiversity Convention (United Nations 1992) have only been targeting cities and their biodiversity since around 2010.

The Global Partnership on Cities and Biodiversity under the Convention on Biological Diversity led to important international events where national and regional policy makers, including city governments, met and cooperated with international organizations such as the International Council for Local Environmental Initiatives (ICLEI) and its Cities Biodiversity Centre (CBC) with representatives of scientific organizations, NGOs and their political or scientific representatives. The aim was to define positions on the way to protect and develop urban biodiversity, to network with each other and especially with national and regional policy representatives, and to agree on steps towards better protection of urban biodiversity.

At such meetings, which usually have hundreds of participants and usually well over 50 country and city representatives, resolutions, declarations and statements on urban biodiversity and its urban management have usually been adopted. They all aim at

- International cooperation
- Improving the transfer of knowledge from research to decision-makers
- Improve planning and design with ecological principles
- more effective outreach and education to improve public acceptance and support for these goals
- Further research in the field of urban biodiversity
- The development of National and Regional Urban Biodiversity Strategies

The focus is on cooperative interaction between politics, management, business and research in international networks and comparison of results achieved. The following events are particularly noteworthy:

- World Summit on Sustainable Development in Johannesburg (2002)
- Curitiba Declaration on Cities und Biodiversity (2007)
- Erfurt Declaration, URBIO (2008)
- Ninth meeting of the Conference of the Parties (COP) to the Convention on Biological Diversity (CBD) in Bonn (2008)
- Durban Commitment (2008)
- Fifth World Urban Forum in Rio de Janeiro (2010)
- Second Curitiba Declaration on Local Authorities und Biodiversity (2010)
- Expo Shanghai (2010)
- Nagoya Declaration (2010)
- Biodiversity Summit for Cities & Subnational Governments, Gangwon Province, Korea (2014)
- Incheon Declaration, Korea (2014)
- 5th Global Biodiversity Summit of Cities &Subnational Governments, Cancun, Mexico (2016)
- The Panama City Declaration: URBIO and IFLA (2016)
- 6th Global Biodiversity Summit of Cities and Subnational Governments, Sharm el-Sheikh, Egypt (2018)
- Virtual World Biodiversity Forum 2021
- 7th Global Biodiversity Summit of Cities and Subnational Governments will be held in parallel to CBD COP 15, scheduled to take place in Kunming, China

The search for indicators to measure and compare urban biodiversity, the commitment of cities and regions to achieve often ambitious biodiversity targets, and strategies to promote "greener cities" have been and continue to be on the agenda.

The Curitiba Declaration (2007) repeats the loss of biodiversity on a global scale already lamented at the World Summit on Sustainable Development in Johannesburg in 2002, although it does not even apply to many cities. Ecosystem services for people, especially for the poor, are directly attributed to biodiversity. Biodiversity thus acquires a general significance for human well-being. Without biodiversity, it is argued, there is a threat of a loss of quality of life. The reference to cities is rather low, which later declarations clearly improved.

The Erfurt Declaration (2008) fundamentally underlined the importance of urban biodiversity for culture, social affairs, climate change and the design of cities. It emphasized the need for further scientific research on urban ecosystems and urban biodiversity, to derive assessments from them, and to understand urban spaces as centres of evolution and adaptation and hotspots of regional biodiversity. Of importance is the emphasis on the close relationship of the concept of urban biodiversity to the "quality-of-life concept":

» "Urban biodiversity is the only biodiversity that many people directly experience. Experiencing urban biodiversity will be the key to halt the loss of global biodiversity, because people are more likely to take action for biodiversity if they have direct contact with nature" (Erfurt Declaration 2008, p. 1).

Long-term monitoring and research on urban biodiversity was also called for here for the first time (Erfurt Declaration, URBIO 2008).

The Durban Commitment (2008) and Second Curitiba Declaration on Local Authorities and Biodiversity (2010) are political declarations by local governments for biodiversity. In Durban, these were 20 cities, including Bonn in Germany. The declaration established a link between biodiversity conservation and poverty reduction, using the concept of ecosystem services, which was subsumed under biodiversity as a key term.

The Second Curitiba Declaration on Local Authorities and Biodiversity (2010) continued the contents of the Nagoya Declaration 2010 and reaffirmed them at the political level (Second Curitiba Declaration on Local Authorities and Biodiversity 2010) (see excursus Nagoya Declaration).

The "Quintana Roo Communiqué on Mainstreaming Local and Subnational Biodiversity Action" was adopted in 2016 at the 5th Global Biodiversity Summit of Cities & Subnational Governments in Cancun, Mexico. It is the most recent international statements for Urban Biodiversity, and was created as a result of the 13th International Biodiversity Conference. Conference of the Parties (COP) the Convention on Biological Diversity (5th Global Biodiversity Summit of Cities und Subnational Governments 2016).

Biodiversity is now clearly used to refer to the general relationship of people **to** nature, and this is seen **as the basis of well-being and economic success:**

» "...people are inextricably linked to, and part of nature, and that this connection is essential to the health, resilience and well-being of our rapidly growing urban communities and their local and regional economies" (5th Global Biodiversity Summit of Cities und Subnational Governments 2016, p. 1)

Cities are now expressing more precisely what their concrete needs are in order to actually promote and preserve biodiversity. They see biodiversity as a prerequisite for sustainable development! To this end, biodiversity should be integrated in all areas and across all sectors. The term *"mainstreaming biodiversity"* is used for this.

Mainstreaming biodiversity (IUCN)

"Biodiversity conservation is thus a pre-condition for achieving sustainable develop-ment. As such, it needs to be integrated into all sectors and across sectors: biodiver-sity needs to be mainstreamed" (IUCN 2019).

More accurate, specific and scientifically sound knowledge, decision-support tools and the need to evaluate these results on urban biodiversity are called for. Horizon-tal *mainstreaming*, *vertical alignment*, cooperative and integrated management, combined with measurable reporting mechanisms at all levels of local government, coordination between sectoral policies and support for ecosystem restoration are called for. Achieving all this would indeed be a breakthrough and would also over-come the previous municipal weaknesses in the enforcement of biodiversity strate-gies.

New partnerships and funding instruments are being sought for this purpose. ICLEI's Local Action for Biodiversity (LAB) and the Singapore Index on Cities' Biodiversity (excursus of the same name) are considered supportive of this (5th Global Biodiversity Summit of Cities and Subnational Governments 2016; IUCN 2019).

The Panama City Declaration 2016 is the outcome of the Urban Biodiversity and Design Network (URBIO) and International Federation of Landscape Architects of the Americas Region (IFLA-AR) conference 'From Cities to Landscape: Design for Health & Biodiversity' 2016 in Panama City. It highlighted health and biodiversity as the focus for sustainable cities (The Panama City Declaration 2016).

The **National Strategy on Biological Diversity in Germany does** not associate biodiversity (in the city) first and foremost with the goal of nature and species con-servation, but sees biodiversity as the basis for ecosystem services (see ▶ Sect. 6.4). Both concepts intermingle, so that it is hardly possible to distinguish between the two anymore. The goal of both biodiversity and ecosystem services is urban nature, "used by people [...] and as a habitat for species" (BmU 2007, p. 42).

The vision for cities is (see ▶ Sect. 7.2):
- Diverse greenery,
- Improved air quality and urban climate,
- Extensive opportunities for recreation, play and experiencing nature, and
- Continue to promote active inner-city development

The goals are:
- To significantly increase the greening of settlements by 2020,
- Publicly accessible green space with a variety of qualities and functions within walking distance of users,
- Preservation and expansion of habitats for endangered species typical of urban areas (for example, bats, chicory, wall ferns),

- Nature experience spaces for healthy mental and physical development of children,
- Expansion of natural areas in inner cities, taking into account the different demands of the various population groups, and
- In many urban areas, more green space can be used by people and serve as a habitat for species.

The study on Natural Capital Germany – TEEB DE (2016) takes a much more sensitive approach to biodiversity (see ▶ Sect. 6.3). On the one hand, it justifiably points to the surprisingly high biodiversity of for example, plant species in cities compared to equally large areas in their rural surroundings, but also to the often related small population sizes and the restriction of rare and endangered species to special habitats (Kowarik 1992; Wittig 2002). The more or less close links between biodiversity and ecosystem services (Fuller et al. 2007; Botzat et al. 2016; Shwartz et al. 2014), which have been proven mainly by scientific studies, need to be pointed out here (see ▶ Sect. 6.1). The authors of TEEB DE 2016, p. 18, only see "opportunities" here with the "ecosystem services approach to simultaneously promote the quality of life of the population and biodiversity in urban areas".

Regional urban biodiversity strategies presuppose either a city network or city-urban regions in close cooperation. This is the case for the Ruhr metropolitan region and the Frankfurt city-region (Biodiversitätsregion Frankfurt, Netzwerk Urbane Biodiversität Ruhrgebiet 2013; BioFrankfurt 2018b) (see ▶ Sect. 6.1).

The task of **municipal biodiversity strategies** (for example, Berlin, SenStadt 2012) could be to support the development of green urban infrastructure that leads to the strengthening of ecosystem services and at the same time promotes biodiversity in different urban ecosystems.

Municipal biodiversity strategies develop green infrastructure, the Green City, as a basis for the quality of life of the population through ecosystem services AND at the same time for the development of biodiversity.

The "Berlin Strategy on Biological Diversity" lists 38 goals in four thematic fields, which are also exemplary for other cities (SenStadt 2012):

Species and habitats
- Preserving biodiversity and taking responsibility for special species
- Alien species are monitored but only regulated if they threaten to significantly affect biodiversity
- FFH habitats maintained in favourable condition
- Conservation and development of specially protected biotopes
- Development of a biotope network
- Improvement of the passability of water bodies
- Aim for water bodies: water quality class II and a significantly higher proportion of near-natural water body sections and riparian zones.
- At least one third of the shorelines of the lakes of the rivers Spree, Dahme and Havel should again have reed beds in good condition
- Sustainable management of groundwater, conservation of groundwater-dependent habitats
- Preserving peatlands as habitats for species typical of peatlands and wetlands

- Agriculture makes important contributions to the design of an attractive and eventful urban landscape (for example, as a provider of leisure and educational services, especially for children and young people, establishment of selected agricultural uses on suitable inner-city open spaces)
- Promotion of near-natural and site-typical mixed forests, preservation of sparse forests and recreationally effective open land habitats
- Forest management continues to be following the FSC standards for the improvement and long-term preservation of the functional and productive capacity

Genetic diversity
- Preserve the genetic diversity of traditional ornamental and useful plants, of traditional livestock and wild species of animals and plants.
- Sustainably strengthen the supply of traditional livestock breeds and crop varieties for use in the field
- Use certified native planting and sowing material in landscaping and landscape management measures
- No risk to biodiversity from genetically modified plants

Urban diversity
- To preserve typical urban species and to secure them in the long term
- Promotion of urban wilderness as a space for experience and as a space for dynamic and largely uncontrolled natural development
- Strengthening the contribution of allotment gardens to the conservation and sustainable use of biodiversity. New forms of urban gardening provide access to biodiversity and anchor this knowledge broadly in society.
- Nature-friendly maintenance of public green spaces and outdoor facilities of public buildings. The aim is to extend this objective to open spaces owned by religious or other institutions.
- Significant increase in the proportion of private open spaces designed close to nature, especially house and front gardens, inner courtyards, facades and roofs.
- Increasing biodiversity on company-owned buildings and company premises through cooperation with Berlin's business community and incentives
- Preservation and development of the street tree population as well as the roadside greenery through qualified maintenance
- Making it possible to experience the biological richness of open landscapes of former traffic areas (for example, after-use of Tempelhof and Tegel airports)

Society
- Standards for biodiversity conservation in public construction and procurement
- Consideration of biodiversity in legal regulations and planning principles
- Consideration of the topic of "biological diversity" in the framework curricula of schools and in the educational programme for children in daycare facilities
- Expansion of the number of places in nature (adventure) and forest kindergartens

- Promotion of environmental education institutions, anchoring of the topic "biological diversity" in environmental information, education and experience offers
- Promotion and use of biodiversity research
- Spaces for experiencing nature, especially in densely populated areas
- Support for initiatives to promote nature-friendly human-nature interactions, especially for socially disadvantaged children and young people
- Companies are increasingly involved in promoting biodiversity research and conservation projects
- Commitment of the economy: Certification and accounting in EMAS (= Eco-Management and Audit Scheme and for example, ISO 14001)
- Companies and credit institutions ensure compliance with international and German environmental standards when investing abroad
- Increase the proportion of imported natural materials and products from environmentally and socially responsible use and production
- Promotion of social commitment to the areas of "nature conservation and environmental education" (SenStadt 2012).

The Nagoya Declaration: Urban Biodiversity Turns Cities into *"Green, Pleasant and Prosperous Places"*

In the International Year of Biodiversity 2010, the 14th expert meeting of the participants of the Convention on Biological Diversity (after Rio) took place in Nagoya. The main subject of the conference was "Urban Biodiversity in the Ecological Network" with two focal points, "Ecosystem Network and Quality of Habitats in and around the Urban Area" and "Networking the Activities of Urban People". The topics covered were:

- Maintenance, conservation and development of urban ecology networks
- Urban life in "harmony with nature"
- Preservation of *"native biodiversity"* (biodiversity of indigenous species)
- Develop a quantitative urban biodiversity assessment system to compare biodiversity between cities and over time
- Close cooperation between the public sector and the private sector

- Increase public awareness of urban biodiversity through environmental education and community participation.

To this end, the following measures were recommended:
- Planning and design of "resilient ecological corridors" in cities and better understanding of the interactions between habitats, corridors and their surroundings
- Climate change mitigation and adaptation
- Comparative studies of different cities
- Linking biodiversity and ecosystem services
- Ecological design should take biodiversity and climate change into account
- Improve the application of research results in planning and design through better cooperation between all stakeholders involved

7

- Attention to the links between the economy and biodiversity
- Better information for policy makers
- Call for the commitment of national governments, regional administrations, funding partners and relevant organizations (Nagoya Declaration 2010) (see ◘ Fig. 7.3)

The Conference of the Parties to the Convention on Biological Diversity (CBD), which followed in 2010, adopted a "Plan of Action on Subnational Governments, Cities and Other Local Authorities for Biodiversity" (2011–2020), which contains the main points mentioned above (Plan of Action 2010).

◘ **Fig. 7.3**   Central Park in Nagoya, Japan. (© Breuste 2010)

City Biodiversity Index (CBI): Biodiversity Comparison for the Convention on Biological Diversity (CBD)

Native Biodiversity the Natural Ecosystems

The global competition for biodiversity, also in cities, has led to the international comparative assessment of not only biodiversity, but also biodiversity in broad complexity in urban areas through an indicator concept. Previous environmental indices mainly included *brown*

issues as indicators, such as clean water, wastewater, energy efficiency, air quality and wastewater management. Biodiversity indicators were mostly applied by including different indices at national level, but not at local level (2005 Environmental Sustainable Index (ESI) and 2008 Environmental Performance Index (EPI)). However, the aim was to combine all biodiversity-related characteristics into one index at the local level.

At the ninth Conference of Participants of the Convention on Biological Diversity (CBD) (COP 9) 2008 in Bonn, Singapore's Minister for National Development, Mah Bow Tan, proposed the development of a City Biodiversity Index to compare cities' achievements in *biodiversity conservation.*

Since 1992, the city-state of Singapore has been one of the pioneers of urban biodiversity conservation worldwide (Davison et al. 2012). The National Parks Board of Singapore, in collaboration with experts from various countries, developed a 'Singapore Index on Cities' Biodiversity or 'Singapore Index' (SI) in two workshops in 2009 and 2010 (Chan and Djoghlaf 2009), in order to provide the CBD Secretariat with a suitable tool for cities in their monitoring and self-assessment of biodiversity conservation for comparative surveys (Chan et al. 2010).

The CBI/SI records 23 indicators, which result in a maximum of 92 points by summing up the scores for the degree of fulfillment. In the first part (profile of the city), localisation, size, ecosystems, species, population size and qualitative information on biodiversity are recorded.

This is followed by three core components

- *Native biodiversity* (10 indicators, covering birds, vascular plants, butterflies and unspecified *other taxonomic groups*)
- *Ecosystem services provided by biodiversity* (4 Indicators)

- *Governance and management of native biodiversity within the city* (9 Indicators)

Thus, in addition to characteristics of a *native biodiversity,* ecosystem services (regulation of water balance, climate, CO_2 sequestration, recreation and environmental education) and management measures are recorded and evaluated. This goes far beyond what is scientifically understood by "biodiversity" and generally excludes non-native species from the biodiversity assessment. As this is an official instrument of the CBD Secretariat, it is also of political importance.

The search for "natural nature" *(natural ecosystems)* in urban areas as an indicator of biodiversity is central and questionable:

» "**Natural ecosystems** contain more species than disturbed or human-made landscapes, hence, the higher the proportion of natural areas to the total city area indicates the biodiversity richness" (Chan et al. 2010, S. 11; Rodericks 2010, p. 3).

As a working definition, *natural areas* are the focus:

» "**Natural areas** comprise predominantly native species and natural ecosystems, which are not, or no longer, or only slightly influenced by human actions, except where such action is intended to conserve or enhance native biodiversity" (Chan et al. 2010, S. 11; Rodericks 2010, p. 3).

» "**Natural ecosystems** are defined as all areas that are natural and not highly disturbed or completely human-made landscapes. Some examples of natural ecosystems are forests, mangroves, freshwater swamps, natural grasslands, streams, lakes, etc. Parks, golf courses, roadside plantings are not considered as natural" (Chan et al. 2010, S. 11, Rodericks 2010, p. 3).

The City Biodiversity Index (CBI) excludes typical urban nature and focuses solely on the remnants of nature in urban areas that are little influenced by humans (*natural ecosystems* and *native biodiversity*). What constitutes urban nature – the *man-made ecosystems* – remains unnoticed (see ◘ Fig. 7.4).

Already in 2010, more than 30 cities worldwide had applied the CBI (Rodericks 2010). Unfortunately, the Secretariat of the Convention on Biological Diversity (CBD), as the administrator of the tool, does not state on its website how many cities have applied the index after 10 years. The collection of the required data is likely to be difficult and hardly repeatable for many municipalities, given that hardly any data sets already exist and are kept up to date.

◘ **Fig. 7.4**   Rehabilitated course of the Isar river in Munich. (© Breuste 2016)

Native Biodiversity Versus Non-native Biodiversity: Biodiversity Only Through Native Species as a Goal for Cities?

Can urban biodiversity be based only on native species? This has already been argued to be impossible in ▶ Sect. 6.1. Many species in cities, often the majority, are not native, but they are not invasive either! The Nagoya Declaration and the City Biodiversity Index (CBI) explicitly assume *native biodiversity* and use the otherwise rarely used term. They thus exclude any biodiversity by non-native species.

The term does not (yet) exists in German. What is meant is biodiversity that is based exclusively on native species, in contrast *(non-native biodiversity)*. The equation of non-native with invasive species is the justification for native biodiversity. However, it is false (Biodiversity Gardening 2018)! Not only "protection" of native species, but also belonging of "native" plants and animals to cultural tradition and belief system as part of identity are given as reasons of preserving *native biodiversity*. This may be true for island ecosystems with a high proportion of endemics, such as Hawaii's Big Island (Department of Land and Natural Resources 2018), but can hardly be applied to cities (Kowarik 2011).

Biodiversity Region BioFrankfurt

Twelve institutions from the fields of research, education and nature conservation, including the University of Frankfurt/M. and the renowned Senckenberg Society, currently support the non-profit association BioFrankfurt. The institutions pool their experience and knowledge to work together to promote the BioFrankfurt biodiversity region and raise public awareness of it. BioFrankfurt was founded as a network in 2004 and established as an association in 2013. Together, they have set themselves the goal of raising awareness of the outstanding importance of biodiversity and its conservation among the media and the general public, bundling the knowledge and experience of individual institutions and promoting biodiversity more efficiently through cooperation and information exchange. This includes targeted educational offerings on the topic of "biodiversity" (with exhibitions, symposia, lectures, guided tours and excursions), information and training offerings for schools and teachers, jointly conducted research projects to gain new insights and exploit valuable synergies, and ongoing dialogue with partners from business and society.

In the joint project "Cities dare wilderness" (2016–2021), three cities, Frankfurt, Hanover and Dessau-Roßlau, are working together to gain more biodiversity through more "urban wilderness" and more understanding of it. BioFrankfurt is a pioneer and communicator (BioFrankfurt 2018b) (see ◘ Fig. 7.5).

7

What we do

* Exhibitions, lectures. Excursions and tours around nature, man and environment

* Action week "Living biodiversity"

* Moderation of the dialogue between science, economy and society for a sustainable living and working space in the Rhine-Main region

* Initiation of international cooperations and research projects

* Organization of conferences

* Biozahl des,jahres: Reports on biodiversity in Frankfurt and the world

* Workshops on biodiversity for schoolchildren of all ages and teachers

* Partner in the joint project "Stadte wagen - Wildnis", called for by the Bundesamt Fur Naturschutz (Federal Agency for Nature Conservation) with funding from the BMU (Federal Ministry for the Environment, Nature Conservation and Nuclear Safety)

* Implementation of the project "Frankfurt dares wilderness - for more wilderness development, nature diversity and nature experience in Frankfurt", demanded by the Frankfurt Airport Foundation for the region.

Aktuelle Informationen zu unseren Aktionen finden Sie unter:

WWW.biofrankfurt.de

WWW.staedte-wagen-wildnis.de

■ **Fig. 7.5**   BioFrankfurt – Together for diversity and sustainability, BioFrankfurt (2018a)

The "Berlin Biodiversity Strategy": "Green Metropolis" Berlin Through Biodiversity

Berlin, as Germany's largest city, sets species and habitats with urban links as a separate focus in its municipal biodiversity strategy. To this end, **measures and indicators** are being developed to **monitor** success (SenStadt 2012). Habitats, ecosystems, animal and plant species and their genetic resources are to be preserved even as the city continues to develop.

The State Commissioner for Nature Conservation and Landscape Management, Prof. Dr. Ingo Kowarik, states: "Biodiversity makes you happy! We already have a high level of biodiversity in many places in the city" (SenStadt 2012, p. 4).

Noted:

— Access to urban nature as a carrier of biodiversity varies greatly in Berlin's neighbourhoods

— Experiencing nature is important for children in the city

— The city is getting denser

— Land uses are changing,

— Climate change requires adaptation

— Funds in the city budget for the development and maintenance of urban nature are becoming increasingly scarce.

The changing framework conditions give rise to a need for a strategy that takes up proven nature conservation objectives, but also goes beyond them (see ◘ Figs. 7.6 and 7.7).

◘ **Fig. 7.6** Berlin's largest park (355 ha) is Tempelhofer Feld, a former airfield site opened in 2010. (© Breuste 2011)

7

■ **Fig. 7.7** Berlin's most recent park (2011/2013) is the Park am Gleisdreieck (26 ha), an evolved wilderness and designed open space, awarded the 2013 Berlin Architecture Prize, the 2014 German Urban Design Special Prize and the 2015 German Landscape Architecture Prize. (© Breuste 2015)

Cities Engage in Competition for Urban Biodiversity: Internationalisation and Local Efforts for Nature for Urban Dwellers

Politically, it is emphasized that the loss of biodiversity is, next to climate change, the greatest global environmental challenge, which cannot be overcome without the active participation of local authorities.

The Federal City of Bonn participated in the process of developing international biodiversity strategies at an early stage. Since 2005, the Mayor of Bonn has chaired the World Mayors' Council on Climate Change WMCCC. Bonn was already part of the pilot group of "Local Action for Biodiversity", a global city project for biodiversity, in 2007 and signed the Curitiba Declaration on Cities and Biodiversity (2007), where the city was represented by its mayor. Bonn hosted the 9th Convention on Biological Diversity (CBD), Conference of the Parties (COP 9) in 2008 and prepared the first **Bonn Biodiversity Report** in 2008 as a status report on local urban biodiversity and the targeted goals. Bonn also participated in the Erfurt and Durban Declaration 2008.

Bonn is the only German city to belong to the Global Action Alliance for Cities and Sub-National Governments for Biodiversity. Within this alliance, the city works very closely with several UN agencies, scientific institutions, city networks and other cities internationally.

The Biodiversity Report of the City of Bonn presents the biodiversity in the city area, strategies for its conservation (biotope protection) and other activities of the City of Bonn for the conservation of biodiversity at local, national and international level. The basis of Bonn's biodiversity strategy is an **intensive analysis of the landscape, supported by scientific research,** which is combined with concrete projects for the conservation of biodiversity.

The biodiversity map for Bonn shows the distribution of areas with high, medium and low importance for urban biodiversity. 51% of the city area is protected areas. The forest areas surrounding the city (approx. 25% of the city area, predominantly near-natural deciduous forests) are protected. Together with floodplains and watercourses, they form the backbone of Bonn's biodiversity. The various species groups are very well documented in Bonn as one of the best botanically and zoologically studied regions in North Rhine-Westphalia.

On the occasion of the International Year of Biodiversity 2010, the German Environmental Aid Association (Deutsche Umwelthilfe) is awarding prizes to cities and municipalities for their special commitment to protecting and promoting local biodiversity. (DUH) awards cities and municipalities for their special commitment to the protection and promotion of local, municipal biodiversity. The "Lebendige Stadt" Foundation, as the sponsor of the **"Federal Capital of Biodiversity"** competition, offered prize money total-ing EUR 50,000 for the winning municipalities, which was to be earmarked for municipal projects to protect biodiversity. The competition was held as part of the international project "Capitals of Biodiversity", which was funded by the European environmental programme LIFE +. The winner of the competition in 2011 was the state capital of Hanover, which was allowed to call itself the "Federal Capital of Biodiversity" due to its special commitment to the preservation of biological diversity. A total of 124 cities and municipalities, including Bonn, submitted almost 900 projects in a total of six categories. A comprehensive questionnaire was used to record the commitment of the cities and municipalities to biodiversity in the areas of "Nature in the City", "Environmental Education and Environmental Justice", "Species and Biotope Protection", "Sustainable Use", "Conception", "Communication and Cooperation" and "Monitoring Local Development". Hannover convinced with a holistic programme for the protection of biodiversity, near-natural silviculture, the water renaturation programme, the promotion of ecological agriculture, the programme "More Nature in the City" and especially a large number of projects that aim to **inspire the population for nature in their vicinity,** for example, environmental education programmes for schools and kindergartens, guided tour programme "Green Hannover", children's forest, forest station Eilenriede with forest tower, school biology centre (Stadt Bonn 2008; Stiftung Lebendige Stadt 2010).

7

Singapore Protects Its Biodiversity with a National Biodiversity Strategy

Singapore, as an island nation with 719.9 km², a population of 5.6 million in 2017, and a very high and growing population density of 7.8 inhabitants/km² (Government of Singapore 2018). It is not only particularly limited, but also particularly reduced in its urban nature stock and under pressure to grow. Just 200 years ago, the island was almost entirely forested (82% tropical forests). Today this is still 23.06% of the area (164 km²), with a slight downward trend. Efforts to preserve the remaining natural remnants in the city are correspondingly intensive. In 1992, the first Singapore Green Plan was drafted, since supplemented by the Singapore Blue Plan. A National Climate Change Strategy and a National Biodiversity Strategy with an Action Plan were adopted and incorporated into the 10-year Concept Plan and 5-year Master Plan for urban development. Singapore has set the protection of its biodiversity as a special task. The "City in a Garden" concept sees the city embedded in a natural environment with coordinated urban nature management and protection of remaining forest remnants. The National Park Board manages four nature reserves (3347 ha), 2269 ha of urban green spaces (59 regional parks and 255 district parks) 2664 ha of street green spaces, including more than 1 million trees, and 1679 ha of open land in use. Since 1992, Singapore has thus been one of the pioneers of urban biodiversity conservation worldwide (Davison et al. 2012) (see ◘ Fig. 7.8).

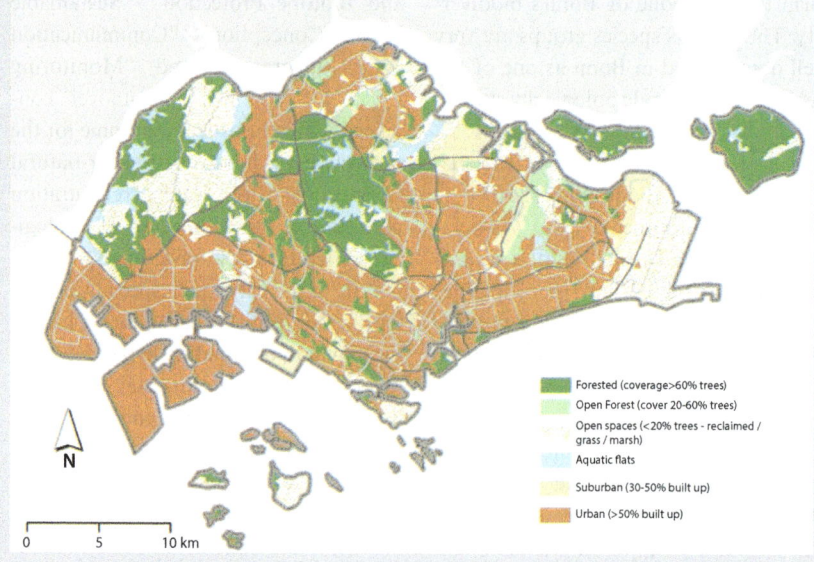

N

	Forested (coverage>60% trees)
	Open Forest (cover 20-60% trees)
	Open spaces (<20% trees - reclaimed / grass / marsh)
	Aquatic flats
	Suburban (30-50% built up)
	Urban (>50% built up)

0   5   10 km

◘ **Fig. 7.8** Urban nature in Singapore. (Design: J. Breuste, drawing: W. Gruber after Davison et al. 2012; Breuste et al. 2016, ◘ Fig. 4.24, p. 122)

7.4 Wild Goes Urban: Wilderness as Part of Urban Nature

While "wilderness" evokes predominantly positive associations, this is usually not the case for "urban wildernesses". The contrast to the lived civil society of the city seems too great. In addition to the "old" wildernesses that may exist as remnants (for example, forests), new types of wildernesses are also emerging in cities that represent an important component of urban nature. Both have wilderness character, but they are clearly different. The special significance of urban wilderness is based on its proximity to the population, the associated cultural ecosystem services and the special opportunities to experience nature here in a different way than in parks and green spaces. Through integrated concepts, acceptance and access to the urban wilderness can be promoted and biodiversity can be directly experienced.

Wilderness as a concept from a social perspective has its origins in the Enlightenment and Romanticism. While Enlightenment and Romanticists discovered wilderness as sublime nature and direct experience of God's work, conservatism in the mid-nineteenth century developed the idea of wilderness as the "fountain of youth" of the "degenerate people" (Trepl 2011).

In the mid-nineteenth century, the influential German conservative Wilhelm Heinrich Riehl (1823–1897) applied the wilderness idea to the modern big city and regarded it as a place of education and a necessary complement to the cultivated land and especially to the civilized society of the big city, which he rejected as far removed from nature (Riehl 1854; Vicenzotti and Trepl 2009). The modern metropolis stands for "over-civilization" and "pathological culture-nature relationships". As a counter-image to the industrial city of modernity, wilderness is based on the conservative critique of civilization of the nineteenth century. Wilderness became a counterworld to the amoral order of culture, to the urbanization of modernity (and postmodernity) (Oelschlaeger 1991; Kirchhoff and Trepl 2009; Kirchhoff and Vicenzotti 2014). Forests and high mountains, later complemented by wetlands, remote from industrialization access, became ideal types of wilderness and the object of conservation efforts (Schwarzer 2007). Spanier (2015) rightly points out the cultural construction of wilderness, its romantic character and current models of thinking of "use-free and human-free" nature, which cannot be transferred to cities in this way.

Wilderness protection today continues to emphasize **"pristine"** and **"untouched"** nature and contrast with human settlement. The US Congress Wilderness Act of 1964 and the International Union for Conservation of Nature and Natural Resources (IUCN) protected area categorization of 2008 are based on this. The US Congress Wilderness Act and other conservation efforts aim to create a "wilderness system" *"... for the permanent good of the whole people, and for other purposes"* (US Congress 1964, p. 1).

7

Wilderness as a protected object and counter-image to the settlement area

» "A wilderness, in **contrast with those areas where man and his own works dominate** the landscape, is hereby recognized as an area where the earth and its community of life are untrammeled by man, where man himself is a visitor who does not remain. … without permanent improvements or human habitation" (US Congress 1964, S. 1).

Today, the concept of wilderness is associated with the search for the "original" in nature:

» "Wilderness is defined as an extensive, **pristine** or slightly modified area that has retained its original character, has largely undisturbed habitat dynamics and biodiversity, in which there are no permanent settlements or other infrastructure with a serious impact, and whose protection and management are designed to maintain its original character" (Wilderness Area IUCN Category Ib, Dudley 2008).

Wilderness can be viewed from a social, environmental sociological, cultural, environmental psychological perspective and the perspective of other disciplines (Ridder 2007; Hofmeister 2009; Lupp et al. 2011; Kowarik 2015). In addition, there is an everyday language use of terms of diverse content (*wilderness is what people perceive as wilderness,* Kowarik 2018, p. 336). **Wilderness from an ecological perspective** does not necessarily translate into the social realm (Kowarik 2018).

Although wilderness continues to be a concept of protecting nature that is remote from cities and (largely) untouched, there is now also a concept of wilderness that includes urban space (Kowarik 2018; Threlfall and Kendal 2017; Diemer et al. 2003; Kowarik and Körner 2005; Jorgensen and Tylecote 2007; Rink 2009; Vicenzotti and Trepl 2009; Jorgensen and Keenan 2012; Stöcker et al. 2014).

New terms such as *urban wilderness* (Diemer et al. 2003; Kowarik 2018), *urban wildscape* (Jorgensen and Keenan 2012), wilderness in urban in-between spaces (Jorgensen and Tylecote 2007), and *urban wild spaces* (Threlfall and Kendal 2017) emerged to express the special nature of the connection between wilderness and the city.

Urban wilderness areas

Urban wildernesses are components of urban nature that develop without significant human intervention. They are not free of human influences, are often small in area and close to populated areas, and can thus also be exposed to pollutant or nutrient inputs and noise. Self-dynamic development processes are deliberately permitted so that natural succession can take place. Such wildernesses are remnants of original pre-urban nature (Nature of the 1st Kind, Kowarik 1992, see ► Sect. 1.3), but also new urban ecosystems on inner-city wastelands after abandonment of use (Nature of the 4th Kind, Kowarik 1992, see ► Sect. 1.3). Urban wilderness can supplement the landscape and biological diversity of a city with additional, different and novel habitats.

The remnants of **pre-urban nature can** be referred to **as old urban wildernesses** *(ancient urban wilderness)* and newly emerging wildernesses on **succession areas as new urban wildernesses** *(novel urban wilderness)* (Kowarik 2018, see also Diemer et al. 2003; Wissel 2016, see ◘ Table 7.1).

Urban wilderness concept

Urban wildernesses also exist without planning and acceptance by the population. However, only through both do they become part of a concept that equally includes the conservation of species and their habitats and the experience of nature in its spontaneous development.

◘ **Table 7.1** Comparison between "wilderness" and "urban wilderness"

Feature	Wilderness	Urban wilderness
Size	Often large areas	Often small areas
Fragmentation	Often uncut	Often cut up
Management intensity	No management	Preferably no maintenance, but often influenced by environmental uses.
Natural process flows	No human influences	Natural processes are controlled and possibly influenced
Importance for biodiversity	Very large	Large, especially in the regional urban environment
Biotope network	Significance in the large-scale biotope network	Often isolated
Importance for environmental education	Large, but often remote in low-population regions	Very large, within reach of a variety of potential users who can gain nature experience here, easy inclusion in environmental education programs for many
Habitat types	Often old forests, rivers, floodplains, mountains	Ancient wildernesses: ancient forests new wildernesses: Succession and spontaneous vegetation areas
Anthropogenic influence	None to low	Little to clearly

Modified after Stöcker et al. (2014)

Urban wilderness as a concept cannot solely consider the wilderness area as a natural object, but must include the human dimension (Cronon 1996; Hoffmann 2010; Kowarik 2018). Potential components of an urban wilderness concept can support the identification of wilderness areas in planning from an ecological perspective, link the concept to biodiversity conservation, and must take into account increasing social demands.

Kowarik (2018) argues for a unified *urban wilderness* concept that combines the social and ecological dimensions of wilderness and both with planning approaches. He proposes a three-dimensional socio-ecological system consisting of ecological, social and planning dimensions (see ◘ Figs. 7.9 and 7.10).

The urban wilderness concept includes:

- Identification of urban wilderness areas as natural elements that meet with acceptance and demand from the population. For this purpose, the existing wilderness areas to be reviewed from an ecological point of view must be identified (supply-side).
- Identify demand requirements for urban wilderness (access, risk management, structure, etc.).
- Physical and mental access to urban wilderness must be provided through planning, interactive acceptance development, and citizen participation (see ◘ Fig. 7.10).

Kowarik and Kendal (2018) take a global perspective on how urban wildernesses can be integrated into concepts of green planning, green infrastructure, urban biodiversity strategies, and sustainable urban development and find positive results. This provides unique but rarely explored opportunities to develop livable cities and reconnect urban dwellers with nature. Threlfall and Kendal (2017) show the special conditions of urban wildernesses as diverse contributors to ecological urban structure from a multidisciplinary perspective (see ◘ Fig. 7.11).

◘ **Fig. 7.9**   Urban wilderness as a socioecological system: challenges in three overarching dimensions. (After Kowarik 2018, p. 337)

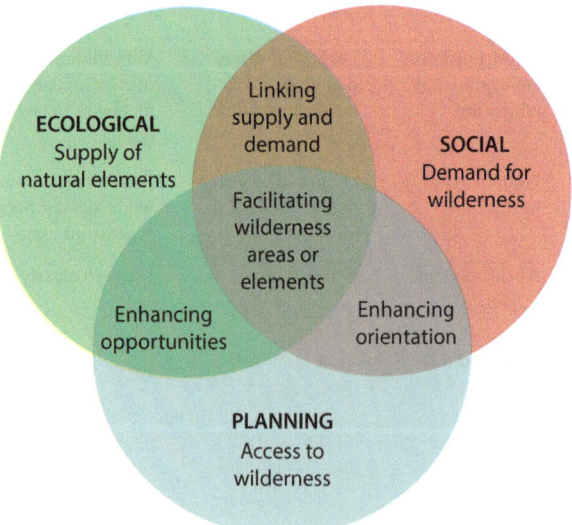

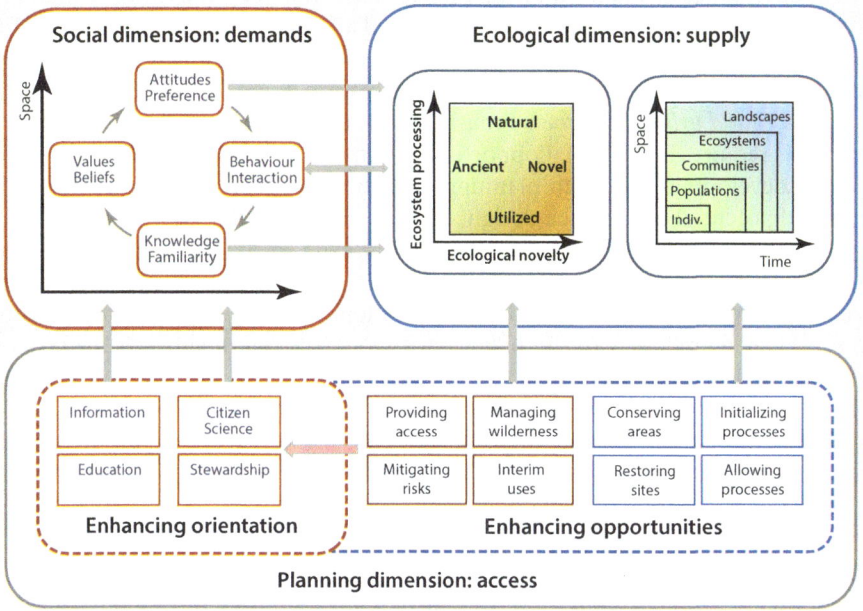

■ **Fig. 7.10** Main components of the ecological and social dimensions of urban wilderness and references to planning practice. (Kowarik 2018, p. 338)

■ **Fig. 7.11** Typology of urban wilderness with anthropogenic influences. (Threlfall and Kendal 2017, p. 349)

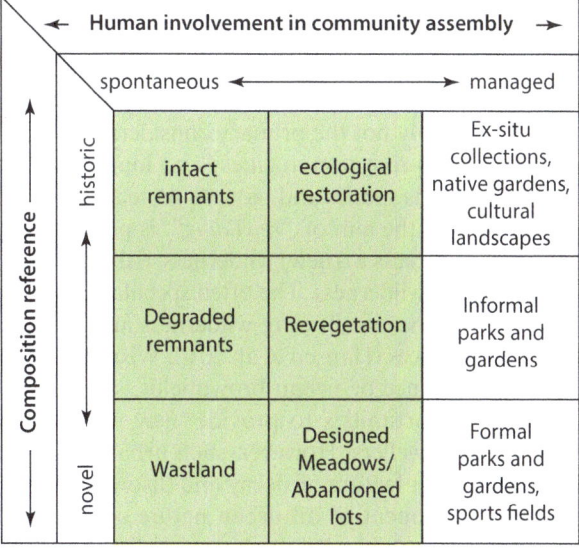

The acceptance of old urban wildernesses *(ancient urban wilderness)*, especially forests, watercourses and lakes, through cultural adaptation as a known and useful part of the "established cultural landscape", is significantly greater than for the unknown newly emerging wildernesses on succession areas. These new urban wildernesses *(novel urban wilderness)* must first be learned about to be accepted and used. This requires educational opportunities, emotional familiarisation and good risk-free development through planning, but above all early involvement of residents and potential users in the design.

New and old urban wildernesses should not exist solely because of their high biodiversity. First and foremost, they are part of the urban green concept, the green infrastructure. This includes appropriation, use and appropriate development. Achieving precisely this in balance with their ecological values is the task of an urban wilderness concept.

Kowarik (2018) argues in biodiversity conservation not to solely include the old urban wildernesses with their primarily native species, but also the new urban wildernesses with their non-native species and their diverse ecosystem services:

» "While many conservation approaches tend to focus on such relict habitats and native species in zn urban settings … (I) argue(s) for a paradigm shift towards considering the whole range of urban ecosystems. Although conservation attitudes may be challenged by the novelty of some urban ecosystems, which are often linked to high numbers of nonnative species, it is promising to consider their associated ecosystem services, social benefits, and possible contribution to biodiversity conservation" (Kowarik 2011, S. 1)

In the German National Strategy on Biological Diversity adopted in 2007 (BmU 2007), the German government set itself the goal of allowing nature to develop undisturbed according to its own laws on 2% of Germany's territory by 2020, and of allowing wilderness to emerge. This certainly includes urban wilderness, even if this was certainly not the primary consideration (see ◘ Table 7.2).

Wilderness protection in cities is no longer the sole aim of nature conservation oriented towards species and biotope protection. Withdrawal of maintenance (management), with the aim of *"rewilding"*, is part of innovative urban green concepts. This includes access to (new) wilderness, risk reduction in use and educational initiatives on urban wilderness. The often speculated risks of invasion by non-native species in the course of allowing wilderness are significantly lower than the expected ecosystem services (Hansen et al. 2012; Wissel 2016; Mathey et al. 2018).

The emergence of urban brownfields, some long term and others short term, opens up opportunities to provide 'new wildernesses' directly in the residential areas of city dwellers. However, such urban nature is often rejected as unknown, unwanted and a feature of decay and disorder by local residents. Preserving it as a valuable component of an urban nature concept for people thus requires acceptance by these people. While this exists for the "old wildernesses", especially old forests, it must first be created for "new wildernesses". Information events and environmental education programmes, starting with children, can make a considerable contribution to this. However, it is also necessary to know the already existing acceptance for different states of spontaneous vegetation development and succes-

◼ Table 7.2 Examples of new urban wildernesses in Berlin and their integration into the green infrastructure concept

Area (size/year of establishment)	Previous use	Protection status	Position in the city	Habitat type	Development Goal	Interventions	Access
Park Hallesche Straße/Möckernstraße (0.7 ha; 1987)	Fallow	Protected landscape feature	City centre	Forest, dominantly non-native. Tree species	Wilderness, nature experience	Route network	Public freely accessible
Park am Gleisdreieck (26 ha; 2013)	Goods station	Park	City centre	Forest, dominantly native. Tree species	Recreation, nature experience, forest	Designed with wild vegetation	Public freely accessible
Park am Nordbahnhof (5.5 ha; 2009)	Train station	Park	City centre	Grassland forest mosaic	Recreation, nature experience	Network of paths, landscaped with wild vegetation	Public freely accessible
Tempelhofer Feld (300 ha; 2010)	Airfield	Park	City centre	Grassland, referendum 2014 for stock	Recreation, bird protection	Mowing	Public freely accessible
Nature Park Südgelände (16.7 ha; 1999)	Goods station	Nature and landscape protection	Inner ring road	Grassland-forest mosaic (native/ non-native)	Wilderness, species conservation, nature experience	Road network, grazing	Public freely accessible
Johannisthal airfield (26 ha; 2002)	Airfield	Nature and landscape protection	Inner ring road	Gassland mosaic	Species protection, recreation	Mowing, path network, landscaped areas	Limited access
Spandau Citadel (13.1 ha; 1959)	Fortress	Landscape protection, FFH	Inner ring road	Forest, caves	Species protection, forest	Care of the hist. Structures	Limited access

(continued)

■ Table 7.2 (continued)

Area (size/year of establishment)	Previous use	Protection status	Position in the city	Habitat type	Development Goal	Interventions	Access
Grünauer Kreuz (34.2 ha; 2004)	Railroad property	Nature protection	Outer ring road	Grassland	Species protection, forest development	Mowing	No public access
Fortress Hahneberg (29.2 ha; 2009)	Fortress	Nature protection, FFH	Outer ring road	Forest, caves	Species protection, forest development	Mowing, maintenance of hist. Structures	Limited access, guided tours
Falkenberger Rieselfelder (60 ha; 1995)	Areas for waste water trickling	Nature protection	Outer ring road	Semi-open woodland, hedgerows, grassland, wetland	Species protection, recreation	Grazing, road network, information system, partial agricultural use	Access on paths
Karolinenhöhe sewage farm (220.4 ha; 1987)	Areas for waste water trickling	Landscape protection	Outer ring road	Hedges, grassland	Recreation	Road network, agricultural use	Public freely accessible

After Kowarik (2018, p. 341)

sion and their reasons. Mathey and Rink (2010), Banse and Mathey (2013) conducted surveys in Dresden and Mathey et al. (2015, 2018) in Dresden and Leipzig on the acceptance and use of succession stages of spontaneous vegetation development, visualized in pictures. The results can be very helpful for wilderness development: New urban wildernesses are accepted for about a quarter of the respondents and worth preserving for a third. Nevertheless, traditional values such as tidiness, cleanliness and safety emerged as dominant, leading to largely traditional perceptions of green space. Urban wilderness is not seen as desirable by most respondents. About a third of respondents perceive new urban wildernesses as disruptions to the cityscape rather than enrichments of urban nature. However, it could be clearly shown that attraction or aversion also depends on the knowledge of vegetation structures, safe accessibility and aesthetic perception. Mathey and Rink (2010) already showed that in a succession of succession stages from

- Open pioneer vegetation, younger than 3 years
- Perennial spontaneous vegetation with perennials and single shrubs (3–10 years),
- Perennial spontaneous vegetation with individual trees over 10 m (10–50 years) and
- Spontaneous woody stands, dense (over 50 years)

Neither the unattractive, open condition a), nor the inaccessible grove condition d), but succession condition c) receives the highest approval ratings (Dresden approx. 1/3, Leipzig approx. 50%) (see ☐ Fig. 7.12).

This mosaic of tall herbaceous and woody vegetation in successional state d), provides aesthetic diversity and variety, visual control, sense of security and accessibility best. The most rejected is the dense spontaneous forest (approximately 50%). However, the fact that this is accepted positively in the Schöneberger

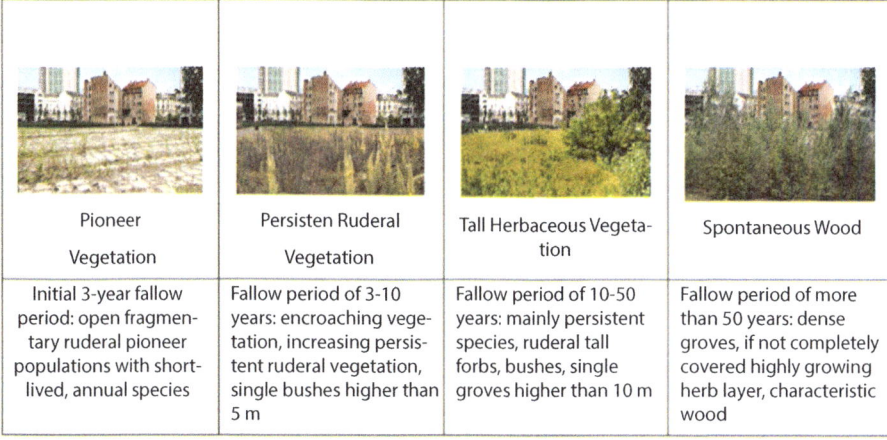

Pioneer Vegetation	Persisten Ruderal Vegetation	Tall Herbaceous Vegetation	Spontaneous Wood
Initial 3-year fallow period: open fragmentary ruderal pioneer populations with short-lived, annual species	Fallow period of 3-10 years: encroaching vegetation, increasing persistent ruderal vegetation, single bushes higher than 5 m	Fallow period of 10-50 years: mainly persistent species, ruderal tall forbs, bushes, single groves higher than 10 m	Fallow period of more than 50 years: dense groves, if not completely covered highly growing herb layer, characteristic wood

☐ **Fig. 7.12** Stages of ecological succession on urban brownfields based on length of fallow period. The characteristic vegetation on the four types of habitat which can develop on urban brownfields during the process of ecological succession are shown and described. (Modified after Mathey et al. 2018, p. 388. Photomontages: E-M Tittel; background photo: J Voltz (drawing: W. Gruber))

Südgelände Nature Park, for example, is certainly related to the attractive path development and explanatory introduction for visitors there. These are factors that can significantly increase acceptance. Structural diversity, together with attractive, safe accessibility and targeted environmental information, prove to be factors that promote attractiveness.

Wild gardens and wild forests open up new experiences of nature in cities that have so far only been offered outside of cities. Bringing them into the city is the promise of urban wilderness concepts. Fulfilling them is currently still a long way off, but it already begun. Idealization of urban wilderness is just as inappropriate as its rejection: wilderness is not the "better", but a "different urban nature" (see ◘ Figs. 7.13, 7.14, 7.15, and 7.16).

◘ **Fig. 7.13** Nutrient-poor grassland a new, conceived, wild street green. (© Breuste 2018)

◘ **Fig. 7.14** "Wild" abandoned garden. (© Breuste 2018)

Fig. 7.15 "Wilderness" without concept - railway facilities of Voestalpine AG in Linz, Austria. (© Breuste 2017)

Fig. 7.16 New industrial woodland wilderness on the site of the Zollverein World Heritage, Gelsenkirchen. (© Breuste 2011)

Project "Cities Dare Wilderness": German Federal Programme on Biological Diversity – Pilot Projects to Develop Acceptance for Urban Wilderness

» "When humans retreat, nature takes over – even in the middle of a big city. What emerge are unusual images of the urban landscape" (Leben.Natur. Vielfalt. Das Bundesprogramm 2018, translated).

The German Federal Ministry for the Environment, Nature Conservation and Nuclear Safety (BmU) and the Federal Agency for Nature Conservation (BfN) are funding the inter- and transdisciplinary project "Cities dare Wilderness" as part of the Federal Programme *Biodiversity*. The aim is to investigate from an ecological and social science perspective how nature develops amid an urban environment designed for human needs.

The project cities of Hanover, Frankfurt am Main and Dessau-Roßlau want to integrate more "wilderness areas" into their urban development and thereby

- Contribute to the conservation and promotion of species and biotope diversity
- Increase the urban quality of life in cities
- Getting people excited about urban wilderness
- Make urban nature accessible and experienceable
- Establish new types of landscapes and test their maintenance and use strategies.

Wilderness thus stands in the project for nature promotion for the benefit of biodiversity and nature awareness. Urban dwellers should "observe, marvel, enjoy and discover". To this end, the cities will let selected open spaces in the city "go wild" in 2016–2021, that is, they will take back their maintenance so that "new wildernesses" of spontaneous development can emerge. Those responsible see the greatest challenge for the project as achieving acceptance among the population, because for many city dwellers "urban wilderness" stands for "chaos, danger and neglect". Together with scientific partners, a series of events will be held to develop acceptance and understanding for wilderness.

Dessau-Roßlau is focusing on establishing and strengthening the wild fruit population, Frankfurt on special environmental education programmes on wilderness for the Frankfurt green belt to inform and sensitise the next generation of nature conservationists, teachers and other multipliers and thus pass on acceptance. Hanover is integrating the project into its overall "More Nature in the City" programme and is focusing on coordinated and reduced maintenance intensity. On 11 existing green spaces, allotment garden fallows and two forest parcels, wilderness development is to be promoted with different focal points in each case and presented to the public in a comprehensible way ("wild islands, wild gardens, post-industrial wilderness, wild forests").

The Tempelhofer Feld in Berlin: Acceptance and Expectations When Urban Nature Meets Wildlife Habitats

After the abandonment of flight operations at the inner-city airport Berlin-Tempelhof in 2008, a large inner-city green space remained for which a new use was sought. In addition to development plans, the idea of a designed park landscape emerged (SSB 2010). The former airfield site with its open grassy areas and asphalt runways is so far as "Tempelhofer Feld" since 2010 Berlin's largest urban park (355 ha) and popular recreation area for about 1 million visitors already in the first 4 weeks. The development of the park landscape Tempelhofer Feld can only take place with the population, which was included in the planning through participation procedures. The expectations of participating citizens also express their values regarding certain natural components and their use-related amenities (see ▣ Table 7.3). In 2014, the Senate Administration confirmed that no construction projects would take place on the site, long an option, and that the entire area would be preserved for the population as a park.

Recreation and nature conservation concepts come together in the park. The inner meadow areas, rare smooth oat meadows and sandy dry grasslands are among the most valuable habitats of nationally highly endangered ground-nesting bird species (for example, wheatear, meadow pipit, fallow pipit, corn bunting and skylark). There are 368 wild plant species in the park. Their habitat is integrated into the park as moderately maintained "wilderness" (see ◼ Fig. 7.17) and can be stably preserved so far. In addition, a rare opportunity for dialogue opens up about "wild urban nature" and its integration into recreation-related use concepts, the development of acceptance and understanding for nature. A habitat has emerged that can develop a new understanding of nature and self-image of people in the city as a basis (SSB 2010).

◼ **Table 7.3** Most frequently expressed expectations of residents and visitors regarding the development of Tempelhofer Feld as park in Berlin, figures in percent

Wishes of the respondents	Citizen survey in the catchment area	Visitor survey
Big trees	92	83
Benches for rest and meeting	91	85
Smaller protected areas for recreation	90	87
WC facilities	89	90
Small water elements	87	77
Large lawns for lying and playing	85	84
Areas with flowering shrubs	78	71
Separated areas for nature conservation	75	82
Gastronomic offers	75	70
Moving terrain with hills and depressions	70	65
Meeting points, communication areas also for picnics	69	75
Flowerbeds	68	57
Play areas for different age groups	65	72
Possibility of observing nature	62	70
Areas for recreational sports	62	66
Naturally designed areas with sunbathing lawns	60	56

Argus GmbH (2009)

◘ Fig. 7.17 Information board in Tempelhofer Park on conditions of use and the habitat of ground-nesting birds (bird symbols in the middle), information board in the park. (© Breuste 2011)

Sensitive Urban Wilderness: Riccarton Bush Christchurch, New Zealand

Christchurch is the largest city in New Zealand's South Island with a population of 367,800 (2015), settled by colonists in an extensive swampy lowland forest area since 1850.

The relationship between urban biodiversity and regional, national con-sciousness is particularly pronounced among the New Zealand population. People see themselves and New Zealand's nature as "native" and unique, and identify directly with it. Its conservation is a direct concern of New Zealanders, who express this in many initiatives and citizen movements and in the media (Meurk and Hall 2006; Stewart et al. 2004; Ignatieva et al. 2008).

Riccartin Bush represents the last remaining woodland remnant of the region's originally widespread swamp stonecrop woodland *(Podocarpaceae)*. The only remaining 7 ha of the original thousands of ha have been preserved and protected since 1914 by the Riccarton Bush Trust and public funding as a protected forest stand in the middle of the city (Chilton 1924). This makes New Zealand's oldest protected area an "urban wilderness".

The forest is the last island of original forest wilderness that has long disappeared in the intensively used agricultural landscape, but happened to be preserved in this urban area. The character trees Kahikatea *(White Pine, Dacrycarpus dacrydioides)*, Totara *(Podocarpus totara)*, Kowhai *(Sophora microphylla)* and Hinau *(Elaeocarpus dentatus)* form a dense stand of trees surrounded by a high protective fence with an access gate *(predator proof* fence*)*, to protect the rare native ground-nesting bird population from invasive small mammals from Australia *Australian Brushtail Possum (Trichosurus vulpecula)* and European hedgehogs and rats.

The urban wilderness is being selectively developed natively through establishment attempts of native animals *(Great spotted Kiwi/Roroa, Apteryx haastii)* and plants (regional tree species Wētā, *Hemideina femorata*) (see ◘ Fig. 7.18).

◘ **Fig. 7.18**   Riccarton Bush, Christchurch, New Zealand, predator proof fence to protect native ground-nesting birds from introduced small mammals (for example, *possums* from Australia). (© Breuste 2006; Breuste et al. 2016, Fig. 7.1, p. 243)

Manicured Urban Nature Versus Wilderness: The Example of Linz, Austria

A study (Breuste and Astner 2018) investigated the use of different nature offerings in the urban environment in the newly built district "solarCity Linz", Austria, in 2005. The protected area "Natura 2000 Traun-Donau-Auen", an "old" woodland wilderness, is directly adjacent with a landscape park of traditional park design embedded in between. Thus, "recreational park" and "wilderness" compete directly with each other. In a survey, 153 residents were asked about their acceptance of, use of, and relationships to these different types of nature in urban settings. As a comparison group, 91 visitors to the urban wilderness area provided information on the same aspects. To support the survey, images were used that depicted a gradient between wilderness and well-maintained recreational space.

The results show that the majority of residents (75%) and visitors (66%) are not aware of the protected area status of the wilderness area. However, the nearby protected area wilderness has high acceptance among the majority of both survey groups.

Wilderness nature as a backdrop is valued by 59%; once "messy" natural processes (for example, fallen trees) are considered, appreciation drops to only 11%. "Order" and trail development in the wilderness is valued by 43%. The landscape park brings it to 91% appreciation. 33% never visit the wilderness despite its close proximity (5 minutes' walk), the disturbed areas even less (58% never). 80% use the landscape park often or very often. Appreciation and actual use thus differ significantly.

Favorite activities in the wilderness area most often cited are walking (28%), observing nature (24%), relaxing (18%), and exercising (6%, mostly jogging). Preferred activities in the wilderness area are walking, relaxing or meeting friends.

It can be seen that nature is generally important for 54% of the respondents. The type of nature that is preferred and that there are reservations about "too much wild nature" depends on factors such as: Visitor quality, quality of infrastructure, perception of safety and accessibility (see ◘ Figs. 7.19, 7.20 and 7.21).

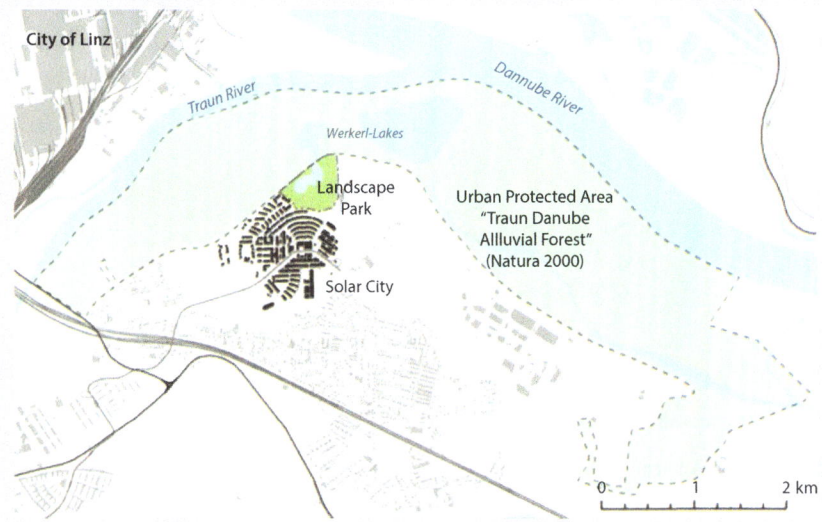

◘ **Fig. 7.19** Traun-Danube floodplain near Linz, Natura 200 site, the relationship between forest – wilderness – landscape park – settlement. (© Drawing: W. Gruber 2016 in Breuste and Astner 2018)

◘ **Fig. 7.20** Landscape park in the urban environment of the Linz district of solarCity, Austria. (© Astner 2014)

7

■ **Fig. 7.21**   Ancient forest wilderness in the urban environment of the Linz district of solarCity, Austria. (© Astner 2014)

Unspectacular Wilderness Becomes Spectacular: Schöneberger Südgelände Nature Park, Berlin

Areas left to their own devices undergo a natural process of succession, which in Central Europe reaches its climax state as a forest.

This process is rarely brought to completion because there is usually not enough time and a new use that has nothing to do with the forest is established on the "fallow land". Thus, we know little about what species composition the new forest will have if 100 or more years are available for it and whether such new forests will differ regionally, which can be expected. There is little experimental land available.

One of these is the Schöneberger Südgelände Nature Park in Berlin. Here, on the 18 ha site of the former Tempelhof marshaling yard, nature has been developing unhindered for 67 years, since the railway facilities at the Anhalter station in Berlin were abandoned in 1952. The herbaceous vegetation, which was still dominant after approx. 30 years, has

already given way to area-dominant forest vegetation after approx. 40 years, in which *Betula pendula* accounts for 23.8% and *Robinia pseudoacacia for* 21.3% of the dominant tree species. But is this already a stable "end state"? To preserve the dry-warm open areas with their species-rich flora and fauna, the forest development is interfered with by mowing on some areas (see ■ Table 7.4).

The Citizens' Initiative Nature Park South Area, founded in 1987, campaigned early on for the preservation of the area regarding its ecological value, which had been proven in the meantime. In 1995, Deutsche Bahn AG transferred the Schöneberger Südgelände to the Senate of the City of Berlin as a compensation area for interventions by its transport facilities elsewhere. A state-owned Grün Berlin GmbH, supported with 1.8 million euros by the Allianz Environmental Foundation, developed the urban

▣ Table 7.4 Table development of vegetation structure in the Nature Park Schöneberger Südgelände in Berlin

(in%)	1981	1991
Herbaceous vegetation	63.5	30.9
Woody vegetation	36.5	69.1

Kowarik and Langer (2005)

▣ Fig. 7.22 Nature Park Schöneberger Südgelände, Berlin. (© Breuste 2011)

wilderness as a forest park, which was opened in 1999 and placed under nature and landscape protection. In 2000, the Nature Park Schöneberger Südgelände became an official German EXPO project. Urban forest wilderness was opened up, made accessible, explained, staged and invitingly presented, and ultimately "accepted" by Berliners and their guests. A high level of biodiversity has now been established there. 366 different species of ferns and flowering plants, 49 species of large fungi, 49 species of birds, 14 species of grasshoppers and crickets, 57 species of spiders and 95 species of bees can be found, including more than 60 endangered species.

2.7 km of steel grid paths lead visitors elevated above the ground next to the old, still existing railway tracks through the urban wilderness. Additional cultural offerings on the nature of the site are intended to bring even more visitors into contact with unfamiliar wilderness in the city in the future. Guided tours on the flora, fauna and history of the area are already trying to develop this contact. Wild nature, culture and recreation combine to create a new attraction for one euro admission (Kowarik and Langer 2005; Senate Department for Urban Development – Communication – Berlin 2011; GrünBerlin GmbH 2013; Cobbers 2001) (see ▣ Fig. 7.22).

New Wilderness as a New Opportunity for More Contact with Nature: Ecological Parks in London

When Max Nicholson combined nature conservation with nature experience in 1976 and the Trust for Urban Ecology (TRUE) realised this idea through an Ecological Park (William Curtis Ecological Park) on an old lorry park near London Bridge, the first park integrating "novel" urban ecosystems was created. New urban wildernesses were given a place in the spectrum of traditional urban nature provision. Stave Hill Ecological Park in London's Docklands followed in 1986–1988, and since then many more have been created with local residents in residential areas on brownfield land. The nine-meter high Stave Hill is made of demolition material from the London docks and presents nature in various stages of natural succession over an area of 2.1 ha. The unspectacular is staged spectacularly and designed to do so. Not all Ecological Parks are simply "urban wildernesses" left to their own devices. But all of them have areas where nature is left to itself to a certain extent and can be observed and also explained. The simple concept of *"enjoy nature"* works with and for the residents. Other *ecological parks* and *urban wildlife habitats* have been added in London and other UK cities since 1988. TRUE now manages Stave Hill, Greenwich Peninsula Ecology Park, Dulwich Upper Wood and Lavender Pond Nature Park alongside. In 2012, TRUE became part of The Conservation Volunteers (TVC) (TVC 2018).

The TRUE Ecological Parks take a new approach to nature conservation in the city by familiarising people with (un)spectacular urban nature. They create new habitats for plants and animals *(habitat for* urban *wildlife),* enable urban ecological research, bring city dwellers, especially children, closer to urban nature through their own experience (environmental education) and demonstrate "creative urban nature conservation" off the beaten track, involving citizens as volunteers in management. London's Ecological Parks on brownfield sites with natural succession and use-based management have proven to be a new idea for adding value to urban nature (Breuste et al. 2016).

7.5 The Green City Concept in European and Global Perspective

In 1994, European cities started an initiative of European Cities and Towns towards sustainability. When a European process for the sustainable development of cities through self-committed actions became concrete in Aalborg in 2004 (European Sustainable Cities & Towns Campaign), some 2500 local and regional administrations in 39 countries and 80 European cities and towns were involved. In Aalborg, Denmark, the approximately 1000 participants of the Fourth Conference of Sustainable Cities & Towns, Aalborg +10, adopted the "Aalborg Commitments" (10 thematic areas). Urban nature, while not a central component, was integral. Goal 3 "Community natural assets: commit to taking full responsibility for protecting and conserving the natural commons and ensuring their equitable distribution" directly addresses urban nature. Scope of Work 3 commits to "promote and enhance biodiversity and expand and maintain protected areas and green spaces." Also "ecologically productive land" is to be preserved "sustainable forestry" promoted, "water, soil and air quality" improved (ESCTS 2004, 2013).

The Leipzig Charter 2007 recommends an "integrated urban development policy" as a process of simultaneous and equitable consideration of the concerns and interests relevant to the development of cities. It focuses on innovation, competitiveness, citizen participation and balance, but does not see urban nature in a significant role for these objectives (Leipzig Charter 2010).

20 European countries are already members of the European Landscape Contractors Association (ELCA). It represents 74,000 companies and 330,000 members. The Green City concept was developed in an initial workshop in 2011.

» "For us the green city is the model of the future, creating urban structures with environments with life-quality" (ECLA 2011, p. 4).

This model should help to solve the identified problems of the coming decades. To this end, it also calls for a new design of urban green space that combines water management, biodiversity, climate adaptation and health.

Four basic principles on the way to a Green City have been identified, for example, by Green City Hungary:

— Reconnecting cities with nature for the benefit of its positive effects
— Urban greening must be an integral part of urban planning from the outset
— Urban greenery is an interdisciplinary concern in planning and execution
— Urban greening should no longer be the neglected part of sustainability efforts (ECLA 2011) (for example, ▣ Fig. 7.23).

As early as 2008, European scientists saw city councils, national governments and the European Commission as having a much stronger responsibility to do more to develop **Urban Greening strategies** with a view to a green city. Among other things, they demand for:

▣ **Fig. 7.23** Intensively used green space at Miradouro de Santa Catarina, Lisbon, Portugal. (© Breuste 2013)

From Municipalities
- Develop urban greening strategies, implement them, review their application, make them a key element of their cities' spatial planning and work cooperatively with other cities to do so
- Urban green strategies should be developed with all planning sectors and broad public participation
- Promote cooperation and public responsibility for green spaces
- Develop local standards of green quality, quantity and accessibility, taking into account different user groups
- Provide adequate funding and training for the planning and maintenance of urban green spaces.

From the European governments
- Promote urban sustainability strategies on a regional and local scale
- Promote understanding that urban nature has ecological functions and provides ecosystem services, and promote appropriate data collection and GIS applications to this end,
- To provide financial resources for urban green spaces and the development of urban green strategies
- To promote the national and European exchange of experience between institutions in the field of green space and urban planning.

From the European commission
- Integrate urban greening strategies into EU policy documents and initiatives to ensure urban sustainability and competitiveness
- Develop, locally adapt, disseminate and provide funding for concrete urban greening strategies that encompass all urban nature types according to the needs of city dwellers, with a focus on biodiversity, air quality, climate change adaptation and health
- Promote research and knowledge transfer on the Green City and increase exchange between green and urban planning institutions in Europe and support this through new European organisational forms (GreenKeys 2018) (see ◘ Figs. 7.24 and 7.25).

The global perspective on sustainability was opened up as early as 1992 in Rio de Janeiro, and concretised for cities by the Habitat I–III conferences.

The Millennium Ecosystem Assessment Report 2005 (WRI 2005) does not treat cities as ecosystems and thus does not address urban nature or a perspective for Green Cities. On a global level, this will only happen again through the Habitat III 2016 conference in Quito and its preparatory meetings.

Habitat III adopted the New Urban Agenda on "sustainable cities and human settlements for all". It is intended to be a shared vision and a political commitment

■ **Fig. 7.24** Thuringian Train Station Park in Halle/Saale, Germany, on former track area with reduced maintenance. (© Breuste 2018)

■ **Fig. 7.25** Hans Donnenberg Park in Salzburg, Austria - reduced mowing of woody borders leads to a significant increase in species diversity. (© Borysiak 2017)

to promote and achieve sustainable urban development (UN 2016). The principles and commitments of the New Urban Agenda do not include a Green City vision. In the area of environment, resource efficiency to increase urban resilience and environmental sustainability are clearly in the foreground. The usual commitments to protect ecological resources and biodiversity remain little concrete and without clear targets and instruments. Instead, risk management and resilience clearly come to the fore. In the working group "Urban Ecosystems and Resource Management", which already adopted a strategy paper in New York in 2015, important key concepts of a Green City such as Ecosystem services (ES), Ecosystem-based Adaptation (EbA), Green infrastructure (GI), also *preserving ecosystem-based management of cities, disaster risk reduction, health and recreation* and even *Citizens need to connect with nature, and benefit from this connection* (p. 6) can be found, but none of this has made it into the vision of the New Urban Agenda. Thus, a global vision of the Green City is sadly missing, which must have taken a back seat to the rest of the urban problem burden, such as poverty, inequality and increasing risks (UN 2015).

The vision of the Green City, therefore, remains first of all local. This is where it belongs and where it can be realised in an exemplary manner. It is an originally European vision that has long since received growing attention, further development, application and support as a Green City all over the world. The Green City provides orientation in an essential area, the relationship of the city to nature. Thus, the vision remains manageable and can be pursued as a guiding idea with individual steps. It is an essential part of what is described as a very complex and multifaceted vision of a sustainable city.

7

Dresden: On the Way to Becoming a Green City with the Guiding Idea "Compact City in an Ecological Network"

The guiding idea of Dresden's urban planning is developing compact urban settlement structures embedded in a network of ecological functional spaces. The existing and multi-structured watercourse system is the spatial basis for Dresden's ecological network. Together with the Elbe, the 400 municipal streams form an almost comprehensive network. It is planned to gradually expand this network together with green spaces to form an ecological network. Its natural subspaces are assigned concrete functions:

- Fresh air supply and healthy urban climate
- Sufficient groundwater recharge
- Flood prevention, water retention and watercourse development
- Recreational spaces for people
- Habitats for plants and animals, migration corridors
- Beauty and uniqueness of the cultural landscape

Watercourse development measures strengthen the ecological network (e.g., brownfield revitalisation). Moderate structural growth takes place in defined spatial cells, stronger structural growth in defined corridors into the surrounding countryside. The ecological network is based on urban nature corridors. The concept is implemented in a development and measures concept (EMK). The 2012 landscape plan of the state capital Dresden is also based on this model. Urban nature is understood as infrastructure and **open spaces as guiding structures for urban development.**

The necessary adaptation to climate change can thus be better managed. It requires more green spaces to mitigate summer heat and allow rainwater to infiltrate, especially during heavy rainfall (REGKLAM 2015; Wende et al. 2014; Breuste et al. 2016) (see ◘ Fig. 7.26).

NET STRUCTURE
- Areas of complex values and functions
- Corridors of functional connections
- Special corridors of functional connections
- Complementary corridors

CELL STRUCTURE
- Inner city
- Built-up areas
- Urban cells of flexible development
- Rural cells

- Administrative urban boundaries
- Waters
- Highways
- Main roads

N

0   4   8 km

Fig. 7.26 Landscape plan Dresden. (Source REGKLAM 2015; design: J. Breuste, drawing: W. Gruber, Breuste et al. 2016, Fig. 6.9, p. 189)

Rebuilding Cities in 'Balance with Nature'

The vision of eco-cities in new urban designs attempts to develop a fundamentally new approach to urban design and life in cities. The perspectives and designs of eco-cities vary. What they have in common is a focus on resource efficiency, energy efficiency, use of renewable energy, reuse of waste products, renaturalization (especially water bodies), and inner-city efficient transportation. While some eco-city designs assume that these and other problems are solvable through more efficient technology (for example, smart cities), others assume a radical rethinking of urban development. Register (2006, p. 5) says: *"Cities need to be rebuilt from their roots.... They need to be reorganized and rebuilt upon ecological principles"* (Register 2006, p. 5). This view assumes a necessary *balance with nature* for cities, which concretely means giving nature greater importance in urban development. Seen in this light, these eco-cities are green cities that *coexist peacefully with nature* (Register 2006, p. 5) (see ▪ Fig. 7.27).

7

◘ Fig. 7.27   Salzburg, Austria, is often considered the ideal city *"in balance with nature"*. (© Breuste 2003)

2050 Nagoya Strategy for Biodiversity: A City Sets Measurable Goals on the Way to a Green City

In the Japanese city of Nagoya, building development has reduced the proportion of green space from 40% in 1967 to 25% in 2008. This was associated with significant impacts on the increasingly poor thermal comfort for residents. This prompted the city of Nagoya to make a far-reaching, forward-looking decision in 2008. Looking ahead to 2050, the "2050 Nagoya Strategy for Biodiversity" was adopted with ambitious goals and declared a municipal policy document. Steps of this strategy are:

- **No further reduction in existing green space,** including private woodland,

- Greening of major streets through tree and shrub plantings,
- Establishment of (landscaped) greenways for primarily pedestrians and bicyclists,
- Support public-private partnership programs for green development,
- Adoption of a Water Cycle Revitalization Plan to increase the 2008 precipitation infiltration rate from 24% to 33% by 2050. Measures to achieve this include new green spaces, green roofs, permeable sealing blankets (Kazmierczak and Carter 2010; City of Nagoya 2008) (see ◘ Fig. 7.28).

■ **Fig. 7.28** Green infrastructure of Nagoya, Japan. (© Breuste 2010)

References

5th Global Biodiversity Summit of Cities & Subnational Governments (2016) Quintana Roo Communiqué on mainstreaming local and subnational biodiversity action. www.nrg4sd.org/.../ biodiversity_summit_for_cities_and_subnati. Accessed 1 June 2018

Arbeitsgruppe Methodik der Biotopkartierung im besiedelten Bereich (1993) Flächendeckende Biotopkartierung im besiedelten Bereich als Grundlage einer am Naturschutz orientierten Planung: Programm für die Bestandsaufnahme, Gliederung und Bewertung des besiedelten Bereichs und dessen Randzonen: Überarbeitete Fassung 1993. Nat Landsch 68(10):491–526

Argus GmbH (2009) Wettbewerb Parklandschaft Tempelhof. Ergebnisse des Besuchermonitorings 2009 – Bürgerbeteiligung zum Wettbewerbsverfahren 2009

Banse J, Mathey J (2013) Wahrnehmung, Akzeptanz und Nutzung von Stadtbrachen. Ergebnisse einer Befragung in ausgewählten Stadtgebieten von Dresden. In: Breuste J, Pauleit S, Pain J (eds) Stadtlandschaft – vielfältige Natur und ungleiche Entwicklung. CONTUREC 5, Darmstadt, S 37–54

Biodiversity Gardening (2018) What is native biodiversity? https://www.biodiversitygardening.com/ what-is-native-biodiversit. Accessed 1 June 2018

BioFrankfurt (2018a) Gemeinsam für Vielfalt und Nachhaltigkeit BioFrankfurt. Flyer, Frankfurt a. M

BioFrankfurt (2018b) https://www.biofrankfurt.de/. Accessed 7 May 2018

Botzat A, Fischer LK, Kowarik I (2016) Unexploited opportunities in understanding liveable and biodiverse cities. A review on urban biodiversity perception and valuation. Glob Environ Chang 39:220–233

Breuste J (1994a) "Urbanisierung" des Naturschutzgedankens: Diskussion von gegenwärtigen Problemen des Stadtnaturschutzes. Naturschutz Landschaftsplanung 26(6):214–220

Breuste J (1994b) Flächennutzung als stadtökologische Steuergröße und Indikator. Geobotan. Kolloquium, Frankfurt a. M. 11:67–81

Breuste J, Astner A (2018) Which kind of nature is liked in urban context? A case study of solarCity Linz, Austria. Mitt Österreichischen Geographischen Ges 158:105–129

Breuste J, Pauleit S, Pain J (eds) (2013) Stadtlandschaft – vielfältige Natur und ungleiche Entwicklung. Schriftenreihe des Kompetenznetzwerkes Stadtökologie. Conturec 5, Darmstadt

Breuste J, Pauleit S, Haase D, Sauerwein M (eds) (2016) Stadtökosysteme. Springer, Berlin

British Ecological Society (2018) Biodiversity conservation. https://www.britishecologicalsociety.org. 'Policy' Policy topics. Accessed 1 June 2018

Bundesamt für Naturschutz (BfN) (ed) (2017) Urbane grüne Infrastruktur. Grundlage für attraktive und zukunftsfähige Städte. Hinweise für die Kommunale Praxis. BfN, Bonn

Bundesinstitut für Bau-, Stadt- und Raumforschung (BBSR) (ed) (2017) Handlungsziele für Stadtgrün und deren empirische Evidenz. Indikatoren, Kenn- und Orientierungswerte. BBSR, Bonn

Bundesministerium für Umwelt, Naturschutz, Bau und Reaktorsicherheit (BmU) (2007) Nationale Strategie zur biologischen Vielfalt. Vom Bundeskabinett am 7. November 2007 beschlossen. Reihe Umweltpolitik. BmU, Berlin

Bundesministerium für Umwelt, Naturschutz, Bau und Reaktorsicherheit (BmU) (2015) Grün in der Stadt – Für eine lebenswerte Zukunft. Grünbuch Stadtgrün. BmU, Berlin

Bundesministerium für Umwelt, Naturschutz, Bau und Reaktorsicherheit (BMUB) (ed) (2017) Grün in der Stadt – Für eine lebenswerte Zukunft. BMUB, Berlin

Chan L, Djoghlaf A (2009) Invitation to help compile an index of biodiversity in cities. Nature 460:33. http://www.nature.com/nature/journal/v460/n7251/full/460033a.html. Accessed 11 Oct 2011

Chan L, Calcaterra E, Elmqvist T, Hillel O, Holman N, Mader A, Werner P (2010) User's manual for the city biodiversity index. Latest version: 27 September 2010. https://www.cbd.int/subnational/partners…/city-biodiversity-index. Accessed 2 June 2018

Chilton C (1924) Riccarton Bush. A remnant of the Kahikatea swamp forest formerly existing in the neighbourhood of Christchurch, New Zealand. The Canterbury Publishing, Christchurch

City of Nagoya (ed) (2008) Biodiversity report. Local action for biodiversity, and ICLEI initiative. ICLEI, Nagoya

Cobbers A (2001) Vor Einfahrt HALT – ein neuer Park mit alten Geschichten. Der Natur-Park Schöneberger Südgelände in Berlin. Jaron, Berlin

Cronon W (1996) The trouble with wilderness: or, getting back to the wrong nature. In: Cronon W (ed) Uncommon ground: rethinking the human place in nature. W. W. Norton, New York, pp 69–90

Curitiba Declaration on Cities and Biodiversity (2007) https://www.cbd.int/doc/…/city/…01/mayors--01-declaration-en.pd. Accessed 1 June 2018

Davison G, Tan R, Lee B (2012) Wild Singapore. John Beaufoy, Oxford

De Groene Stad (NL) (2018) De Groene Stad. www.degroenestad.nl. Accessed 5 Jan 2018

Department of Land and Natural Resources (2018) Issue 6: conservation of native biodiversity. www.dlnr.hawaii.gov/forestry/files/2013/09/SWARS-Issue-6.pdf. Accessed 1 June 2018

Die Grüne Stadt (2018) Die Grüne Stadt. www.die-gruene-stadt.de. Accessed 5 Jan 2018

Diemer M, Held M, Hofmeister S (2003) Urban wilderness in Central Europe. Rewilding at the urban fringe. Int J Wilderness 9(3):7–11

Dover JW (2015) Green infrastructure. Incorporating plants and enhancing biodiversity in buildings and urban environments. Earthscan und Routledge, London

Dudley N (Hrsg) (2008) Guidelines for applying protected area management categories. IUCN, Gland. https://www.iucn.org/…areas/…areas…/category-ib-wilderness-are. Accessed 3 June 2018

Durban Commitment (2008) www.joondalup.wa.gov.au/files/…/2008/Attach2brf080708.pdf. Accessed 1 June 2018

Erfurt Declaration, URBIO 2008 (2008) https://www.fh-erfurt.de/urbio/…/ErfurtDeclaration_Eng.php. Accessed June 1 2018

Europäische Kommission, Generaldirektion Umwelt (ed) (2009) Zielrichtung: eine grüne Infrastruktur in Europa. Natura 2000: Newsletter "Nature and Biodiversity" of the European Commission 27:3–7

European Landscape Contractors Association (ELCA) (ed) (2011) ELCA research workshop green city Europe for a better life in European Cities. www.green-city.de. Accessed 5 Jan 2018

European Sustainable Cities & Towns Compaign (ESCTC) (2004) Die Aalborg Commitments. www.ccre.org/docs/Aalborg03_05_deutsch.pdf. Accessed 7 Jan 2018

European Sustainable Cities & Towns Compaign (ESCTC) (2013) European sustainable cities. www.sustainablecities.eu. Accessed 12 Jan 2014

Forman R (2014) Urban ecology. Science of cities. Cambridge University Press, Cambridge

Fuller RA, Irvine KN, Devine-Wright P, Warren PH, Gaston KJ (2007) Psychological benefits of greenspace increase with biodiversity. Biol Lett 3(4):390–394

Government of Singapore (2018) Singapore Department of Statistics (DOS). https://www.singstat.gov.sg/. Accessed 4 Dec 2018

GreenKeys (2018) Recommendations for new urban green policies and an agenda for future action. www.greenkeys-project.net; greenkeys@ioer.de. Accessed 7 Jan 2018

GrünBerlin GmbH (Hrsg) (2013) Natur-Park Schöneberger Südgelände. http://www.gruen-berlin.de/parks-gaerten/natur-park-suedgelaende/. Accessed 21 Dec 2013

Hansen R, Heidebach M, Kuchler F, Pauleit S (2012) Brachflächen im Spannungsfeld zwischen Naturschutz und (baulicher) Wiedernutzung. BfN-Skripten 324, Bonn

Hoffmann M (2010) Urbane Wildnis aus Sicht der Nutzer. Wahrnehmung und Bewertung vegetationsbestandener städtischer Brachflächen. Dissertation Mathematisch-Wissenschaftlichen Fakultät II Humboldt-Universität zu Berlin, Berlin

Hofmeister S (2009) Natures running wild: a social-ecological perspective on wilderness. Nat Cult 4(3):293–315

Ignatieva ME, Meurk C, van Roon M, Simcock R, Stewart G (2008) How to put nature into our neigbourhoods. Application of Low Impact Urban Design and Development (LIUDD) principles, with a biodiversity focus, for New Zealand developers and homeowners, vol 35. Landcare Research Science. Manaaki Whenua, Christchurch

IUCN, International Union for Conservation of Nature (2019) Mainstreaming biodiversity. https://www.iucn.org/theme/global…/mainstreaming-biodiversity. Zugegriffen 5 Jan 2019

Jedicke E (2016) Biodiversitätsschutz. In: Riedel W, Lange H, Jedicke E, Reinke M (eds) Landschaftsplanung. Springer Reference Naturwissenschaften. Springer Spektrum, Berlin

Jorgensen A, Keenan R (eds) (2012) Urban wildscapes. Routledge, Oxon

Jorgensen A, Tylecote M (2007) Ambivalent landscapes – wilderness in the urban interstices. Landsc Res 32(4):443–462

Kazmierczak A, Carter J (2010) Adaptation to climate change using green and blue infrastructure. A database of case studies. Manchester. https://www.escholar.manchester.ac.uk/uk-ac-man--scw:128518. Accessed7 Jan 2018

Kirchhoff T, Trepl L (eds) (2009) Vieldeutige Natur. Landschaft, Wildnis und Ökosystem als kulturgeschichtliche Phänomene. Transcript, Bielefeld

Kirchhoff T, Vicenzotti V (2014) A historical and systematic survey of European perceptions of wilderness. Environ Values 23(4):443–464

Kowarik I (1992) Berücksichtigung von nichteinheimischen Pflanzenarten, von "Kulturflüchtlingen" sowie von Pflanzenvorkommen auf Sekundärstandorten bei der Aufstellung Roter Listen. Schriftenr Vegetationskunde 23:175–190

Kowarik I (2011) Novel urban ecosystems, biodiversity and conservation. Environ Pollut 159(8–9):1974–1983

Kowarik I (2015) Wildnis in urbanen Räumen. Erscheinungsformen, Chancen und Herausforderungen. Natur und Landschaft online; Natur und Landschaft Jahrgang 2015. https://www.natur-und--landschaft.de/de/news/wildnis-in-urbanen-raumen. Accessed 17 Aug 2018

Kowarik I (2018) Urban wilderness: supply, demand, and access. Urban For Urban Green 29:36–347

Kowarik I, Kendal D (2018) The contribution of wild urban ecosystems to liveable cities. Urban For Urban Green 29:334–335

Kowarik I, Körner S (eds) (2005) Wild urban woodlands. New perspectives for urban forestry. Springer, Berlin

Kowarik I, Langer A (2005) Natur-Park Südgelände: linking conservation and recreation in an abandoned raiyard in Berlin. In: Kowarik I, Körner S (eds) Wild urban woodlands. New perspectives for urban forestry. Springer, Heidelberg, S 287–299

Leben.Natur.Vielfalt. Das Bundesprogramm (2018) Städte wagen Wildnis. https://www.staedte-wagen-wildnis.de/das-projekt.html. Accessed 18 Aug 2018

Leipzig Charta (2010) Leipzig-Charta zur nachhaltigen europäischen Stadt 2007. Informationen zur Raumentwicklung 4:315–319

Lupp G, Höchtl F, Wende W (2011) Wilderness – a designation for central European landscapes? Land Use Policy 28(3):594–603

Mathey J, Rink D (2010) Urban wastelands – a chance for biodiversity in cities? Ecological aspects, social perceptions and acceptance of wilderness by residents. In: Müller N, Werner P, Kelcey JG (eds) Urban biodiversity and design. Conservation science and practice series. Wiley-Blackwell, Oxford, pp 406–424

Mathey J, Röbler S, Banse J, Lehmann I, Bräuer A (2015) Brownfields as an element of Green Brownfields as an element of green infrastructure for implementing ecosystem services into urban areas. J Urban Plan Dev 141(3):A4015001. https://doi.org/10.1061/(asce)up.1943-5444.0000275

Mathey J, Arndt T, Banse J, Rink D (2018) Public perception of spontaneous vegetation on brownfields in urban areas – results from surveys in Dresden and Leipzig (Germany). Urban For Urban Green 29:384–392

Meurk CD, Hall GMJ (2006) Options for enhancing forest biodiversity across New Zealand's managed landscapes based on ecosystem modelling and spacial design. N Z J Ecol 30:131–146

Nagoya Declaration, Urbio 2010 (2010). https://www.cbd.int/.../doc/NagoyaDeclaration-URBIO-2010.pdf. Accessed 1 Jan 2018

Naturkapital Deutschland – TEEB DE (2016) Ökosystemleistungen in der Stadt – Gesundheit schützen und Lebensqualität erhöhen. In: von Kowarik I, Bartz R, Brenck M (eds) Technische Universität Berlin, Helmholtz-Zentrum für Umweltforschung – UFZ. TEEB DE, Berlin

Naumann S, McKenna D, Kaphengst T, Pieterse M, Rayment M (2011) Design, implementation and cost elements of Green infrastructure projects. Final report to the European Commission, DG Environment, Contract no. 070307/2010/577182/ETU/F.1. Ecologic Institute and GHK Consulting, Brüssel

Neßhöfer C, Kugel C, Schniewind I (2012) Ökosystemleistungen im Europäischen Kontext: EU Biodiversitätsstrategie 2020 und "Grüne Infrastruktur". In: Hansjürgens B, Neßhöver C, Schniewind I (eds) Der Nutzen von Ökonomie und Ökosystemleistungen für die Naturschutzpraxis. Workshop I: Einführung und Grundlagen, vol 318. BfN-Skripte, S 22–27

Netzwerk Urbane Biodiversität Ruhrgebiet (2013) Urbane Biodiversität – ein Postionspapier. www.urbane-biodiversitaet.de/.../Urbane_Biodiversitaet_Positionspapier.pdf. Accessed 16 April 2018

Oelschlaeger M (1991) The idea of wilderness: from prehistory to the age of ecology. Yale University Press, New Haven

Plan of Action on Subnational Governments, Cities and Other Local Authorities for Biodiversity (2010) https://www.cbd.int/decision/cop/default.shtml?id=12288 Accessed 1 June 2018

Register R (2006) Ecocities: rebuilding cities in balance with nature, Revised edn. New Society Pub, Gabriola Island

REGKLAM Regionales Klimaanpassungsprogramm für die Modellregion Dresden (2015) Grüne und kompakte Städte. www.regklam.de/...programm/.../gruene-und-kompakte-staedte/?tx. Accessed 15 Apr 2015

Ridder B (2007) The naturalness versus wildness debate: ambiguity, inconsistency, and unattainable objectivity. Restor Ecol 15(1):8–12

Riehl WH (1854) Naturgeschichte des Volkes als Grundlage einer deutschen Social-Politik, Bd 1. Cotta'scher Verlag, Stuttgart

Rink D (2009) Wilderness: the nature of urban shrinkage? The debate on urban restructuring and restoration in Eastern Germany. Nat Cult 4(3):275–292

Rodericks S (2010) Singapore city biodiversity index. TEEBweb.org. Accessed 2 June 2018

Sandström UF (2002) Green infrastructure planning in urban Sweden. Plan Pract Res 17(4):373–385

Schrijnen PM (2000) Infrastructure networks and red-green patterns in city regions. Landsc Urban Plan 48:191–204

Schwarzer M (2007) Wald und Hochgebirge als Idealtypen von Wildnis. Eine kulturhistorische und phänomenologische Untersuchung vor dem Hintergrund der Wildnisdebatte in Naturschutz und Landschaftsplanung. Diploma Thesis, Chair of Landscape Ecology, Technical University Munich, München

Second Curitiba Declaration on Local Authorities and Biodiversity (2010) www.biodic.go.jp/biodiversity/activity/international/…/urbio1.pdf. Accessed 1 June 2018

Senatsverwaltung für Stadtentwicklung – Kommunikation – Berlin (Hrsg) (2011) Natur-Park Schöneberger Südgelände. Faltblatt, Senatsverwaltung für Stadtentwicklung, Berlin

Senatsverwaltung für Stadtentwicklung Berlin (SSB) (2010) Ideenfreiheit Tempelhof. Auf dem Weg zur Stadt von morgen. www.stadtentwicklung.berlin.de/…/tempelhof/…/ideenfreiheit_tempelhof. Accessed 4 Jan 2014

Senstadt – Senatsverwaltung für Stadtentwicklung und Umwelt Berlin (ed) (2012) Berliner Strategie zur Biologischen Vielfalt. Begründung, Themenfelder und strategische Ziele. http://www.stadtentwicklung.berlin.de/natur_gruen/naturschutz/downloads/publikationen/biologische_vielfalt_strategie.pdf. Accessed 12 Nov 2011

Shwartz A, Turbé A, Simon L, Julliard R (2014) Enhancing urban biodiversity and its influence on city-dwellers: an experiment. Biol Conserv 171:82–90

Spanier H (2015) Zur kulturellen Konstruiertheit von Wildnis. Nat Landsch online; Nat Landsch 90 (2015):09/10. Accessed 17 Aug 2018

Stadt Bonn (2008) Biodiversitätsbericht Bonn. Zusammenfassung. www.bonn.de›…› Internationaler Konferenzstandort. Accessed 3 June 2018

Stewart GH, Ignatieva ME, Ignatieva ME, Meurk CD, Earl RD (2004) The re-emergence of indigenous forest in an urban environment, Christchurch, New Zealand. Urban For Urban Green 2:149–158

Stiftung Lebendige Stadt (2010) "Bundeshauptstadt der Biodiversität" – Auszeichnung für Anstrengungen im Artenschutz. Hannover ist Bundeshauptstadt der Biodiversität. www.lebendige-stadt.de/web/goto.asp?sid=204. Accessed 2 June 2018

Stöcker U, Suntken S, Wissel S (2014) A new relationship between city and wilderness. A case for wilder urban nature. Deutsche Umwelthilfe e. V. (ed), Berlin. www.duh.de. Zugegriffen: 4. Juni 2018

Sukopp H, Sukopp U (1987) Leitlinien für den Naturschutz in Städten Zentraleuropas. In: Miyawaki A, Bogenrieder A, Bogenrieder A, Bogenrieder A, Bogenrieder A, Okuda S, White J, White J (eds) Vegetation ecology and creation of new environments. Tokai University Press, Tokyo, pp 347–355

Sukopp H, Weiler S (1986) Biotopkartierung im besiedelten Bereich der Bundesrepublik Deutschland. Landsch Stadt 18(1):25–38

The Conservation Volunteers (TVC) (2018) The community volunteering charity. https://www.tcv.org.uk/. Accessed 18 June 2018

The Panama City Declaration: URBIO and IFLA (2016) urbionetwork.org/data/documents/PanamaCityDeclaration.pdf. Accessed 18 June 2018

Threlfall CG, Kendal D (2017) The distinct ecological and social roles that wild spaces play in urban ecosystems. Urban For Urban Green. https://doi.org/10.1016/j.ufug. 2017.05.12. www.sciencedirect.com/science/article/pii/…/pdf?md5…pid. Accessed 5 June 2018

Trepl L (2011) Die Idee der Landschaft. Transcript, Bielefeld

Tzoulas K, Korpela K, Venn S, Yli-Pelikonen V, Kazmierczak Niemela J, James P (2007) Promoting ecosystem and human health in urban areas using green infrastructure: a literatur review. Landsc Urban Plan 81:167–178

United Nations (1992) Convention on biological diversity (deutsch: Übereinkommen über die biologische Vielfalt), Rio de Janeiro. Übersetzung Bundesmin. F. Umwelt). www.dgvn.de/fileadmin/user…/UEbereinkommen_ueber_biologische_Vielfalt.pdf. Accessed 17 Apr 2018

United Nations (UN) (2015) Habitat III Issued Papers. 16 Urban ecosystems and resource management. New York. www.habitat3.org. Accessed 18 June 2018

United Nations (UN) (2016) Habitat III. Neue Urbane Agenda. www.habitat3.org. Zugegriffen: 18 Juni 2018

US Congress (1964) Wilderness Act. Public Law 88–577. 88th Congress, Second Session, Act of September 3, 16 U.S.C., S. 1131–1136. http://www.wilderness.net/NWPS/documents//publiclaws/PDF/16_USC_1131-1136.pdf. Accessed 3 June 2018

van der Ryn S, Cowan S (1996) Ecological design. Island Press, Washington, DC

Vicenzotti V, Trepl L (2009) City as wilderness: the wilderness metaphor from Wilhelm Heinrich Riehl to contemporary urban designers. Landsc Res 34(4):379–396

Walmsley A (2006) Greenways: multiplying and diversifying in the 21st century. Landsc Urban Plan 76:252–290

Wende W, Rößler S, Krüger T (eds) (2014) Grundlagen für eine klimawandelangepasste Stadt- und Freiraumplanung. Publikationsreihe des BMBF-geförderten Projektes REGKLAM – Regionales Klimaanpassungsprogramm für die Modellregion Dresden, 6th edn. Dresden, IÖR

Wissel S (2016) Perspektiven für Wildnis in der Stadt. Naturentwicklung in urbanen Räumen zulassen und kommunizieren. Deutsche Umwelthilfe (ed), Berlin 27 s. www.duh.de. Accessed 3 June 2018

Wittig R (2002) Siedlungsvegetation. Ulmer, Stuttgart

World Ressources Institut (WRI) (2005) Millennium ecosystem assessment. Washington. http://www.millenniumassessment.org/documents/document.356.aspx.pdf. Accessed 18 June 2018

7

What Ways Are There to a Green City?

Contents

8.1   The First Steps – 317

8.2   Maintain, Gain and Connect Space for Urban Nature – 321

8.3   Making Urban Nature Accessible to All – 328

8.4   Enhancing the Benefits of All Types of Urban Nature – 333

8.5   Making Urban Nature a Space for Experiencing Nature and a Learning Place – 337

8.6   Protecting and Using Urban Nature – 346

8.7   Using Urban Nature for Climate Moderation – 359

© Springer-Verlag GmbH Germany, part of Springer Nature 2022
J. Breuste, *The Green City*,
https://doi.org/10.1007/978-3-662-63976-4_8

8.8   Solving Problems and
       Reducing Risks with Urban
       Nature – Nature-based
       Solutions (NbS) – 362

8.9   Providing Guidance Through
       Good Examples – 367

       References – 379

The paths to the Green City can be diverse and develop in the local and regional context. This chapter describes in eight points the "main routes" of the "road map" to the Green City. The effort towards a Green City will always start with first steps to gain a perspective on the tasks to be tackled. It will always be a matter of preserving space for urban nature, gaining new space and networking it. Making urban nature accessible to all also means tapping existing urban nature potential. In general, the aim is to increase the benefits of urban nature for the citizens of the city and to include all types of nature. To this end, it is necessary to turn urban nature into a space for experiencing nature and a place for learning about nature. The classic conflict between "protecting" and "using" nature must be eliminated in the city in favour of a nature conservation concept that focuses on *protection through use*.

It is also becoming increasingly clear that urban nature brings a wide range of benefits for urban dwellers, more than could be calculated from saved financial expenditure, and is capable of making a significant contribution to the moderation of climate change in cities through wise and targeted planning. This makes it possible to use urban nature to help solve problems and reduce risks (*nature-based solutions*). Ultimately, it is always good examples that have been implemented that encourage us to follow them and take the path to the Green City a little further.

8.1 The First Steps

The paths to a Green City differ depending on the region, natural conditions, culture and tradition, social opportunities and other factors. Cities with good preconditions, for example, with already many green spaces, functioning, well-equipped and assertive administration, committed citizens, state support, etc., will be able to follow different paths than those where the start of local community involvement comes from citizens in their neighbourhood or is an initial committed project. Thus, the paths to the Green City cannot be generalized. Germany (see ► Sects. 7.1 and 7.2) has been setting clear policy directions towards the Green City since 2017 (BMUB 2017; BBSR 2017). Elsewhere, efforts will be regional or only local, providing good examples for more attention in the country or beyond.

The problem situation in other cities, especially outside Europe, is quite different and a green awareness and a Green Agenda are only beginning to develop. Other more pressing problems, such as the social situation, housing, employment, available drinking water, health care, education, good road connections, are higher on the local agenda. However, it is certain that people there also benefit from urban nature and want to see it integrated into their living environment, as a green space for social interaction, as a play space for their children or as a garden. It would be presumptuous to regard urban nature there as a deferrable "first world" luxury. The great appreciation for urban nature can also be found everywhere where other problems clearly occupy the city administrations more. Here, much more attention needs to be paid to local initiatives and the urban nature that still exist. Recognizing this and launching initiatives are the first steps on the way to a Green City. These steps will primarily come "from below", that is, from the citizens.

Urban nature in the form of green and blue infrastructure is an equally important infrastructure for the functioning of a liveable city as other technical infrastructures (see ▶ Sect. 7.2). This is increasingly being recognised by decision-makers. The appreciation of urban nature is also growing in society, not least through media attention. International projects for the development or restoration of different types of urban nature can be found everywhere. This can be done through individual projects, but also strategically as a development concept. It is becoming increasingly accepted that the development of cities should also be thought of and planned based on their natural conditions. This reduces the risks associated with natural processes (for example, landslides, flooding along watercourses and in drainage channels) and allows the urban population to participate to a large extent in the benefits of urban green spaces. Planning is an appropriate means to this end. However, the integration of urban nature into urban planning is not done efficiently and strategically everywhere. In many regions of the world, urban planning is either marginal or has little influence on urban development. Planning in a European sense cannot therefore, be relied upon everywhere.

In Germany, there is a call for integrated planning for urban green spaces, combining regional, landscape and green space plans. Here, the conditions for a path to the Green City are comparatively favourable. In many other countries, the path will probably have to be taken from below through many individual local measures.

Both planning and action, however, begin by thinking of the city from its nature side in balance with other aspects. Modernist architects have already begun to do this:

» "Modern Life demands, and is waiting for a new kind of plan, both for the house and the city." (Le Corbusier) (Exhibition in the Le Courbusier Centre in Chandigarh, India, 2016)

The first steps always include the urban dwellers, determine their interests, preferences, user habits and preferences regarding the interaction with nature. This should lead to a significant improvement in the information base of the urban population regarding the urban nature that surrounds them. Often there is less knowledge about this from the media, than about special nature highlights elsewhere. Another important task is to build up or develop the emotional attachment of the inhabitants of a city to "their urban nature". This can be tackled very well by city administrations in cooperation with NGOs or knowledge mediators (universities, technical colleges, adult education centres, schools or similar institutions in other countries) as a longer-term programme. *Better information about urban nature is the first step, better involvement in decision-making the second.*

Urban planning with the goal of the Green City must also adapt to the needs of urban dwellers, even if these do not always coincide with the scientifically justified goals being pursued. For the Green City, compromises have to be made and educational work has to be done. *In the long term, nothing can be implemented against the citizens, but only with them.* Well-intentioned dirigiste measures or surprising "nature-gratifications" from administrations in residential areas will tend

to be less successful than cooperative action with residents (Landeshauptstadt Dresden 2014; Rusche et al. 2015; Penn-Bressel 2018) (see ■ Figs. 8.1 and 8.2).

The aim should be to create awareness and acceptance for the requirements of Green Cities among the population, to develop strategies and concepts for this in the city administrations and to start with a first project for this purpose, thus attracting attention.

◻ **Fig. 8.2** **Urban** nature trail station with stamp house in Incheon, South Korea. (© Breuste 2014)

8

Experts Develop an International Research Agenda for the Green City

In 2009, as a result of intensive research cooperation, a European team of experts identified five scientific research priorities for an international research agenda for the Green City and developed 35 research questions. These are aimed at developing urban nature and green spaces even better as a key contributor to better living conditions in European cities. This international research agenda first draws attention to where there are deficits in knowledge and research in relevant research areas (ecosystem services, changes in urban development, competing uses, user demands and offers through urban nature, etc.). The identified issues are:

1. How can ecosystem services be more precisely determined and quantified?
2. How can (new) urban nature bring benefits to urban areas with environmental and social problems?
3. How can quality, quantity and form be determined for urban nature elements in relation to different urban structures in order to provide optimal ecosystem services?

4. How can the direct and indirect effects of expected climate change on urban nature be determined in relation to urban quality of life?
5. How resilient is designed urban nature (including tree cover) to the effects of climate change and how can this resilience be improved?
6. How can the supply (quantity and quality of ecosystem services) and management of urban waters be improved through linkages with urban nature (James et al. 2009)?

8.2 Maintain, Gain and Connect Space for Urban Nature

In cities and their surrounding areas, space for urban nature is in particular competition with other uses with strong, assertive and well-founded interests (building development, infrastructure, etc.). Due to urban development, more and more urban nature is being lost in cities, but especially on the outskirts in the urban environment. In most cases, this important agricultural land is important for the supply of the urban population, but also as habitat, climate compensation space and often also as rural recreation space.

Worldwide, population and area are growing in the majority of cities. The growth of these two characteristics is by no means correlative. In many cities in the developed industrialized countries, urban land use is growing much faster than the population. This is evidence of growing land consumption per inhabitant, a characteristic of prosperity. Germany aims to limit the daily land consumption for settlement and transport purposes to 30 ha by 2020. This new use of settlement areas also includes a growing proportion of horticultural urban nature (nature of the third kind, Kowarik 1992, see ▶ Sect. 1.3). In the balance, however, a previously agriculturally used area (nature of the second type, Kowarik 1992, see ▶ Sect. 1.3) is usually lost beforehand. Thus, a conversion from agricultural urban nature predominantly to building and infrastructure (sealing) and to horticultural urban nature also takes place. The urban nature balance is negative overall.

In some countries, land consumption for settlement and transport purposes is even more rapid than in Germany. This is due either to even higher values of land redesignation per inhabitant or to a particularly strong increase in population, mainly as a result of migration.

The result is areal cities of enormous spatial scale that are developing either according to plan (for example, USA, Australia) or completely unplanned (for example, Latin America, India, Pakistan, Nigeria and many others) (Habibi and Asadi 2011; Taubenböck et al. 2012, 2015; Breuste et al. 2016, see ◘ Figs. 8.3 and 8.4).

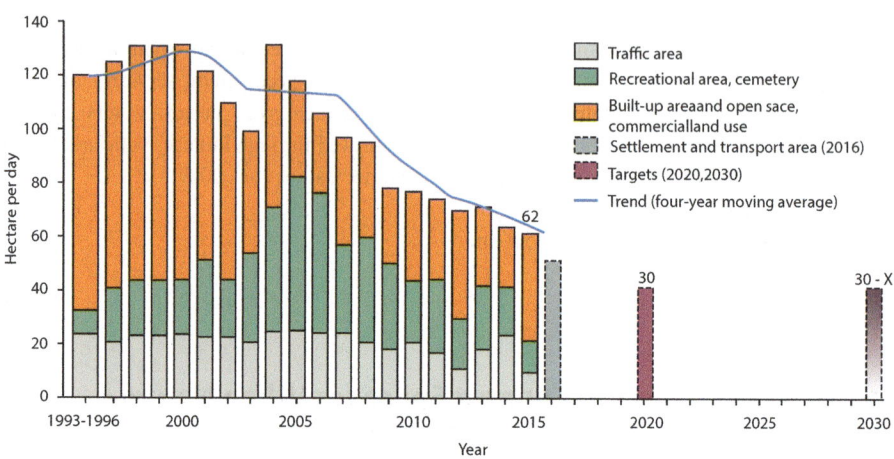

Fig. 8.3 Increase in settlement and transport area in Germany according to calculations by the Federal Environment Agency. (Statistisches Bundesamt 2017)

The Sprawl City

A "Sprawl City" is characterised by its large spatial extent, often with a low population and building density. Urban districts (residential, commercial) grow in a planned or unplanned manner into the undeveloped space of the urban hinterland, often beyond the administrative city boundaries. Inefficient use of resources, impairment of ecological functionality, reduction of biodiversity and ecosystem services are accompanying symptoms (Breuste et al. 2016).

It remains to be seen whether shrinking cities, whose population and settlement area in use are declining, will permanently gain more space for urban nature. In most cases, these abandoned areas are only "green" for a limited period, but are primarily land awaiting construction for intended new development. New wildernesses (see ▶ Sect. 7.4) are also created in spatially limited shrinking phenomena in cities where growth is taking place elsewhere (Schwarz and Sturm 2008) (see ▪ Figs. 8.5 and 8.6).

In Detroit, it is estimated that about one-third of the total city area lies fallow. On a large part of this, new urban nature is emerging unplanned (see ▪ Fig. 8.7).

Land consumption also takes place within cities up to extreme building densities (for example, in large cities of developing countries such as Mumbai in India or La Paz in Bolivia). The result is the reduction of almost all urban nature, which only occurs sporadically (for example, in a few green spaces, cemeteries, etc.). This is associated with the loss of all ecosystem services in the immediate living environment of people, an increase in natural risks due to thermal and hydrological processes and an extensive alienation from nature. In order to improve the living conditions of the mostly disadvantaged and marginalised population living there, often in informal settlements, this extreme building density must be counteracted

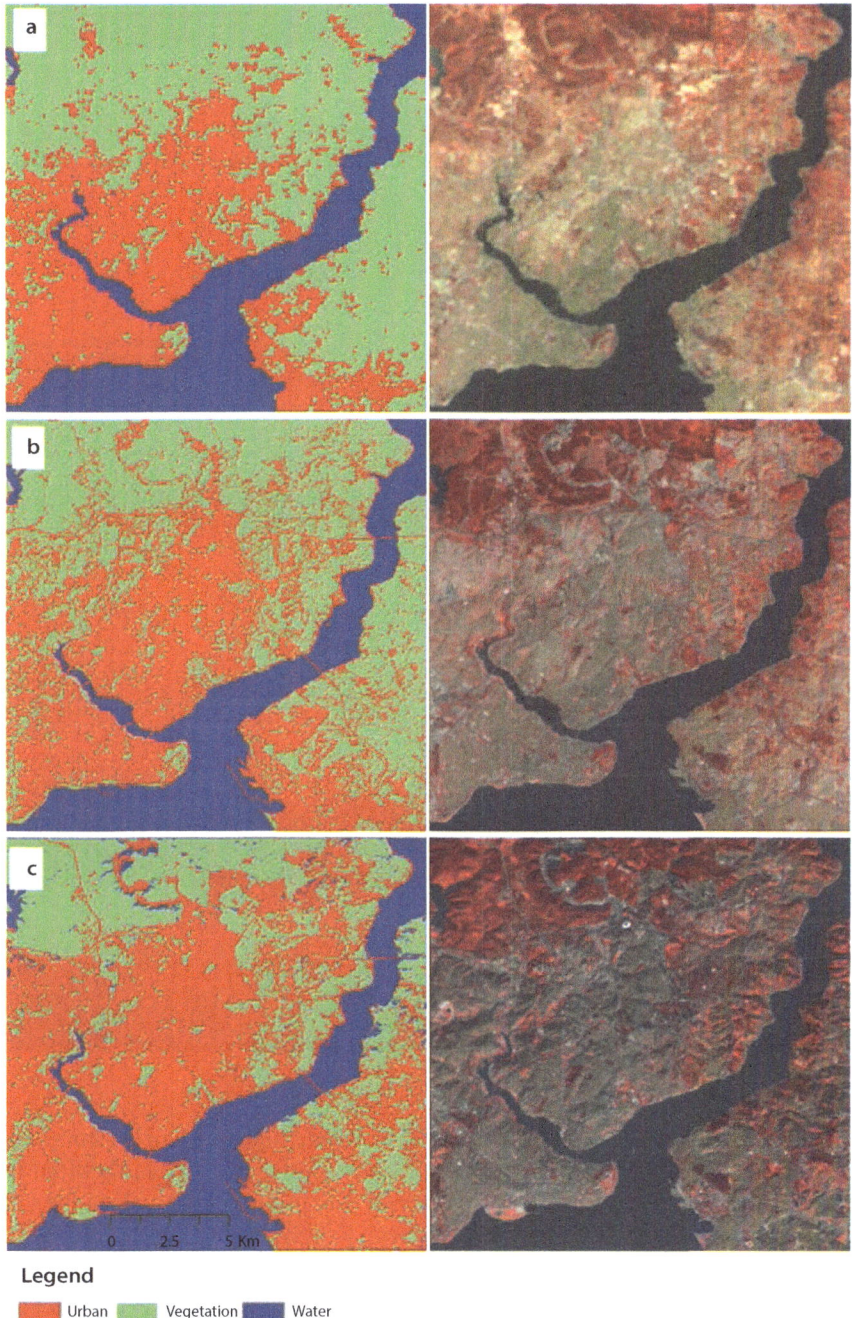

Legend

Urban   Vegetation   Water

◘ **Fig. 8.4** Urban development by urban sprawl in Istanbul Metropolitan Area, Turkey. **a** Classification and Landsat Global Land Survey 1975 Mosaic; **b** Classification and Landsat 5 TM TOA reflectance 1990 Mosaic; **c** Classification and Landsat 8 TOA reflectance 2015 Mosaic. (Taubenböck et al. 2012)

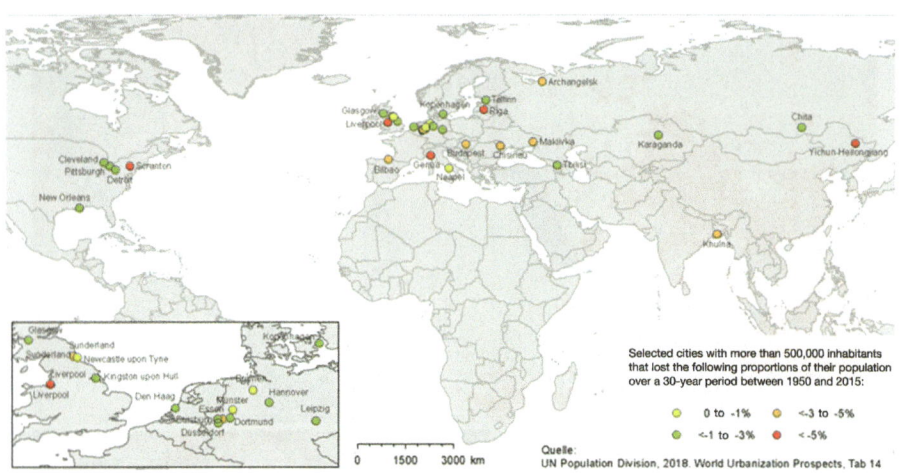

Selected cities with more than 500,000 inhabitants that lost the following proportions of their population over a 30-year period between 1950 and 2015:

- ○ 0 to -1%   ○ <-3 to -5%
- ○ <-1 to -3%   ● <-5%

Quelle:
UN Population Division, 2018. World Urbanization Prospects, Tab 14

■ **Fig. 8.5** Shrinking cities, selection 1950–2000. (Design: J. Breuste, drawing: W. Gruber according to Klett-Verlag 2018 and various other sources)

8

■ **Fig. 8.6** New urban nature as wilderness on abandoned industrial sites in Chemnitz, Saxony. (© Breuste 2017)

by the city authorities. This is only possible in a functioning urban system, under controlled, plannable conditions with effective executive power. This is often not the case in many cities in developing countries. UN programmes such as the UN-Habitat programme City Development Strategy (CDS) on urban development and the Urban Management Programme (UMP) on urban management are intended to strengthen governance and planning capacity in such cities (Ribbeck 2007) (see ■ Fig. 8.8).

Subsequent "interventions" in the extremely dense built substance are usually impossible, so that only additions of urban nature in the accessible surrounding

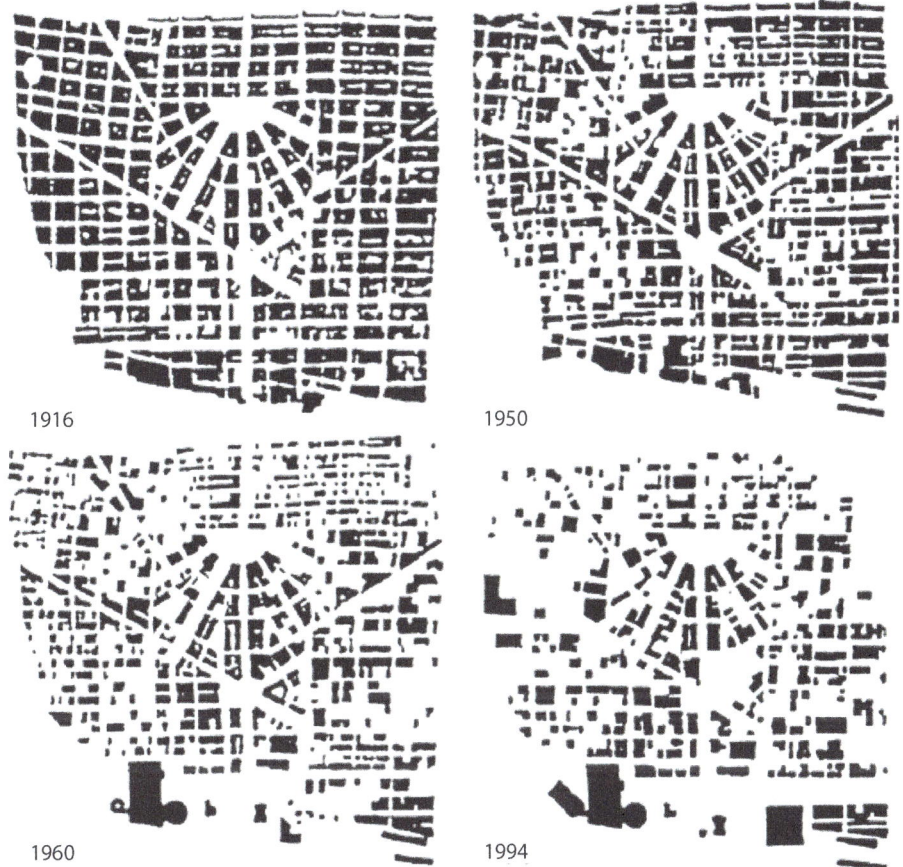

1916

1950

1960

1994

◘ Fig. 8.7 Shrinkage of built-up area and growth of open space in Detroit 1916–1994. (Unger 1995, drawing: W. Gruber)

space are possible. Some cities, such as Medellín in Colombia, have already achieved success with this.

The formation of a **green-blue network** of urban nature areas as **infrastructure** is often based on existing watercourses. The aim here is not solely and primarily to link biotopes to form a habitat network (biotope network), but above all to link urban nature elements in a green way for use by urban dwellers (see ▶ Sect. 7.2). Everyone should have access to this network within a short distance of their place of residence, and it should provide connections to many parts of the city on foot or by bicycle. Networking of urban natural elements also reduces the size of "development and sealing cells" in their area and reduces the formation of heat islands. Dresden has made this principle its long-term, municipal urban development strategy: "Compact city in an ecological network" (Landeshauptstadt Dresden 2014; Rusche et al. 2015, see ▶ Sect. 7.5).

8

◘ **Fig. 8.8** La Paz, Bolivia, Macrodistrito Maximiliano Paredes, the only remaining green is found in the Cementerio General (central cemetery, centre of image). (© Breuste 2015)

The Aim Should Be

To preserve, create and connect as much space as possible for all types of urban nature.

The 30 ha Target in Germany

Despite a stagnating population, the daily land consummation in 2015 was still 62 ha. The goal in Germany is a general reduction in the consumption of resources, as well as the consumption of land for development, that is, also of nature. The target of not converting more than 30 ha of land per day (!) into settlement and transport areas does not stand for a specific state of energy consumption, transport, biodiversity or other ecologically relevant factors, but is based on a general societal objective of the Wuppertal Institute in the 1990s. At that time, the Wuppertal Institute called for reducing the amount of resource consumption (1993–1996: 120 ha area/day) to a quarter of the value at that time. Undeveloped, unspoiled and undissected open spaces in the outskirts of towns were thus to be protected. Such areas, if they are agricultural land, are generally used more intensively today and are often less habitat for plants and animals than natural areas within cities. Nevertheless, this agricultural, undeveloped land is a valuable asset in itself, simply because of its potential for development, which also includes nature (NABU 2017; Umweltbundesamt 2018).

Arvi Park, Medellín, Colombia

The Arví Regional Park in Greater Medellín, Colombia, has been connected to the city's metro and bus network by Metrocable, a cable car, since 2010, and is easily accessible from the densely built-up city of Medellín. It is a natural park that combines the last remaining natural forest (1760 ha) and cultural landscape. The city nature park is the largest and most important in Colombia. The local population benefits from this now easily accessible natural resource through ecotourism, recreational use and nature learning opportunities. In an innovative concept, nature conservation is combined with recreation and contact with nature for all visitors to create sustainable nature tourism. The special concept consists in the mediation of nature, especially for urban children and adults (nature learning places) and was made possible by the joint efforts of the regional administration and private initiatives. Instead of 4 m² in the past, each inhabitant of Medellín now has 12 m² in the balance (Arvi Park 2018) (see ◻ Fig. 8.9).

◻ **Fig. 8.9** Arvi Park Medellín, Colombia, nature conservation, recreation, nature information and involvement of the population (here marketing of local products). (© Breuste 2014)

"Double Internal Development" in Germany

The "double internal development" pursued in Germany seeks to give preference to "internal development" (in existing buildings) over outer development (on the outskirts of towns outside closed buildings) and, in this development, to preserve and develop urban nature in the town in the residential environment of the town's inhabitants.

Double internal development means developing land reserves in existing settlements not only structurally, but also as urban green space. This is intended to protect the landscape in the urban hinterland from further development and at the same time to develop the inner settlement area with efficient and usable green space. This is a particular challenge in densely populated urban areas. As a programme, dual inner development forms an interface between urban development, open space planning and nature conservation (DIFU 2013).

8.3 Making Urban Nature Accessible to All

8

Haase et al. (2017), using examples from Europe, increasingly assess Green City strategies as an aspect of urban renewal, urban development and revitalisation processes. They see these primarily as economically justified efforts to make cities more attractive, aimed particularly at middle- and higher-income populations. Neighborhoods of low-income populations are less of a focus of greening activities. As a result, existing and new social inequalities in the amenities and development of urban nature stabilize. Successes in efforts to create a Green City do not benefit all urban residents equally. This phenomenon has been observed worldwide for a long time. In Europe, however, expectations of social inclusion and equal opportunities to use publicly funded assets such as urban greening are higher (Barbosa et al. 2007).

Not to be forgotten are the services provided to all urban dwellers by urban greenery in green city centres or their additions, also a process that can be observed worldwide. One example is the green coastal belt Cinta Costera in the dynamic metropolis of Panama City, completed in 2009 on newly reclaimed land at a cost of US$189 million (see ■ Figs. 8.10 and 8.11).

Urban nature is also unevenly distributed between cities, but also within cities, due to specificities of location, historical development and culture (see ▶ Sect. 8.3). Scandinavian cities often have extensive green and blue infrastructure, which Mediterranean cities usually lack. Lower levels of urban nature then also lead to lower levels of provisioning of their ecosystem services.

Within cities, the goal of "green justice" is part of the environmental justice to be strived for. What is meant is that public goods, such as urban greenery as nature on public land, should benefit all urban dwellers equally. This requires free access to this urban nature. However, those who have to travel long distances to reach green spaces, for example, are disadvantaged by limited opportunities for use (see ▶ Sect. 8.5).

■ **Fig. 8.10**   Cinta Costera, linear park in the center of Panama City. (© Breuste 2012)

■ **Fig. 8.11**   Cinta Costera, a park for all in the center of Panama City. (© Breuste 2012)

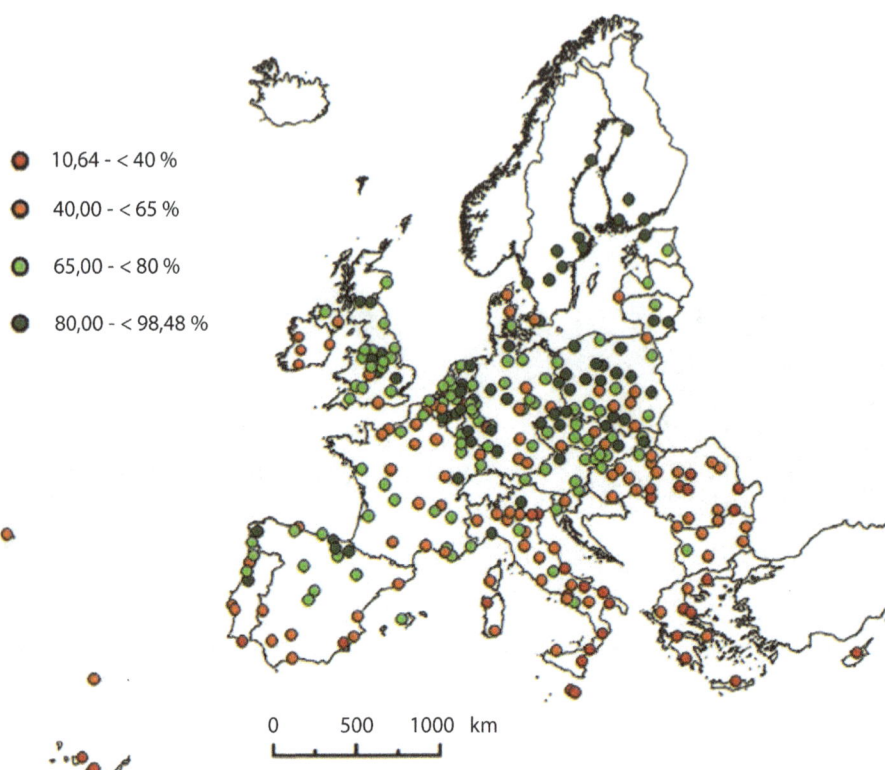

● 10,64 - < 40 %

● 40,00 - < 65 %

○ 65,00 - < 80 %

● 80,00 - < 98,48 %

0   500   1000 km

◻ **Fig. 8.12** Coverage rate of urban population with green spaces of at least 2 ha within 500 m of the place of residence in European cities (in percent). (After Kabisch et al. (2016), drawing: W. Gruber)

In only half of the European cities has a majority of urban dwellers access to public urban green spaces within a 500 m distance to their place of residence (Kabisch et al. 2016) (see ◻ Fig. 8.12).

Grunewald et al. (2016 and 2017) calculate the accessibility of public green spaces for the residential population in Germany. In the 182 German cities with more than 50,000 inhabitants studied, 74.3% of the resident population can reach green spaces and water bodies (≥1 ha) within a distance of 300 m as the crow flies (= 500 m walking distance) as well as green spaces and water bodies (≥10 ha) within a distance of 700 m as the crow flies (= 1000 m walking distance). These are very positive supply values, even in a European comparison (see ▶ Sect. 8.3) (see ◻ Fig. 8.13).

In Germany, the introduction of nationwide minimum standards for a demand-based green space supply is suggested (BBSR 2017). There are hardly any data on this for countries outside Europe. A comparison in the sense of reflecting "better" or "worse" conditions worldwide is out of the question,

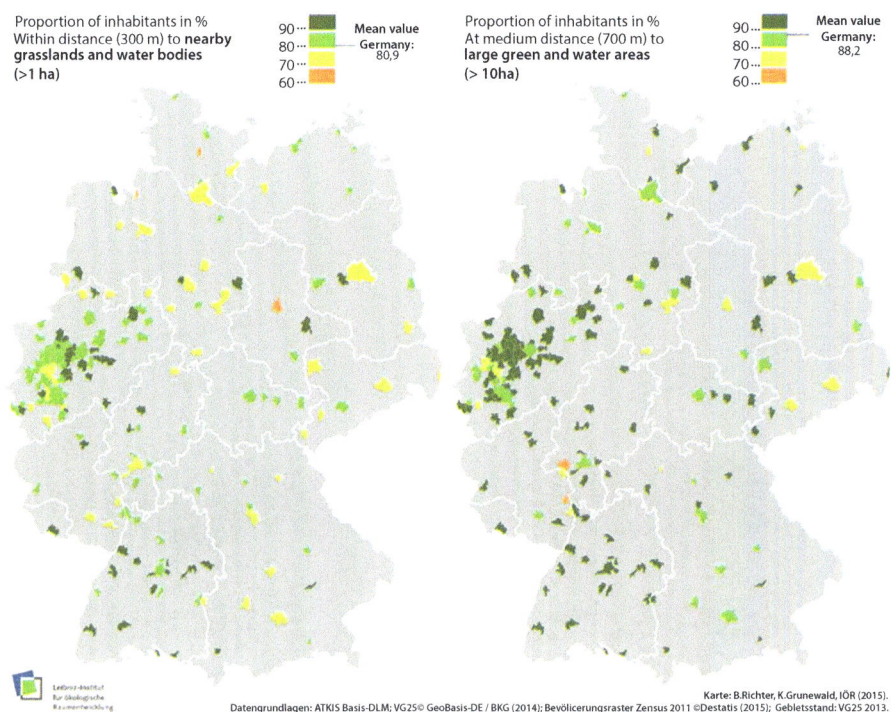

Proportion of inhabitants in %
Within distance (300 m) to **nearby grasslands and water bodies** (>1 ha)

90···
80···
70···
60···

Mean value Germany: 80,9

Proportion of inhabitants in %
At medium distance (700 m) to **large green and water areas** (> 10ha)

90···
80···
70···
60···

Mean value Germany: 88,2

Leibniz-Institut für ökologische Raumentwicklung

Karte: B.Richter, K.Grunewald, IÖR (2015).
Datengrundlagen: ATKIS Basis-DLM; VG25© GeoBasis-DE / BKG (2014); Bevölkerungsraster Zensus 2011 ©Destatis (2015); Gebietsstand: VG25 2013.

Fig. 8.13 Accessibility of public green spaces for the residential population in German large and medium-sized cities (≥50,000 inhabitants), left for nearby smaller (minimal 1 ha, 300 m walking distance), right for larger green spaces (bigger than 10 ha, 1000 m walking distance, including water bodies). (Grunewald et al. 2016 and 2017)

because natural conditions and cultural traditions would be too different. Private green spaces can also replace public green spaces to a certain extent, which must be taken into account. It is clear, however, that the informal settlements and urban districts of the poor in Africa, Asia and Latin America have a high population density and almost no green space, which means that there is no green justice here. What is more, whole sections of the population, often majorities, are excluded from contact with urban nature at all in cities in developing countries (Ernstson 2013) (see ◘ Fig. 8.14).

The German Federal Institute for Research on Building, Urban Affairs and Spatial Development (BBSR) recommends action targets for public green spaces with a minimum size of 1 ha for everyday recreation. They should be located within walking distance (300 m as the crow flies). Green spaces with a minimum size of 10 ha should be located within moderate walking distance (700 m as the crow flies) for local and weekend recreation. The combination of the accessibility of green spaces with minimum sizes should enable optimal use (BBSR 2017).

Fig. 8.14 Low public green provision in extensive neighbourhoods of Buenos Aires, Argentina. (© Breuste 2015)

The Aim Should Be

To enable all urban dwellerso in a short distance to their place of residence to access to sufficiently large areas of urban nature.

Nordelta, Buenos Aires – A Green Island of the Wealthy

Urban nature as private nature is always found where it represents a value for wealthy city dwellers and acts as an object of representation. The possession of green spaces alone identifies city dwellers as privileged. Green is an object of status. Green residential areas with homes and gardens differ from densely built multi-storey building blocks with considerably less (communal) greenery worldwide.

Often, *gated communities* are green islands in the sea of built-up areas in cities or on their edges.

Founded in 1999, the Nordelta waterscape, an exclusive residential development on the outskirts of Buenos Aires, Argentina, created through land fill and landscaping, is home to approximately 25,000 residents (2014) who live undisturbed a park landscape aways the polluted and densely built-up city of Buenos Aires (Nordelta 2015; see ▫ Fig. 8.15).

◘ Fig. 8.15 Nordelta, Buenos Aires, Argentina, an upper-class residential area in a newly created green waterscape in the delta of Paraná river. (© Breuste 2015)

8.4 Enhancing the Benefits of All Types of Urban Nature

The Aim Should Be

To apply the concept of ecosystem services in municipal decision-making, using as a starting point with an overview of local urban nature, its service characteristics and service demands. This inventory should be continuously updated, available for public information and include all types of nature. With its help, urban nature can be preserved and increased where it is most useful.

Applying the concept of ecosystem services means understanding a city's natural assets as its wealth. When this understanding is present or emerging, this wealth can be protected, nurtured and increased. The concept of ecosystem services is designed to help cities around the world preserve and increase their natural wealth. The principle is simple: what benefits me, I preserve and increase. This puts the benefit side in the foreground. Urban planning, urban budgeting and urban services can all be considered. Economic viability and social benefit can be arguments for achieving municipal political commitments.

The concept of ecosystem services can support cities if ecosystem services are included from the outset in decisions on benefit trade-offs. However, in many cities, ecosystem services are not yet or only inadequately included in decisions. The reason for this is often ignorance of the services or a lack of quantifiability in comparison with quantified economic use alternatives. Conservation of ecosystem services often only results when the urban population benefits directly, would be directly affected by a loss and does not accept it. This applies, for example, to urban nature that provides established and sought-after services as a recreational space, for beautifying the cityscape or as a religious site (for example, shrine forests in Japan, the central city park, the nature recreation area on the city land, the rows of trees along rivers) (see ▶ Sect. 3.6). Other urban nature that receives little or only minor attention has less political and administrative support. Its value is often not recognised and, in the face of competitive pressure, it can often be easily removed and replaced by other uses. Strengthening acceptance for urban nature also means informing people, making them aware of the benefits of each individual natural area and raising awareness of the values associated with urban nature among the population (*bottom-up approach*). A strong civil society with various possibilities to represent its interests is helpful or even a prerequisite for the success of this approach (TEEB 2011) (see ▶ Sect. 7.5).

Overall, it is also important to work towards ensuring that the concept of ecosystem services can also be incorporated into formalized planning and decision-making processes. Even in states and cities with developed and established planning systems, this is usually not yet the case (*top-down approach*). The consideration of which of the two approaches, *bottom-up* or *top-town*, or whether both approaches can be applied cooperatively, falls locally.

The municipality of Cape Town, South Africa, for example, has decided to continuously assess the ecosystem services of its urban nature, including monetary valuations (Total Economic Value/TEV approach). This is an immense task that requires value awareness, human resources, experience and budgets. Cape Town aims to set an example for other cities (De Wit et al. 2009; De Wit and van Zyl 2011) (see ▣ Figs. 8.16 and 8.17).

Performance characteristics of urban nature depend not only on the type of urban nature, but also on its quality. The quality of urban nature as a service provider is necessary for its targeted development. This can be achieved by making smaller areas attractive to recreational users, for example, by improving their facilities and cleanliness, and by increasing their attractiveness through networking with others, for example, through attractive pathways. Urban wildernesses can gain importance as nature experience areas if a simple infrastructure of paths enables

▣ **Fig. 8.16**   Cape Town, South Africa, with Table Mountain National Park. (© Breuste 2006)

☐ Fig. 8.17   Cape Town, South Africa, city centre. (© Breuste 2006)

their use. Wetlands that are in danger of drying out can be restored to a valuable, species-rich wetland through appropriate measures. There are many examples of how existing urban nature can be improved in its qualities as a service provider for city dwellers. This is all the more important as urban nature often cannot simply be expanded over a wide area, but can often improve its performance characteristics (see ▶ Sect. 8.5).

The protection and preservation of urban trees is of particular importance. Tree-lined streets and squares where summer temperatures are reduced, small woods, parks and urban forests are among the most attractive and efficient natural elements of cities (see ▶ Sect. 5.2).

Urban water areas often still offer diverse potentials in terms of services if they are made accessible and naturalised in the riparian area. As connecting elements, flowing waters are particularly suitable for forming a green network in the city. Much more attention can be paid to this than has been the case to date. In the future, they can be an essential part of the green-blue infrastructure connecting city districts (see ▶ Sect. 5.4, see ☐ Fig. 8.18).

In Germany, gardens, especially allotment gardens, have repeatedly fallen victim to the pressure of competition from other uses and have been pushed to the outskirts of the city out of the densely built, multi-storey residential areas where they represent a valuable local recreational supplementary space. Keeping them in residential areas and thinking of them as part of the residential-garden combination will be an important task on the way to the Green City. In some countries, for example, Portugal and Spain, new green spaces have already been successfully established (see ▶ Sect. 5.3).

■ **Fig. 8.18**   Karl Heine Canal
with bank vegetation, Leipzig.
(© Breuste 2011)

Agricultural areas in and around cities are the category of nature (nature of the second type, Kowarik 1992, see ▶ Sect. 1.3) that is most in decline and least protected. They are the urban development space accessed in urban development and expansion. However, with proper management and accessibility through usable trails, it is apparent that this category of nature can also provide valuable ecosystem services (for example, food production for the city, recreational space, habitat).

Wetlands are generally in decline in cities. They are species-rich habitats and particularly suitable as nature experience areas. Their preservation and renaturation should be given urgent attention in cities. As watercourses, they are often also the basis of an existing or possibly to be created green-blue network (see ▶ Sects. 6.4 and 7.2).

Building greening (horizontal and vertical) is still applied far too rarely in cities as part of the green infrastructure. There are often technical and economic reservations against it. Also, the benefit from greening buildings should be concrete, measurable goal. Above all, vertical greening and green roofs could be applied much more frequently to new buildings than has been the case to date (see ■ Fig. 8.19). Green roofs are often not accessible to users, that is, they cannot enjoy the benefits of vegetation, as accessibility is a prerequisite for this. Green roofs do not have to consist only of wall pepper vegetation, but can certainly integrate performance greenery (shrubs, smaller trees) and turn roofs into attractive places to stay with natural elements in densely built-up neighbourhoods. Public buildings could be pioneers and examples for this. The greening of the Warsaw University Library is an illustrative example of this (see ■ Fig. 8.20, see ▶ Sect. 8.4) (The Green City 2015).

337   **8**

8.5 · Making Urban Nature a Space for Experiencing Nature...

Fig. 8.19 Vertical greening at the Les Halles supermarket in Avignon, France. (© Breuste 2014)

Fig. 8.20 Roof of Warsaw University Library greened as a usable park, Poland. (© Breuste 2011)

8.5 Making Urban Nature a Space for Experiencing Nature and a Learning Place

The Aim Should Be

To make contact with nature in the urban daily life of children and young people and to stop the existing alienation from nature. To this end, spaces for experiencing nature, especially for children and young people, should be maintained and expanded. Green Learning Spaces should be available for adults formally and informally first for children and young people.

Alienation from nature has become a characteristic feature of many developed, urban societies, and this even though never has more information about nature been available to all people than with today's media support. The difference between what is "natural" and uninfluenced by humans and what is "natural", based on natural processes but heavily influenced by humans, is something most people are not aware of. Plant and animal life surrounds us, but has disappeared from the everyday experience of many urbanites. There are glaring deficits in knowledge, contact and experience, especially in developed, urban societies and there especially among the majority of the population living in cities. City life alienates from nature. The 2017 studies on nature awareness by the Federal Ministry for the Environment, Nature Conservation (BMU) and Nuclear Safety and the Federal Agency for Nature Conservation in Germany make this clear (see ▶ Sect. 6.3):

Fifty-eight percent of the 2065 representative respondents do not know what biodiversity is, although 53% say they feel personally responsible for biodiversity conservation. Only 36% are very convinced of the decline of biodiversity. Three-quarters (75%) of all Germans do not have a high awareness of biodiversity. Social awareness of the importance of biodiversity has hardly changed in recent years, despite an immense flood of information. Nature no longer occupies a significant position in the lifeworlds of entire milieus (for example, middle-class and precarious) (BMU, BFN 2018, see ▶ Sect. 6.3).

Nature is only rarely addressed in the experience of urban dwellers, in the city, for example, through the media. Nature is localised as something that exists outside the cities, and responsibility for it is sought from politics. Social and individual awareness, knowledge of nature, attitudes and behaviour towards experiencing, using and valuing nature depend on social milieus, lifestyles and value orientations (BMU, BFN 2018, see ▶ Sects. 8.6 and 8.7).

However, the main potential space for experiencing nature is urban nature, as this is the everyday nature experienced by more than three-quarters of people in Germany. So far, this potential is far from being exploited. The entire city is a space for experiencing nature. In it, special nature experience and adventure spaces can be specifically set up and designated. In the United Kingdom, a wide range of positive experiences were already made in this respect in the 1980s and 1990s, and corresponding urban nature areas were set up specifically for the use of children and young people in cities such as London (Johnston 1990) (see ▶ Sect. 8.6).

Children and young people are a special target group for improving the declining contact with nature. Since 1997, the "Jugendreport Natur" (*Youth Report on Nature*) has been documenting the change in young people's relationship with nature at irregular intervals. For this purpose, each time between 1200 and 3000 secondary school students of all school types in Germany were asked about their experiences, attitudes and knowledge concerning nature. Apart from various publications, the results are compiled on the website ▶ http://www.natursoziologie.de/NS/alltagsreport-natur/jugendreport-natur.html. In general, it can be stated: Nature is increasingly disappearing from the everyday lives of young people. In just

339   **8**

8.5 · Making Urban Nature a Space for Experiencing Nature...

10 years, the number of those who enjoy visiting nature has halved to less than 20%. Nature sociologist Brämer (2006; Brämer and Koll 2017), head of the study, calls this alienation from nature "nature oblivion". Young people in Germany are increasingly at home involving in the media society and less and less in nature. They usually perpetuate a romantic image of nature. The use of nature, for example, for hunting or the timber industry, is perceived as reprehensible, forests with dead wood as untidy. Sustainability is hardly understood. Traditional environmental education does little to change the nature-related everyday behaviour of young people. In contrast, those young people who spend a lot of time in the forest acquire significantly more nature competence (Brämer and Koll 2017).

Urban nature can and should fulfil two tasks at the same time: It can be an emotional place to experience nature and at the same time a place to learn about nature and its processes. This promotes the healthy development of children and young people, their personal responsibility, creativity, risk competence, social competence as well as their linguistic, motor and scientific skills. Last but not least, dealing with urban nature leads to the formation of general attitudes towards nature that can also be applied outside the cities. This requires the preservation and targeted localisation of urban nature in the residential and living environment. Formal and informal green learning spaces, urban wildernesses, should be preserved and promoted in their diversity (Naturkapital Deutschland – TEEB DE 2016) (see ▶ Sect. 7.4; ◘ Fig. 8.21).

The concept of nature experience spaces is based on experiencing nature with all senses. Although the concept is rather emotionally oriented, emotion and cognition interact in experiencing nature (Reidl et al. 2005) (◘ Fig. 8.22). However, the

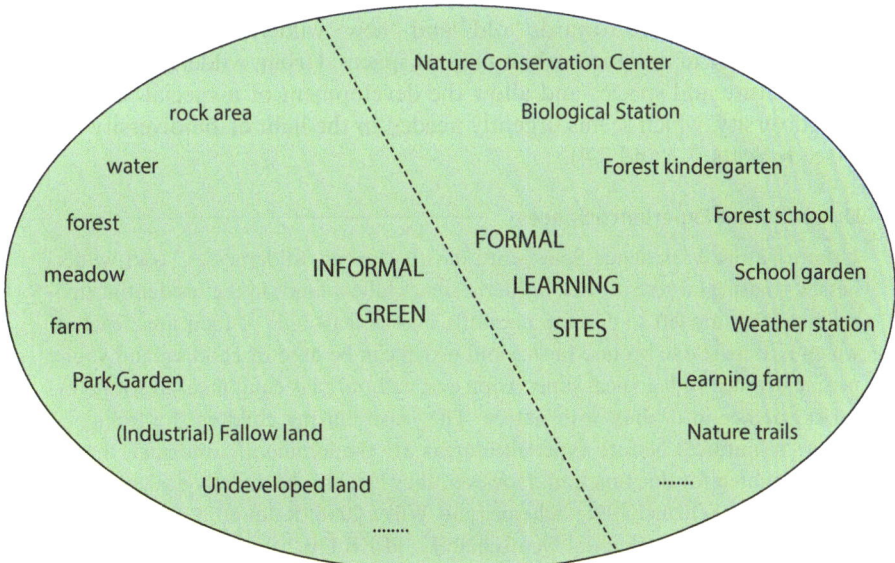

◘ **Fig. 8.21** Concept of Green Learning Places. (Modified after Naturkapital Deutschland – TEEB DE (2016), drawing: W. Gruber)

Fig. 8.22 Concept of experiencing nature. (After Janssen (1988), drawing: W. Gruber)

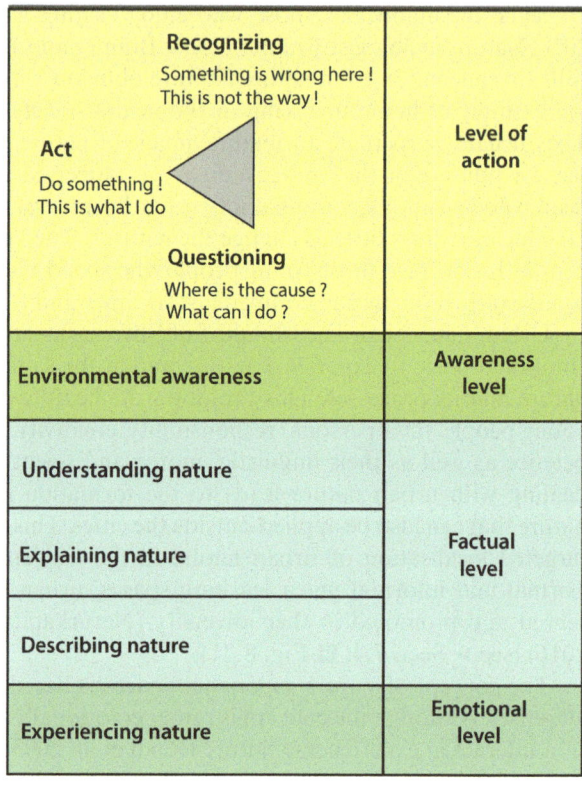

8

concept is often oriented towards "old" and "new" wildernesses, but can include nature experience in designed urban natural spaces. Urban wildernesses are often rich in structure and species and allow the development of a special relationship with biodiversity, which seems urgently needed in the light of biodiversity knowledge (see ▶ Sects. 7.3 and 7.4).

> **Urban Nature Experience Spaces**
>
> Urban Nature Experience Spaces are mostly "urban wildernesses" (old wildernesses = natural forests or new wildernesses = fallow land) in the residential environment that are left to develop naturally over at least half of their area, each of which is at least one hectare in size and which can be used by children and young people without pedagogical supervision and without play equipment to gain playful experience in dealing with nature. They give playing children the feeling of being "in nature". Nature experience areas are predominantly informal, that is, not designated for this function. However, they can also be directly designated for this purpose (Schemel 1998; Schemel and Wilke 2008; Reidl et al. 2005; Stopka and Rank 2013, Naturkapital Deutschland – TEEB DE 2016).

341 **8**

8.5 · Making Urban Nature a Space for Experiencing Nature...

■ **Fig. 8.23** Children experience and learn about nature independently by experimenting and playing, soil quarry edge near the Silberhöhe residential area in Halle/Saale. (© Breuste 2008)

■ **Fig. 8.24** Experiencing nature with a smartphone in the Bailu Wan Wetland, part of the Green Ring in Chengdu, China. (© Breuste 2013)

Children use urban nature less and less independently, but in a regulated and controlled way. Educational institutions provide nature learning through nature experience at organised (formal) green learning places (■ Fig. 8.23). Nature learning places or green learning places are considered important because children in cities have fewer and fewer opportunities for contact with nature and increasingly and dominantly use indoor spaces in their leisure time. Contact with nature plays only a subordinate role for children in cities today. As a result, the numerous proven positive effects of contact with nature in children's development no longer apply. Nature learning places can be components of urban nature, the urban forest, the urban park, the urban water body, the garden or the "new urban wilderness". Children are open to experiencing all types of nature and using them as learning places (Schemel and Wilke 2008) (■ Figs. 8.24 and 8.25). Complementary to green learning places for children and young people, green learning places for adults can be important multipliers. These include learning sites at universities, learning gardens,

Fig. 8.25 Children learn about nature through play, especially on the water, under guidance and supervision, Nordelta, Argentina. (© Breuste 2015)

Fig. 8.26 Bird feeding is popular with urban dwellers – wild bird feeding site on the Vltava River in Prague, Czech Republic. (© Breuste 2017)

teaching and research stations, etc. For adults, too, some urban municipalities try to impart knowledge about their urban nature through boards, information brochures or nature trails. However, the success of these measures is usually not monitored (see ◘ Fig. 8.26).

Green Places of Learning

Educational institutions organise the formal experience of nature by children and young people and include nature learning at "green learning places" in their programmes. In addition to these formal forms of green learning places, any form of urban nature can informally become a green learning place (Naturkapital Deutschland – TEEB DE 2016).

Nature Experience and Nature Learning in Linz, Austria

Half of Linz's urban area is urban nature. This includes the Botanical Garden, the attached Arboretum, the many city parks, urban woodlands such as the extensive Wasserwald Forest, the Schiltenberg, the Traunauen floodplain and Lake Pichlinger. The Natural History Station (*Naturkundliche Station*), which has been attached to the Botanical Garden since 2005, successfully strives to preserve and develop urban nature and is an important local representative for nature conservation and urban ecology in the city of Linz. It was founded as early as 1953. Important tasks here include regular population monitoring of selected animal species (for example, bird species, amphibians, reptiles, dragonflies), nature and species conservation practice, expert service and public relations work.

The station's public relations work is exemplary and addresses all groups of the population, from children to the elderly. It includes activities such as nature walks, bird-watching excursions, children's activities, specialist lectures, exhibitions on urban nature conservation, urban ecology and nature and species conservation in the city.

The city of Linz has a supra-regional impact through its popular scientific journal ÖKO.L – Zeitschrift für Ökologie, Natur- und Umweltschutz (*Journal for Ecology, Nature and Environmental Protection*), which has been published by the Natural History Station since 1979. The journal's concerns are the popular scientific treatment of topics in the field of urban ecology, nature and environmental protection in the city and its surroundings, and the dissemination of the idea of environmental protection in the sense of environmental education. In 2018, ÖKO.L's anniversary year was marked by the Linz exhibition "Bewusst für Natur!" ("Conscious for nature!"). "40 Jahre ÖKO.L – Zeitschrift der Naturkundlichen Station" (Magistrat Linz 2018) (see ◘ Figs. 8.27 and 8.28).

8

Fig. 8.27 ÖKO.L – Zeitschrift für Ökologie, Natur- und Umweltschutz – for 40 years (1979) one of the leading popular science journals on the subject of "urban nature", published by the Municipality of Linz, Austria. (Magistrat Linz 2018)

345   **8**

8.5 · Making Urban Nature a Space for Experiencing Nature...

🔹 **Fig. 8.28** ÖKO.L – Cover page of issue 1, 2003 of ÖKO.L – Zeitschrift für Ökologie, Natur- und Umweltschutz. (Magistrat Linz 2018)

8.6 Protecting and Using Urban Nature

To make urban nature usable and accessible to all urban dwellers while protecting it in an appropriate, differentiated manner.

The Green City can only emerge if urban nature is valued, protected and developed. A broad understanding of different nature, natural processes and their sensitivity or robustness is a prerequisite for this. Urban nature should not have the character of a "reserve" and be "visited" like a Zoological Garden, but should be used. This requires widespread ideas about the non-destructive usability and resilience of various parts of urban nature. This applies just as much to the lawns of a park as to the woodland on the edge of the city (see ▪ Fig. 8.29).

Cities have often already provided protection from destruction for parts of urban nature. To this end, they apply state laws or enact regulations that are intended in particular to regulate the use of this urban nature. However, these protective provisions are usually oriented towards a few natural elements that are considered to be of high value (for example, rare species and their habitats). Their compliance is rarely systematically monitored. Urban protected areas for nature

▪ **Fig. 8.29** Guidance for visitors on environmental behaviour – note on grass conservation in a park in Suzhou, China. (© Breuste 2006)

are thus often "islands of protection", with no connecting link between them. They are often protected with the intention of reducing or even excluding human use. Different categories of protection serve this purpose (see ▶ Sect. 7.3).

Nature Conservation in the City

The exclusion of human use of nature should be the rare exception in cities. Nature conservation is best fulfilled in cities through responsible use of nature, based on fundamental knowledge about nature.

Only accepted nature can be protected for the urban dweller, unaccepted nature only against them. Protecting nature in the city from the urban dweller should be the rare exception of special justification (for example, protection is not possible elsewhere) (Breuste 1994).

The issue of acceptance is of particular importance. Achieving acceptance among the urban population for nature and nature conservation measures is thus also one of the efforts of institutionalized urban nature conservation (see ◘ Fig. 8.30).

Protecting nature and using nature must be combined today far more than in the past, especially in cities. This can also be seen as an opportunity, as many pro-

◘ **Fig. 8.30** Nature conservation through use and maintenance – information board in the Lainzer Tiergarten, a 2450 ha nature reserve in Vienna, Austria. (© Breuste 2018)

tected areas are either located in the vicinity of cities and are already used for local recreation or are even located in or directly adjacent to cities (Landy 2018). This even applies to several national parks that are now or specifically part of urban nature, for example, national parks such as Table Mountain NP in Cape Town, Nairobi NP in Nairobi, Sanjay Gandhi NP in Mumbai, Tijuca NP in Rio de Janeiro, Tyresta NP in Stockholm, Danube Floodplains NP near Vienna, etc.

Whereas 10 years ago a quarter of the world's protected areas were located within a 17 km radius of cities, today this figure is already significantly higher. The reason for this is the growth of cities into the wider hinterland and the resulting new adjacency of protected areas and urban structures. This does not always lead to possible positive partnerships, but also to conflicts. However, the new neighbourhoods have the potential to bring people closer to nature in their living environment "city" and thereby enable their own experiences of nature. This acquaintance can lead to a greater appreciation of nature. Protecting and using can thus enter into a positive partnership.

Urban Protected Areas (Urban Nature Reserves)

The International Union for Conservation of Nature (IUCN) defines protected areas as:

» *"A clearly defined geographical space, recognised, dedicated and managed, through legal or other effective means, to achieve the long-term conservation of nature with associated ecosystem services and cultural values."* (Trzyna 2014b)

This includes six categories in public, private or combined management:

» *"Ia Strict nature reserve; Ib Wilderness area; II National park; III Natural monument or feature; IV Habitat/species management area; V Protected landscape or seascape; VI Protected areas with sustainable use of natural resources."* (Trzyna 2014b)

Protected Areas are made by their location to Urban Protected Areas, when they are *"situated in or at the edge of larger population centres"* (Trzyna 2014b, S. xi).

Since Protected Areas are exclusively about *"conservation of nature"*, the translation with nature reserves is justified without referring to a specific category of protected area (for example, NSG in Germany).

Urban nature reserves are particularly characterised by:
- A large number of daily users, greater ethnic and social diversity than in remote protected areas
- Diverse relationships with different actors in the city
- Their acute threat from urban development
- Particular concern about crime, vandalism, litter, light pollution and noise nuisance
- Particular affected by urban edge effects, such as frequent and more severe fires, air and water pollution, and the spread of invasive species (Trzyna 2014b, p. xi).

Trzyna (2014b, p. 3) rightly emphasizes:

» *"The international conservation movement traditionally has concentrated on protecting large, remote areas that have relatively intact natural ecosystems. It has given a lot less attention to urban places and urban people."*

Urban nature reserves
— Are also a habitat for some globally endangered species that can only be preserved here
— Are part of the tourism concept of the cities (for example, Cape Town, Rio de Janeiro, Stockholm)
— Can decisively improve the quality of life of urban residents, create jobs and contribute to municipal stability (Trzyna 2014a)

This means that urban nature conservation is faced with new and, in comparison with nature conservation outside cities, additional and in some cases even different, extended tasks.

These include, for example:
— Development of new nature relations,
— Development of familiarity with and belonging to an area ("sense of home"),
— Opening up opportunities for non-regulated children's play,
— Recreational opportunities in nature,
— Offer learning, research and teaching spaces on nature,
— Contribution to environmental protection and the landscape (water balance, reduction of water pollution, cleaning from air pollution, reduction of noise),
— Bioindication of environmental changes and pressures, and
— Opening opportunities for ecological research (see also Sukopp and Weiler 1986; Trepl 1991; Breuste 1994; Breuste et al. 2016; Landy 2018).

Thus, urban nature conservation is about the interface of natural science and social science approaches. The goals of use are different for different user groups. Management must adapt to this, but not lose sight of the goal of preserving diverse nature (Plachter 1991).

It makes sense to determine adapted, settlement-specific nature conservation goals for different urban areas.

Concerning Auhagen and Sukopp (1983), Plachter (1991) develops "settlement-specific nature conservation objectives" and measures, with particular reference to zooecological aspects (p. 137):
— Preservation and restoration of continuous green corridors,
— Reduction of the degree of sealing,
— Avoidance of new buildings in areas that are of particular importance as dispersal axes for animals and plants, such as watercourse banks or forest edge areas,
— Strict adherence to the principle of intervention minimization in all construction measures
— Reduction in maintenance intensity on a significant proportion of public land,
— Targeted protection of high age habitats,

— Development of old tree stands. Restoration measures should only be carried out for compelling reasons of traffic safety,

— Regeneration of watercourses including their riparian zones by reopening piped sections and creating wide, natural riparian zones,

— Protection and, if necessary, regeneration of diverse green belts, and

— Preservation and, if necessary, redevelopment of habitats such as small water bodies, avenues, orchards, farm gardens, ruderal meadows.

How much wilderness we tolerate in cities or even deliberately develop in the system of nature reserves will be a task to be solved locally (see ▶ Sects. 6.3 and ▶ 7.4). To see wilderness development as the only sensible way of urban nature conservation (Hard 1998, p. 41) will probably be the exception rather than the rule. However, applying an urban sense of order everywhere should also be reconsidered.

The better integration of nature into cities is not without conflict. Urban dwellers, often unfamiliar with nature, have fears about nature that are sometimes unfounded but shared by many. They suspect:

— Spread of pathogens,

— Insect calamities,

— Spread of poisonous animal species (for example, spiders, snakes, etc.),

— Spread of "uncleanliness" through natural development processes, and

— Disturbance of social peace (expectation that natural areas in the city could be retreats for criminals or fringe groups with little tolerated behaviour).

It is certain that with more nature, wild animals and wild plants are also becoming established in cities, sometimes in unusual places. This does not always happen in agreement with the city dwellers, who perceive wild animals in particular as a threat or simply as a troublemaker. This is certainly true for bears in Brasov (Romania), leopards in Mumbai (India), wild boars in Berlin, monitor lizards in Bangkok or monkeys in for example, Indian cities. Other wild animals often remain largely invisible and unrecognised in cities, but have long been part of our urban habitat (for example, foxes in Zurich or Stockholm) (see ◘ Figs. 8.31 and 8.32).

Often urban nature reserves are little used, especially if they are less known and promoted, less well made accessible, less well equipped with infrastructure or difficult to reach, or if there are safety concerns. All of these barriers can be worked on to improve them if they have been identified by managers and the aim is to increase the attractiveness of the areas. Sometimes this is not the aim of management, especially when high visitor numbers are suspected to pose threats to species populations and habitat quality by official nature conservation. In most cases, however, these reductions in the quality of nature reserves are, with good manage-

🔲 **Fig. 8.31** Wildlife is part of the urban habitat. Monkey in Nainital, India. (© Breuste 2018)

🔲 **Fig. 8.32** Feral pigs in the waste areas near settlements at Mehrauli Archeological Park in Delhi, India. (© Breuste 2018)

ment, avoidable. While in remote nature reserves a reduction of visitors may also be an objective, this should not apply to urban nature reserves. Here, the use and development of nature contact should be absolutely paramount. Good management can achieve this even while preserving habitats through appropriate visitor management (Breuste et al. 2016).

30 Recommendations for the Development and Management of *Urban Protected Areas* (After Trzyna 2014b)

Urban nature reserves and urban dwellers

1. Create access for all, especially for all ethnic and disadvantaged groups
2. Develop a sense of local ownership of these natural areas
3. Involve volunteers and supporters in management
4. Communicate considerately
5. Demonstrate, facilitate and support good environmental behaviour
6. Clarify and support the benefits of contact with nature and good eating habits for health
7. Avoid throwing away waste
8. Preventing criminal behaviour and not tolerating it
9. Avoid contacts and conflicts with wild animals and thus prevent the spread of infectious diseases
10. Control poaching
11. Control the spread of invasive species (animals and plants)

Urban nature reserves in spatial context

1. Promote links with other nature conservation areas
2. Integrating nature into built-up areas and thus reducing the contrast between "nature" and "urban"
3. Control encroachments on urban protected areas
4. Continuously monitor water bodies and intervene where necessary to shape them
5. Restrict wildfire
6. Reduce the influence of noise and artificial lighting at night, pay attention to research on electromagnetic radiation

Municipal nature reserves and institutions

1. Cooperation with institutions and organisations with similar statutory tasks
2. Cooperation with institutions and organisations representing similar interests
3. Develop a network of supporters and allies
4. Work with universities that train managers for urban protected areas, use these areas for scientific research and special learning.
5. Learning from the experiences of others through collaboration, paying careful attention to structures, processes and the natural object in the process.

Promote, create and improve urban nature reserves

1. Promote and defend urban protected areas
2. Prioritize urban protected areas for national and global protection
3. Creation and extension of urban protected areas
4. Support rules and cultures of interaction that express the differences between urban and other protected areas
5. Recognize that political skills are critical to success, strengthen them, and build political capital
6. Tapping supportive financing from a variety of sources
7. Take advantage of international organizations and exchange programs
8. Improving protected areas through research and evaluation

Principles of Urban Nature Conservation

Principles of urban conservation may include:

1. Priority areas for environmental protection and nature conservation
2. Zonally differentiated focal points of nature conservation and landscape management
3. Consideration of nature development in the downtown area,
4. Historical continuity
5. Preservation of large contiguous open spaces
6. Networking of open spaces
7. Preservation of site differences
8. Differentiated intensities of use
9. Preservation of the diversity of typical elements of the urban landscape
10. Elimination of all avoidable interventions in nature and landscape
11. Functional integration of structures into ecosystems
12. Creation of numerous air exchange paths
13. Protection of all spheres of life (Sukopp and Sukopp 1987, pp. 351–354).

Aravalli Biodiversity Park, Delhi, India

The Aravalli Biodiversity Park in Delhi represents a piece of reclaimed wilderness in the middle of Delhi on an area of 153.7 ha. It was opened on World Environment Day 2010 by making an area of secondary dry forest that had been unused for some time accessible to visitors through infrastructure, fencing and management, and presenting it as a nature reserve. This also prevented other uses from spreading on the unused land.

The park is a biodiversity hotspot in Delhi with 175 species of birds alone. Its emergence is due to private initiatives to restore lost nature in the city. To re-establish native nature, extensive management is required against the spread of *Prosopis juliflora*, a woody species from Mexico that is spreading invasively in subtropical Africa, Asia and Australia and already occupies large areas.

The park provides visitors with a good insight into the flora and fauna of the Aravali hill ranges, which have structurally enclosed the major city of Delhi in recent decades, making it part of the city (Bansal 2017) (see ◘ Fig. 8.33).

◘ Fig. 8.33 Aravali Biodiversity Park in Delhi, India. (© Breuste 2018)

8

Văcăreşti Nature Park – A Wilderness as a Protected Area in the Middle of Bucharest, Romania

The Văcăreşti Nature Park is a 190 ha site in the middle of Bucharest, Romania. It was originally a marshland that was drained and dammed during the Ceausescu era. The intention was to use it as a future dammed water area, for example, for Olympic sailing regattas. However, the water supply required for this purpose from the Argeş River via Lake Mihăileşti was never sufficient. Nor could the growing costs be met. The plans were therefore abandoned. What was left behind was an unfinished project, a huge unused area in the middle of the city, which was left to itself as a partial wetland and became a new "urban wilderness". Informal uses by anglers, hunters, social fringe groups and thus a widespread rejection among the population towards the Văcăreşti area soon set in.

In 2003, the responsible Romanian Ministry of Environment transferred the rights of use for 49 years to a private organization (Royal Romanian Corporation) for 6 million US$ in order to develop a sports and cultural complex with an investment of 1 billion US$. This would have been the end of the natural area. This project also failed.

In 2014, Văcăreşti was provisionally protected by the Romanian government and in 2016 it was declared Văcăreşti Nature Reserve (Parkul Natural Văcăreşti). This was done perhaps in the absence of other major development ideas, perhaps also in recognition of the nature conservation value of the area and under the influence of growing interest in the exceptional and special natural development in the middle of the city.

The Văcăreşti Nature Park is not only the largest urban park in Romania, but also the first nature reserve in the middle of Bucharest. It is also of international importance, as it has now developed into an ecosystem rich in species and worthy of protection, thanks to decades of independent natural development. It is managed by the Asociaşia Parcul Naţional Văcăreşti (Vacaresti National Park Association). The development plans aim to follow the example of the London Wetland Centre (see ▶ Sect. 5.4.2). To this end, international cooperation is taking place with environmental institutions such as Wetland Link International and the Wildfowl and Wetland Trust. The aim is to create a structurally and species-rich "nature oasis" with species-rich habitats that invites residents and guests of the Romanian capital to interact with nature (Manea et al. 2016; Imperialtransylvania 2018) (see ◻ Figs. 8.34 and 8.35).

◻ **Fig. 8.34** Vacaresti Nature Park in the middle of Bucharest, Romania. (© Breuste 2017)

Fig. 8.35 Symbol of the Vacaresti Nature Park in Bucharest, Romania. (© Breuste 2017)

Tokai Forest, Cape Town, South Africa – Moderating the Conflict Between Restoration and Recreational Use

Tokai Forest is a forest area in the southern suburbs of Cape Town. It is also part of Table Mountain National Park, a national park in the city. The forest area was originally established by gardener Joseph Lister in 1885 as an arboretum for exotic tree species to determine which species were suitable for forestry naturalisation under the natural conditions of the Cape. This was an effort to at least partially replace the Afromontane forest of the Table Mountain massif lost to overexploitation, but also to reduce timber imports through own timber production. The current 1500 or so trees, now over 100 years old, include about 150 non-native species, such as oaks, cypresses, pines and California redwoods. The Tokai Forest is a problem area for the conservation of the National Park to preserve the native fynbos vegetation against the encroachment of exotic species that threaten the vegetation stock worth protecting. On the other hand, Tokai Forest is a popular area among Capetonians for camping, picnicking, hiking, horseback riding and mountain biking. The conservation management came up with the idea to develop biodiversity by clearing the non-native forest stand and re-developing it with native fynbos vegetation. This was quickly abandoned in favour of retaining the exotic forest for its cultural ecosystem services, although this is contrary to the general strategy of species and habitat conservation of the Nature Conservancy Management. Recreational use is clearly the most important function here (Trzyna 2014b) (see ◘ Figs. 8.36 and 8.37).

◘ **Fig. 8.36** Tokai Forest, Cape Town, South Africa. (© Breuste 2008)

◘ **Fig. 8.37** Tokai Forest, barbecue site, Cape Town, South Africa. (© Breuste 2008)

Parque Nacional da Tijuca, Rio de Janeiro, Brazil – Hotspot of Global Biodiversity and Urban National Park

The Tijuca National Park is one of the largest urban national parks in the world, covering 39.72 km². It is the natural trademark of Rio de Janeiro, whose green natural scenery it forms.

The Tijuca National Park (*Parque Nacional da Tijuca*) is located within the urban area of Rio de Janeiro and consists of (secondary) Atlantic Rainforest (*Mata Atlântica*), one of the most threatened forest ecosystems in the world.

The area served as a coffee plantation for a long time and from the middle of the nineteenth century onwards its forest structure was deliberately renaturalised with trees, some of which were foreign to the region. The Brazilian Emperor Dom Pedro II gave the order for this in 1861. A few years later he had a network of paths and lookout points laid out for recreational use. In 1961, the forest was declared a national park. In 2012, it became part of the larger UNESCO World Heritage Cultural Landscape (UNESCO Carioca Landscapes) and the Mata Atlântica Biosphere Reserve (area 1660 km²).

The national park is jointly managed by the city and the national government. It has about 2.5 million visitors a year, who can visit the park for free and reach it easily by public transport. About 1000 km of hiking trails open it up in the interior.

Several roads lead through the forest to the viewpoints, hiking trails lead to caves and waterfalls. Striking peaks are the Pedra da Gávea and the 1022 m high Pico da Tijuca. Rio's much-visited landmark, Corcovado, is also located in the park (Mc Neely 2001; Sociedade dos Amigos Parque Nacional da Tijuca 2004; Trzyna 2014a, b) (see ◘ Fig. 8.38).

◘ **Fig. 8.38**   Parque Nacional da Tijuca, Rio de Janeiro, Brazil, Vista Chinesa. (© Jana Breuste 2009)

8.7 Using Urban Nature for Climate Moderation

The Aim Should Be

To achieve a targeted increase in the proportion of green spaces in general and especially in vulnerable and heat-stressed urban districts in the medium term in order to counter the consequences of climate change. The stock of urban nature, especially shady trees, should therefore be preserved at all costs and, if possible, expanded.

The expected climate changes are already noticeable in cities as "real laboratories" of general climate change. On average, cities have a warmer and drier climate than their surroundings. They are also affected by extreme climate events. For city dwellers, summer temperature extremes are the most stressful aspect of the urban climate. These will continue to increase in the future. Here, climate moderation to reduce temperature is a desirable goal to avert health risks to particularly sensitive parts of the population. In the medium to long term, cities can take structural measures to avoid or reduce overheating effects and use urban nature for moderation in a targeted manner or optimise existing urban nature for this purpose. It should be noted, however, that the optimisation of temperature moderation through urban nature generally requires dense vegetation stands with a lot of functional biomass. This regulating ecosystem service can often only be provided at the expense of other ecosystem services (for example, recreation and other cultural services). The degree of temperature moderation can therefore be determined in the best possible (possibly not maximum) way, taking these effects into account. Temperature moderation generally occurs where urban nature is localized (for example, tree cover, street, parkland, forest). Only in special localizations (no obstacles to the dispersion of cold air) can an extension of the moderation effect into the surrounding area be expected (see ► Chap. 5, see ◘ Figs. 8.38, 8.39, and 8.40).

◘ **Fig. 8.39** **Necessary** tree irrigation in Parque General San Martin, Mendoza, Argentina. (© Breuste 2000)

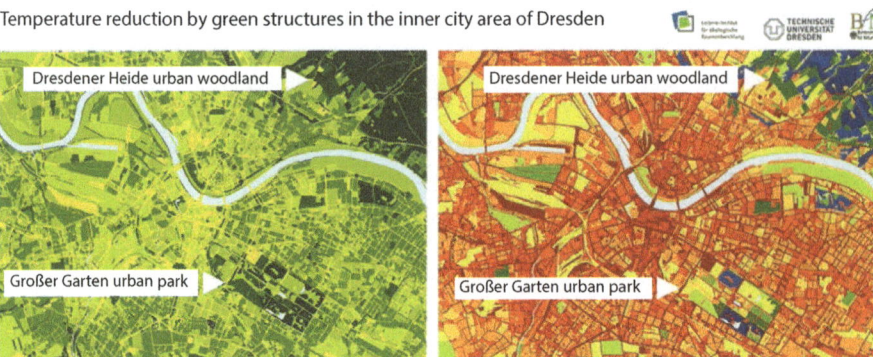

□ Fig. 8.40 Distribution of green volume and associated maximum cooling effects on a radiant summer day in Dresden. Visible are the cooling effects of the urban park Großer Garten and the Dresdener Heide (in K = Kelvin). (Source: Modified after Mathey et al. (2011a))

The temperature-reducing effect of vegetation stands is usually greatest at night, when it can hardly or not at all be claimed by users of the areas. However, urban vegetation is also itself exposed to the consequences of climate change. Drought and heat during the growing season limit the competitiveness of many native species. Targeted plant selection and maintenance measures for water supply during dry periods can counteract this (Roloff et al. 2010).

Key elements of climate moderation in urban communities through urban nature are:

— Analysis of the urban climate in climate function maps
— A small-scale and structurally rich open space system in the inner area, supplemented by undeveloped cold air paths in the peripheral areas
— The larger the green volume, the higher the achievable cooling effect with low air exchange ratios as a rule
— Shady trees and larger lawns usually provide both nighttime cooling and daytime heat load mitigation (University of Manchester 2006; Pauleit 2011; Mathey et al. 2011a, b; Norton et al. 2015) (see □ Fig. 8.41).

In order to effectively plan and implement climate change adaptation measures, the following are required

(a) Identification of overheated areas (urban heat islands)
(b) Identification of exposure due to residence of people and
(c) Determination of sensitivity/vulnerability of exponents to heat (Norton et al. 2015; see □ Fig. 8.42).

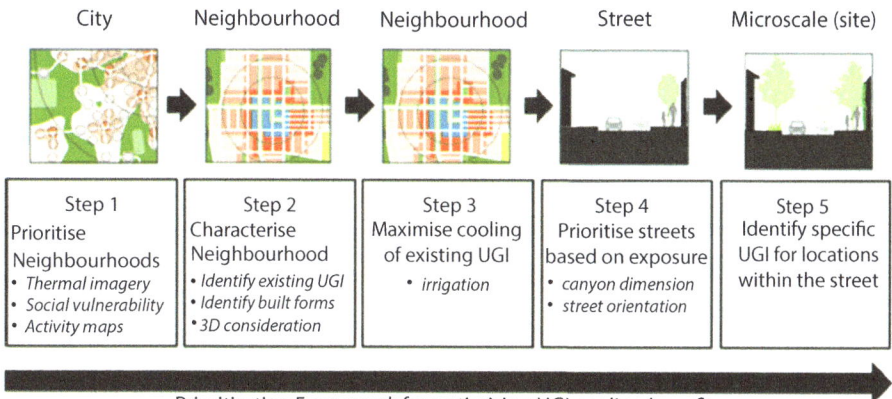

◘ Fig. 8.41   Climate moderation with urban nature in a multi-stage process. (Norton et al. 2015)

◘ Fig. 8.42   Concept of vulnerability. (Strongly simplified after Rittel et al. (2011), drawing: W. Gruber)

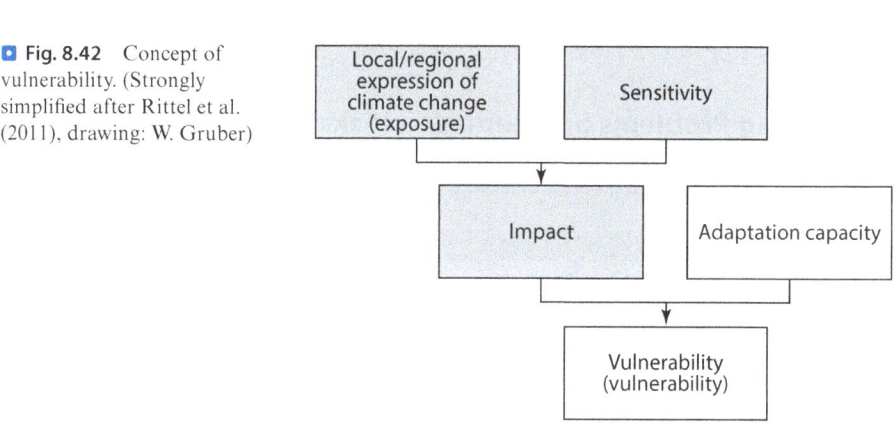

Vulnerable exponents in this consideration are population segments that suffer particularly from heat stress, people with health problems, small children and older people. They are more likely not to expose themselves to heat stress by not visiting open spaces outside their homes during this time. Alternatives, however, would be cooler open spaces in the immediate residential environment. This can be used to determine climatically sensitive residential areas in cities and to take appropriate measures for temperature moderation, including through urban nature (Henseke and Breuste 2015; see ◘ Fig. 8.43).

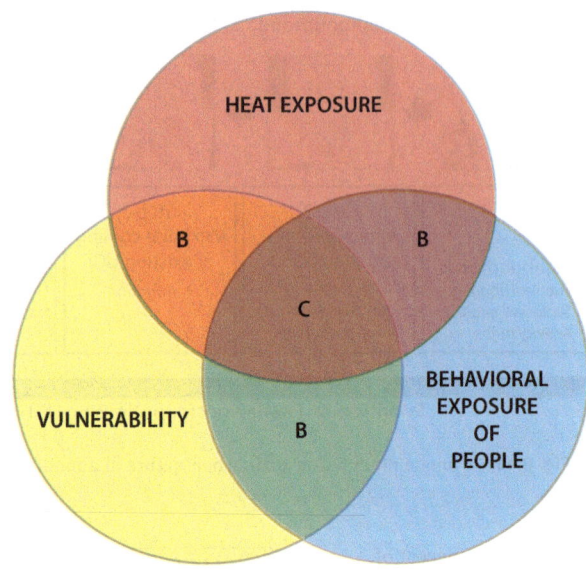

Fig. 8.43 Factors for identifying the priority for temperature moderation of a neighbourhood or a building block. (according to Norton et al. 2015 priority scale: C = high, B = medium, A = moderate, drawing: W. Gruber)

8

8.8 Solving Problems and Reducing Risks with Urban Nature – Nature-based Solutions (NbS)

The Aim Should Be

To work with urban nature to solve urban problems and mitigate risks to improve the urban habitat for urban dwellers (Nature-based Solutions, NbS).

In the context of urban development, the use of technical approaches is usually sought to solve problems and mitigate risks. The concept of the smart city seems to correspond well to this (see ▶ Sect. 8.9). However, environmental engineering has also long offered to use nature to solve problems. However, technology as a problem solver is familiar, predictable, established and accepted, also generally in society and outside the engineering community. This is often the reason for preferring technical solutions. However, the debate on ecosystem services also makes offers for solutions outside of technical interventions or complementary to them. Nature can also take on targeted tasks. Natural processes and structures are used for this purpose. This is offered outside cities and increasingly also in cities. Financial considerations and a broader view of urban planners have also made "Nature-based Solutions" (NbS), a term for such considerations and approaches, possible in cities for about 20 years. This is also related to the increasingly widespread recognition of the fundamental contribution of ecosystems to human well-being in general, a new perspective on many phenomena (Dudley et al. 2010).

From the point of view of the rather passive use of advantages through nature, it was only a small step to increase and optimize these advantages through the targeted design of nature. This view is particularly well suited to cities, where almost the entire urban nature is already designed by people anyway. It is only more important to combine this design with problem-solving strategies and to optimize it. This is already being sought with the concept of ecosystem services. The concept of Nature-Based Solutions (NBS) currently serves primarily as a comprehensible means of communication for politicians and decision-makers with a large "overlap" with the concept of ecosystem services. In scientific publications, the NbS designation is increasingly used (for example, MacKinnon et al. 2011; MacKinnon and Hickey 2009; Eggermont et al. 2015; Maes and Jacobs 2015; Kabisch et al. 2016, 2017). The NbS concept first started with approaches on a global scale, is already widely used on a regional scale (for example, Europe) (ESMERALDA project, ▶ www.esmeralda-project.eu/) and is now increasingly applied in urban settings. The International Union for Conservation of Nature (IUCN) and the European Commission have already incorporated it into policy and governance documents (for example, United Nations Framework Convention on Climate Change (UNFCCC) COP 15, Horizon 2020 Research and Innovation Programme).

Nature-based Solutions (NbS)

Nature-based Solutions (NbS) are defined by the IUCN as

» *"actions to protect, sustainably manage, and restore natural or modified ecosystems that address societal challenges effectively and adaptively, simultaneously providing human well-being and biodiversity benefits."* (International Union for Conservation of Nature 2016)

The European Commission's definition is similar:

» *"Living solutions inspired by, continuously supported by and using nature designed to address various societal challenges in a resource efficient and adaptable manner and to provide simultaneously economic, social and environmental benefits."* (Maes and Jacobs 2015, S. 121)

Modified ecosystems, such as urban ecosystems, are explicitly included as performers. The goals of the problem solutions are
(a) Improvement of living conditions in cities (*human* well-being)
(b) Preservation of biodiversity also in cities (*biodiversity benefits*) (see ◙ Fig. 8.44).

Societal challenges in cities should be parried in an efficient way to generate combined economic, social and environmental benefits.

If NbS are to solve problems in cities, they should also be able to be named concretely. Such problems are typically:
— Noise abatement
— Reduction of air pollution

©IUCN

■ **Fig. 8.44** Concept of Nature-based Solutions (NbS) as a framework term for ecosystem-based approaches. (IUCN 2016)

- Mitigating the effects of climate change (especially temperature reductions)
- Flood risk mitigation
- Promotion of quality of life and recreation
- Conservation and restoration of wetlands and drylands
- Protection of forests and preservation of energy and food resources and local income
- Preservation of biodiversity
- Coastal protection through natural vegetation (for example, mangrove forests)
- Development of green infrastructure through landscaped buildings, trees and area greenery

These aspects are of course also found in the concept of ecosystem services. However, the NbS concept is specifically about replacing or supplementing technical solutions and thereby saving costs and increasing efficiency.

Three main types of NbS also in cities can be determined:

- Sustainable use of existing nature (for example, urban agriculture)
- Restoration of nature through improved management (for example, urban wetlands);
- Design of new ecosystems (for example, green buildings)

Principles of Nature-based Solutions

- Conservation of nature as a service provider (sole service provider or in combination with other problem solutions)
- Incorporation of traditional local and scientific knowledge in the natural and cultural context
- Enabling social benefits in a fair and balanced way, transparently and with broad participation by stakeholders
- Maintaining biological and cultural diversity and the capacity of ecosystems to develop
- Application in landscape scale
- Recognize the balance between a few immediately achievable economic benefits and the long-term impact of ecosystem services

- Integral part of political aspirations, general design measures and specific problem solutions
- Not a substitute for nature conservation measures, but their promotion (however, not all nature conservation measures are also NbS)
- One among several problem-solving options and can be applied together with them
- Support cultural and social components and values
- Specificity into a nature and problem context
- Formats can be both restoration, sustainable management and extending benefits from nature to people (Cohen-Shacham et al. 2016; IUCN 2016; ICLEI 2017)

Restoration of Natural Ecosystems of Urban Wetlands and Waters

Mangrove Rehabilitation Program in Haikou, China

In 2015, the city government, together with a private investor, began a conservation program to restore natural coastal protection in the river marshes of the city of Haikou, China, by replanting previously eliminated mangrove forests (Mangrove Conservation Program). The project also fulfills environmental

education tasks through an educational trial. The area is also used as a recreational area, made accessible by suitable infrastructure (see ◘ Fig. 8.45).

Green Rivers – Urban River Restoration to Reduce Flood Risk

The major river floods in Central Europe in 2002, 2006 and 2013 have led to a significant increase in sensitivity to natural processes and have stimulated planners, decision-makers and scientists to look for new solutions to reduce risks involving natural processes in addition to technical solutions to problems.

The concept of river restoration, which has long been recommended and applied from the point of view of biodiversity, is now also being increasingly used for flood protection. This includes the removal of constricting technical river boundaries, where this is not impossible due to established uses, and the "widening" of river floodplains to preserve retention areas. This is a (re) adaptation to "pulsating", processual river dynamics within the scope of the still existing, urban often limited possibilities, to at least come closer to the goal of *strechable floodplains* and *green rivers*. However, the prevention of further, new, particularly structural uses in the dynamic river floodplains, in order to avoid exposure to the risk in the first place, has failed to materialise. The political will to do so was lacking everywhere because of the shortage of building land and seemingly insurmountable constraints in already densely built-up cities.

These developments – both the flood events as occasions and the actions towards NbS as reactions – are not only found in Central Europe, but worldwide.

Together with design measures, further management must be applied

- To restore people's awareness of nature and natural processes, especially in cities
- Disseminate knowledge about dynamic ecosystems
- To promote acceptance for nature design measures
- Develop an overall plan for new "urban waterscapes" in cities where flood protection, biodiversity and recreation are no longer opposites (IUCN 2016; Cohen-Shacham et al. 2016; see ◘ Fig. 8.46).

◘ **Fig. 8.45** Restoration of mangrove forest under the Mangrove Conservation Program in Haikou, China. (© Breuste 2017)

◘ Fig. 8.46 Flooding of the Salzach River on June 2, 2013 in Salzburg, Austria. (© Breuste 2013)

8.9 Providing Guidance Through Good Examples

The Aim Should Be

To lead the way with good examples locally, in neighbourhoods or urban development plans and new urban designs, proving that the Green City is achievable. It is predominantly the city of today, supplemented and improved by urban nature, which we develop and maintain for the benefit of the city's inhabitants.

Orientation for municipal decision-makers on the way to the Green City is not easy to find. The ideal as a concept does not exist; approaches require adaptation everywhere. The Green Cities are the existing cities of the present, which mostly have to go through a restructuring process slowly in small steps. They will experience this restructuring process with many different requirements anyway. It is only a matter of making urban nature an important co-designer in this process.

Building new "eco-cities" is something that only a few companies can tackle at the moment. China is leading the way. In the eco-city concept, urban nature plays an important role, but often not the decisive one. The eco-city as a green city can therefore also be built from scratch. However, this will remain the exception.

In Europe, where the guiding principle is followed by planning, which is also successful down to the municipal level, the orientation is towards the compact, green, loosened-up image of the European urban tradition. In China and South Korea, an assertive state makes specifications and creates, at least in examples, presentable realities that correspond to the image of "ecological harmony" in society and the city. Real urban nature preferences with clear goals for this are mostly missing.

In rich oil states on the Arabian Peninsula, people are experimenting with energy, water and resource efficiency at the highest possible technological level in the construction of new cities with a lavish and water-consuming "green facade".

In the USA, the Green City model is developing into reality in small neighbourhood steps. Despite good approaches (for example, Seattle, Portland), there are no examples of Green City projects that have been implemented.

In Latin America, innovative examples are being realised, often through the commitment of small groups and individual municipalities. However, a model for the ecological (green) Latin American city does not exist.

In Africa, political and economic elites mostly realise postmodern urban showpieces, far removed from green cities in terms of content. This is due not least to social conditions, but also to the widespread reservations there about urban nature. They have only just proudly made themselves independent of the risky nature outside the cities and participated in modernity through urbanisation. Only in the Western Cape (South Africa) is the unique importance of biodiversity (Cape Flora Kingdom) in an urban and cultural landscape environment perceived as a special task and challenge, also or especially for cities.

Eco-city models started just over 100 years ago as a response to social problems concentrated in cities. The cities were supposed to solve them in an exemplary way. Nature and space for nature in the daily routine of life were absolutely part of the garden city model, for which the "cultivated garden" became a symbol. At the beginning of the last century, the garden city became the garden city movement, the relationship to space and nature its guiding principle (Howard 1898). The ideal design "City of the Future" by the German garden city pioneer Theodor Fritsch had half of the city consist of urban nature (urban forest) (Fritsch 1896, see ► Sects. 2.3 and 2.6).

The architecture of modernity (Athens Charter) makes man the measure of urban design, to whose needs, functionally separated, urban nature must also adapt (Le Corbusier et al. 1943).

Richard Registers, American pioneer of the eco-city idea (Register 1987), sees the future eco-city as a city in which people rediscover the lost balance with nature and make it the guiding principle. Such a city should *coexist peacefully* with nature, an ideal whose realization was developed as a goal in small steps (Register 2006).

Eco-city in the Narrower Sense

An eco-city (Ecocity Builders 2013) should be
- A healthy city based on self-regulating, resilient structures and functions of natural ecosystems and living organisms
- A spatial unit that includes its inhabitants and their ecological influences and is part of its natural environment

The eco-city is understood as an ecosystem. The relationship of the city to the (animate and inanimate) nature of its space and the newly created urban nature are central objects of action. Urban nature offers indispensable ecosystem services, the valorisation of which is the aim (Breuste et al. 2016).

The "ecological urban development" in Europe sees as a guiding principle (especially 1970s–1990s) a city that does not harm but promotes the physical and mental health of humans, that does not pollute or destroy its surrounding country-side, and that promotes the development of all kinds of nature in the city (for example, Wittig et al. 1995).

The eco-city and the smart city (since around 2000) are far larger models than the green city, which can be understood as a building block of these models. Increasingly central at present is the model of the Smart City. It underlines the expectation that existing urban problems are primarily technological and can be solved through the better efficiency of technologies. This includes high-tech developments in smart grids, intelligent transport and IT networks in energy and service provision (Albino et al. 2015; Breuste et al. 2016; Vanolo 2016). However, it is important to remember that cities are primarily social-ecological systems that use technology to meet the needs of their inhabitants (McPhearson et al. 2016).

Smart City

Smart in the context of Smart City means: efficient, technologically advanced, green and socially inclusive. The focus is on technology-based changes and innovations in urban spaces. The term has been used since the 2000s in urban marketing, by large technology corporations and by various actors in politics, business, administration and urban planning. Factors of a Smart City are: Smart Economy (economy), Smart People (population), Smart Governance (administration), Smart Mobility (mobility), Smart Environment (environment) and Smart Living (life). City dwellers should become part of the technical infrastructure of a city. Under *"smart environment"* are treated with indicators: *attractiveness of natural conditions, pollution, environmental protection and sustainable resource management* (Giffinger et al. 2007; Albino et al. 2015; Breuste et al. 2016; Vanolo 2016; TU Wien 2018).

Giffinger et al. (2007) point out the following factors for smart cities in the area of *"smart environment"*:

- The attractiveness of natural conditions (indicators: hours of sunshine and proportion of green space)
- Pollution (indicators: ozone smog, particulate matter pollution, fatal chronic lower respiratory diseases per inhabitant)
- Environmental protection (indicators: individual efforts to participate in nature conservation activities, opportunities for nature conservation) and
- Sustainable resource management (indicators efficient use of water and energy).

Artmann et al. (2019a) try to complement the Smart City with much more content of urban nature and its services to a Smart Green City, to extend both to the Smart Compact Green City and thus provide better guidance.

Objectives are identified in four areas:

■ **Smart built environment:**
– Integration of green infrastructure
– Green quality
– The proximity of green elements to residential areas
– Green Network, connectivity

■ **Smart multifunctionality**
– Social considerations
– Economic considerations
– Ecological considerations

■ **Smart government**
– Strategic management (processes and principles, cooperation, etc.)
– Reflective management (monitoring, targets)
– Multi-scale management

■ **Smart governance**
– Interdisciplinary cooperation for green integration
– Transdisciplinary cooperation for quality of green infrastructure

The Compact Green City Model is being developed as an alternative to the city dominated by urban sprawl and is recommended for planning purposes (Artmann et al. 2019b; see ◘ Figs. 8.47 and 8.48).

┌─ **Smart Compact Green City** ──────────────────────────────

The concept of the Smart Compact Green City complements the Smart City concept by (a) compactness of the urban structure and limitation of land growth as resource and land-use efficiency and (b) preservation and development of urban nature as urban green infrastructure and service provider. The concept can be understood as an indicator-based target system (Artmann et al. 2019a).

Urban greenery and nature are part of the concepts of eco-, smart- and future cities, sometimes with more attention, sometimes with less importance. In any case, the aspect of urban nature is only a partial aspect of holistic urban development concepts and thus also easily falls into the background. This is a good reason to represent the concept of the Green City offensively in sample areas and integrated into other concepts. Good examples provide orientation.

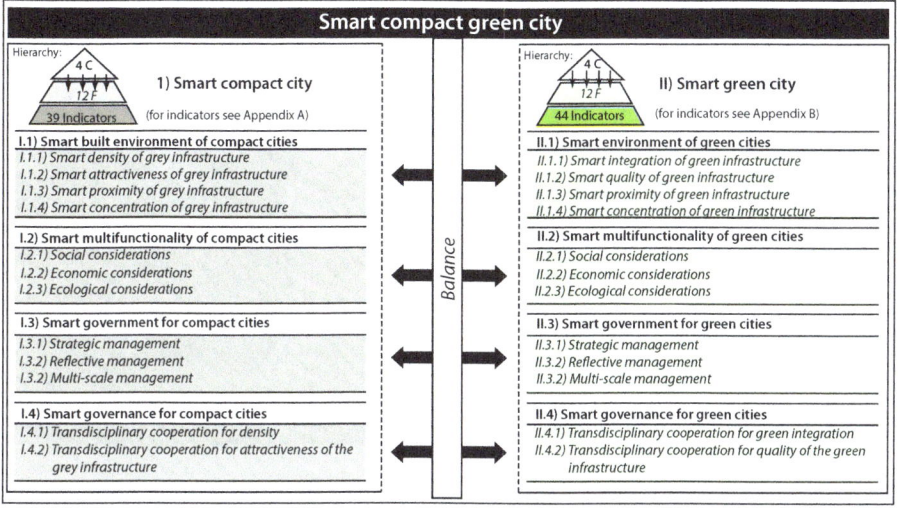

© Artmann & Kohler, IOER 2017

◘ **Fig. 8.47** Smart Compact Green City as a concept. (Artmann et al. 2019a)

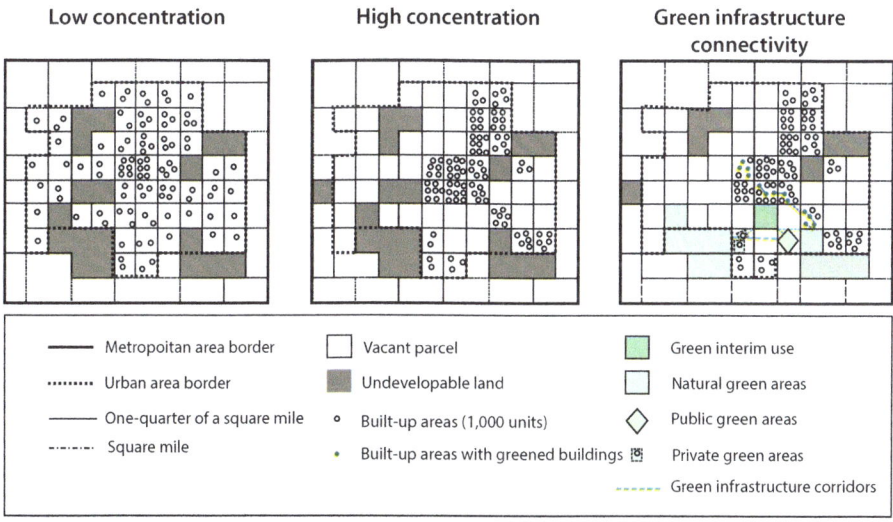

◘ **Fig. 8.48** Scheme of concentration (compactness) and green infrastructure connectivity. (Artmann et al. 2019a, based on the visualization of concentration by Galster et al. 2001)

Urban nature can encounter us in very different forms. In any case, it will seek its own way if we do not pursue a concrete goal with it. Using the services of urban nature in all its forms without destroying it should be the goal of future urban development (see ◘ Figs. 8.49, 8.50, and 8.51).

🔹 **Fig. 8.49**   Living in an orchard, Garden Home, Nature Resort, Thaton, Thailand. (© Breuste 2018)

🔹 **Fig. 8.50**   Urban nature also envelops buildings by itself – if it is allowed to do so, Freilassing, Bavaria. (© Breuste 2018)

🔹 **Fig. 8.51**   Note on environmental protection in a park in Nainital, India. (© Breuste 2018)

8

City of the Future without a Green Concept? – Neom, Saudi Arabia

The lure of creating a future city from scratch without all the legacies of an already existing city is great, assuming far-reaching ambitions. The Saudi royal family has these ambitions and expresses them impressively with the project "Future City Neom".

A mega future city is to be built on an area of 26,500 km² in the Arabian Desert, bordering Jordan and Egypt. It is to have 468 km of coastline and rise from the coast to an altitude of 2500. There has never been anything comparable before: "*The world's most ambitious project: An entirely new land, purpose-built for a new way of living*".

Can the Saudi Arabian urban desert dream also set examples as a Green City?

The areas of energy and water, mobility, biotechnology, food, advanced production and manufacturing, media, entertainment, technological and digital sciences, tourism and sports are included. As an outlook on the way of life of future generations, however, there is only a vague reference in the project to "*lush green spaces*". Since these will require a lot of energy and water in this arid region, it would be good to know what innovative things are planned here. Information on this is lacking so far. It would certainly be interesting because many cities in hot and arid regions suffer from the fact that they can only maintain and develop green spaces at great expense with a lot of irrigation (Neom 2018; see ◘ Fig. 8.52).

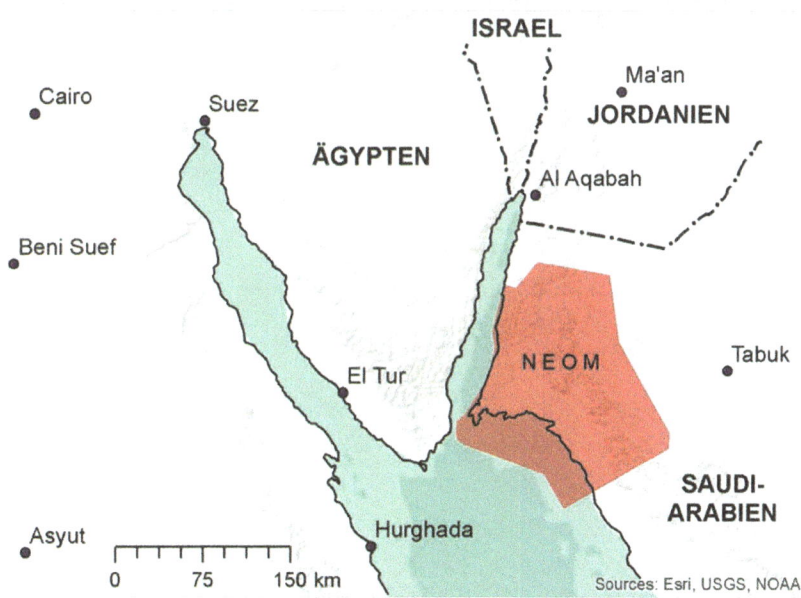

◘ **Fig. 8.52**   Neom City – Planned location of the future city. (Neom 2018, drawing: W. Gruber)

Compact Smart City New Songdo, Incheon, South Korea

New Songdo City is being built 65 km from Seoul in three construction phases from 2003 to 2020 on a 600 ha polder area in the wadden sea. When completed, 60–65,000 people are expected to live here. In 2012, there were 22,000, mostly from the upper middle class. The master plan for the planned city was drawn up by the architectural group Kohn Pedersen Fox (KPF, ► https://www.kpf.com/). In addition to primarily economic benefits, the eco-city is intended to offer living, working, and recreation in a sustainably oriented community with internationally recognized, certified environmental standards for construction and operation and environmentally friendly individual and public transportation. The planned city is committed to the Compact Smart City concept. Residents are involved in continuous data collection. Individual consumption data, access data, etc. are collected in the apartments. This should enable energy optimisation and savings of 30%. Video surveillance of public spaces up to the houses, smart cards with multifunction such as public transport use, health care, apartment access, banking services, etc. are part of the city concept (Alusi et al. 2011). The cost of the project is privately funded and is about US$35 billion, with 40% of the land being parks, green and recreational areas. In 2009, the 40 ha Central Park, modelled on New York, was inaugurated (see ◘ Figs. 8.53 and 8.54).

◘ **Fig. 8.53** Central Park of New Songdo City, Compact Smart City, South Korea. (© Breuste 2014)

> **Fig. 8.54** Symbol of New Songdo City, Compact Smart City, Südkorea, Central Park. (© Breuste 2014)

Ecological Wetland Belt Chengdu – Six Lakes and Eight Wetlands

Chengdu, the capital of the Chinese province of Sichuan with a population of 14 million, is the most important economic centre in western China alongside Chongqing. In 2006, Chengdu ranked fourth among China's most livable cities (China Daily). Chengdu's green belt completely encircles the city through a 200 km² interconnected system of rivers, lakes and wetlands along the ring road around the city. Six lakes and eight wetlands are interconnected. The goal of the Green Belt, in addition to recreation, is to increase biodiversity and reduce air pollution. The model is the 2200-year-old Shu irrigation system in nearby Dujiangyan, which early on made the city a thriving agricultural center protected from floods. Ambitiously, the city is thus also striving for a balance between industrial and ecological development.

The highly ambitious green belt project began in 2012. It will actually surround the city with a wide, green ring. The project has expanded the urban water area from 1.8% to 6% of the total city area, more than tripling it. No other Chinese city is so committed to developing green-blue infrastructure as a development stimulus: *"Building wetlands on such a large scale is uncommon, especially when many other cities are reclaiming land from lakes"* (Li and Peng 2015).

According to the city's green development plan, land in a 500-meter belt on both sides of the ring road around the city was used for the green belt. In 2013, the 160-ha Jincheng Lake Park opened with 42% of the green belt's water areas. The 20,000 ha Bailu Wan Wetland is part of the green belt to the south and is a popular area for recreation and nature observation (Go Chengdu 2014; Li and Peng 2015; China Daily 2017; see ■ Figs. 8.55, 8.56, and 8.57).

8

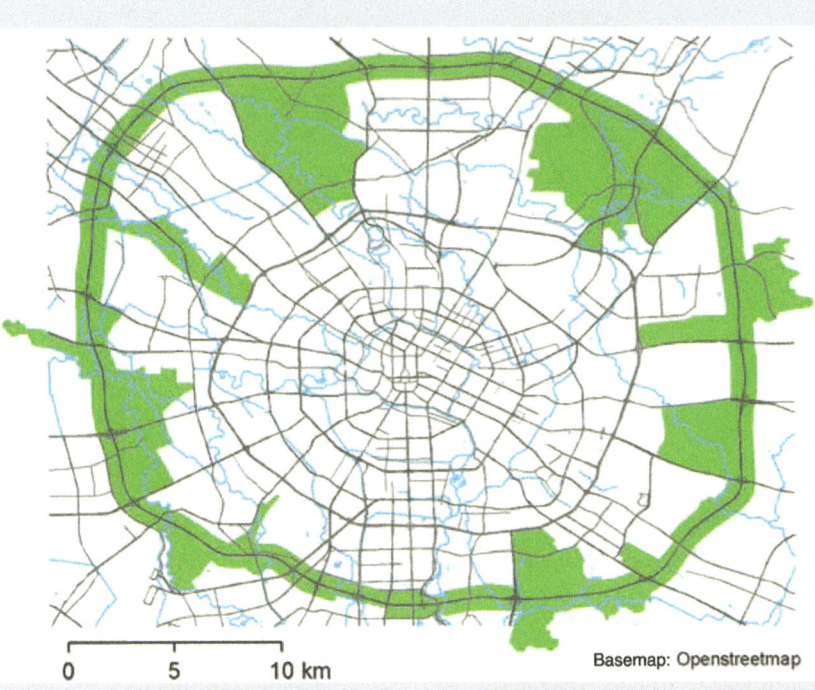

Basemap: Openstreetmap

0   5   10 km

◘ **Fig. 8.55**   Chengdu, China, Green Ring. (Design: J. Breuste, Drawing: W. Gruber)

◘ **Fig. 8.56**   Bailu
Wan Wetland, part of
the Green Ring in
Chengdu, China.
(© Breuste 2013)

■ **Fig. 8.57** Bailu Wan Wetland in Chengdu, China, information board on natural features. (© Breuste 2013)

Bang Krachao – The Green Island in the Bangkok Metropolis

Strictly speaking, Bang Krachao does not belong administratively to Bangkok at all. It is part of the independent Phra Pradeang district of Samut Prakan province in the south of the city. However, Bang Krachao is completely enclosed by the metropolis and its urban structures. It is separated from the city center only by the river, the Mae Nam Chao Phraya, which flows through Bangkok. Except by water, it is accessible only by an access road.

Bang Krachao is an approx. 16 km² large river island in the Mae Nam Chao Phraya. It was originally inhabited by fishermen. The population and use structure is still village-like, even though tourism and nature conservation have long since taken hold. Strict land use regulations protect Bangkok's green island (often called the "green lung") from development by owner-occupied homes and office buildings. About 800 families still live on Bang Krachao on about 400 ha of land from agriculture with other supplementary income. A total of about 40,000 inhabitants live here.

The island is characterized by a loose, agricultural cultivated forest of peasant plots intersected by canals, where the local population also still lives in a rural way, but also increasingly benefits from the day tourism of the capital. Due to the dense green cover and extensive cultivation, the island is rich in wild plant and animal species. Biodiversity is showcased to visitors at Sri Nakhon Khuean Khan Park. In Suan Pa Ket Nom Klao Community Forest, a natural forest has been created since 2007 on 10.16 ha by residents under the guidance of the Royal Forest Department with native species on fallow land, including native crops and including information on them and marketing of their fruits. Eco-tourism in the town is expected to improve the living conditions in a sustainable way.

In 2006, Bang Kachao was named by Time magazine as the "*best urban oasis*" in its "Best of Asia" series. It is now regularly visited by nature lovers, cyclists and tourists and represents a real counter-world to the centre of the Bangkok metropolis, only 6 km away (Marshall 2006). A 3-year project launched in 2016 by the Royal Forest Department, Kasetsart University and the oil company PTT aims to ensure that 60% of the area remains free of new development, mainly construction (Easey 2017). The aspiration is to prevent the land from being sold off to investors through rules and restrictions, but also by ensuring that the rural population can live a sufficient life with urban benefits in a rural environment. Only this will secure the green island in Bangkok in the long run. The example of Bang Krachao shows that biodiversity, agricultural use and modern urban life can be preserved together, at least in "green islands" in the city, if all efforts can also be linked to the interests of the inhabitants and they participate in them (see ◘ Fig. 8.58).

◘ **Fig. 8.58** Bang Krachao – the village in the city of Bangkok, Thailand, poster in Bang Krachao

References

Albino V, Berardi U, Dangelico RM (2015) Smart cities: definitions, dimensions, performance, and initiatives. J Urban Technol 22(1):3–21

Alusi A, Eccles RG, Edmondson AC, Zuzul T (2011) Sustainable cities: oxymoron or the shape of the future? Working paper 11–062, Harvard Business School. https://www.hbs.edu/faculty/Publication%20Files/11-062.pdf. Accessed 5 Sept 2018

Artmann M, Kohler M, Meinel G, Gan J, Ioja J-C (2019a) How smart growth and green infrastructure can mutually support each other – a conceptual framework for compact and green cities. Ecol Indic 96:10–22

Artmann M, Inostroza L, Fan P (2019b) Urban sprawl, compact urban development and green cities. How much do we know, how much do we agree? Ecol Indic 96(2):3–9

Arvi Park (eds) (2018) Arvi Park. https://parquearvi.org/en/. Accessed 30 Aug 2018

Auhagen A, Sukopp H (1983) Ziel, Begründungen und Methoden des Naturschutzes im Rahmen der Stadtentwicklung von Berlin. Natur Landschaft 58(1):9–15

Bansal M (2017) Evaluating the impact of ecological restoration on the bird community of Aravalli Biodiversity Park, Gurugram. M.Sc. dissertation, School of Life Sciences, Jawaharlal Nehru University, New Delhi

Barbosa O, Tratalos JA, Armsworth PR, Davies RG, Fuller RA, Johnson P, Gaston KJ (2007) Who benefits from access to green space? A case study from Sheffield, UK. Landsc Urban Plan 83:187–195

Brämer R (2006) Natur obskur. Wie Jugendliche heute Natur erfahren. Oekom, München

Brämer R, Koll H (2017) Siebter Jugendreport Natur 2016. Grundauswertung: (1) Schwerpunkt Wald. https://www.natursoziologie.de/files/jrn2016-grundauswertung-19_1704111205.pdf. Accessed 1 Sept 2018

Breuste J (1994) "Urbanisierung" des Naturschutzgedankens: Diskussion von gegenwärtigen Problemen des Stadtnaturschutzes. Naturschutz und Landschaftsplanung 26(6):214–220

Breuste J, Pauleit S, Haase D, Sauerwein M (eds) (2016) Stadtökosysteme. Springer, Berlin

Bundesinstitut für Bau-, Stadt- und Raumforschung (BBSR) (Hrsg) (2017) Handlungsziele für Stadtgrün und deren empirische Evidenz. Indikatoren, Kenn- und Orientierungswerte. BBSR, Bonn

Bundesministerium für Umwelt, Naturschutz, Bau und Reaktorsicherheit (BMUB) (Hrsg) (2017) Grün in der Stadt – Für eine lebenswerte Zukunft. BMUB, Berlin

Bundesministerium für Umwelt, Naturschutz und nukleare Sicherheit (BMU), Bundesamt für Naturschutz (BfN) (Hrsg) (2018) Naturbewusstsein 2017. Bevölkerungsumfrage zu Natur und biologischer Vielfalt. Berlin, Bonn

China Daily (2017) Chengdu paints the town green. www.chinadaily.com.cn/m/chengdu/.../content_29115994.htm. Accessed 3 Sept 2018

Cohen-Shacham E, Walters G, Janzen C, Maginnis S (eds) (2016) Nature-based solutions to address global societal challenges. IUCN, Gland

Corbusier L, Giraudoux J, de Villeneuve J (1943) La charte d'Athénes. Plon, Paris

De Wit M, van Zyl H (2011) Assessing the natural assets of Cape Town, South Africa: key lesson for practitioners in other cities. https://www.researchgate.net/.../284724522_Investing_in_Natural_Assets_A_business_c. Accessed 2 Sept 2018

De Wit M, van Zyl H, Crookes D, Blignaut J, Jayiya T, Goiset V, Mahumani B (2009) Investing in natural assets. A business case for the environment in the city of Cape Town'. Cape Town. https://www.researchgate.net/.../284724522_Investing_in_Natural_Assets_A_business_c... Accessed 2 Sept 2018

Deutsches Institut für Urbanistik (DIFU) (2013) Doppelte Innenentwicklung. Strategien, Konzepte und Kriterien im Spannungsfeld von Städtebau, Freiraumplanung und Naturschutz. Difu-Berichte 4/2013. https://difu.de/publikationen/difu-berichte-42013/doppelte-innenentwicklung.html. Zugegriffen 30 Sept 2018

Dudley N, Stolton S, Belokurov A, Krueger L, Lopoukhine N, MacKinnon K, Sandwith T, Sekhran N (Hsrg) (2010) Natural solutions: protected areas helping people cope with climate change.

IUCNWCPA, TNC, UNDP, WCS, The World Bank and WWF, Gland, Switzerland, Washington, DC and New York

Easey R (2017) The battle to save Bangkok's "green lung". Agence France-Presse. https://www.yahoo. com, https://yahoo.com/news/battle-save-bangkoks-green-lu. Accessed 28 Oct 2018

Ecocity Builders (2013) Ecocity. http://www.ecocitybuilders.org/. Accessed 28 Dec 2013

Eggermont H, Balian E, Manuel J, Azevedo N, Beumer V, Brodin T, Claudet J, Fady B, Grube M, Keune H, Lamarque P, Reuter K, Smith M, van Ham C, Weisser WW, Le Roux X (2015) Nature-based solutions: new influence for environmental management and research in Europe. GAIA 24(4):243–248

Ernstson H (2013) The social production of ecosystem services: a framework for studying environmental justice and ecological complexity in urbanized landscapes. Landsc Urban Plan 109:7–17

Fritsch T (1896) Die Stadt der Zukunft. Fritsch, Leipzig

Galster G, Hanson R, Ratcliffe M, Wolman H, Coleman S, Freihage J (2001) Wrestling sprawl to the ground: definition and measuring an elusive concept. Hous Policy Debate 12(4):681–717

GEO (2017) Die Grüne Revolution. Wie die Natur unsere Städte erobert, 9

Giffinger R, Fertner C, Kramar H, Kalasek R, Pichler-Milanovi N, Meijers E (2007) Smart cities. Ranking of European medium-sized cities. Wien, Hrsg. Centre of Regional Science, TU Wien. www.smart-cities.eu/download/smart_cities_final_report.pdf. Accessed 3 Sept 2018

Go Chengdu (ed) (2014) Bailuwan ecological wetland park. www.gochengdu.cn/.../bailuwan-ecological-wetland-park-a283.html. Accessed 3 Sept 2018

Grunewald K, Richter B, Meinel G, Herold H, Syrbe R-U (2016) Vorschlag bundesweiter Indikatoren zur Erreichbarkeit öffentlicher Grünflächen Bewertung der Ökosystemleistung "Erholung in der Stadt". Naturschutz und Landschaftsplanung 48(7):218–226

Grunewald K, Richter B, Meinel G, Herold H, Syrbe R-U (2017) Proposal of indicators regarding the provision and accessibility of green spaces for assessing the ecosystem service "recreation in the city" in Germany. Int J Biodivers Sci Ecosyst Serv & Manage, Spec Issue: Ecosyst Serv Nexus Thinking 13(2):26–39

Haase D, Kabisch S, Haase A, Andersson E, Banzhaf E, Baró F, Brenck M, Fischer LK, Frantzeskaki N, Kabisch N, Krellenberg K, Kremer P, Kronenberg J, Larondelle N, Mathey J, Pauleit S, Ring I, Rink D, Schwarz N, Wolff M (2017) Greening cities – to be socially inclusive? About the alleged paradox of society and ecology in cities. Habitat Int 64:41–48

Habibi S, Asadi N (2011) Causes, results and methods of controlling urban sprawl. Procedia Eng 21:133–141

Hard G (1998) Ruderalvegetation. Ökologie und Ethnoökologie. Ästhetik und Schutz (= Notizbuch 49 der Kasseler Schule). Universiy Kassel, Kassel

Henseke A, Breuste J (2015) Climate change sensitive residential areas and their adaptation capacities by urban green changes: case study of Linz, Austria. J Urban Plan Dev 141(3):A5014007

Howard E (1898) To-morrow: a peaceful path to social reform, Cambridge library collection. Cambridge University Press, London

ICLEI-Local Governments for Sustainability (2017) Nature-based solutions for sustainable urban development. ICLEI Briefing Sheet. http://www.iclei.org/briefingsheets.htm. Accessed 4 Sept 2018

Imperialtransylvania (2018) Văcăreşti natural park. www.imperialtransilvania.com/.../vacaresti-natural-park-paradise. Accessed 18 Oct 2018

IUCN-International Union for Conservation of Nature (2016). Defining Nature-based Solutions (NbS). https://www.iucn.org/. Accessed 4 Sept 2018

James P, Tzoulas K, Adams MD, Barber A, Box J, Breuste J, Elmqvist T, Frith M, Gordon C, Greening KL, Handley J, Hawort S, Kazmierczak AE, Johnston M, Korpela K, Moretti M, Niemelä J, Pauleit S, Roe MH, Sadler JP, Wardthompson C (2009) Towards an integrated understanding of green space in the European built environment. Urban For Urban Green 8(2):65–75

Janssen W (1988) Naturerleben. Unterr Biol 137(9):2–7

Johnston J (1990) Nature areas for city people. A guide to the successful establishment of community wildlife sites, Bd 14. Ecology handbook. The London Ecology Unit, London

Kabisch N, Strohbach M, Haase D, Kronenberg J (2016) Urban green spaces availability in European cities. Ecol Indic 70:586–596

Kabisch N, Korn H, Stadler J, Bonn A (eds) (2017) Nature-based solutions to climate change in urban areas – linkages between science, policy, and practice (= Theory and practice of urban sustainability transitions). Springer, Heidelberg

Klett-Verlag (Hrsg) (2018) Die Zukunft unserer Städte? Fundamente Kursthemen Städtische Räume im Wandel, Schülerbuch, Oberstufe, pp 160–171. ISBN: 978-3-623-29440-7. https://www2.klett.de/sixcms/media.php/229/29260X-8702.pdf. Accessed 30 Aug 2018

Kowarik I (1992) Berücksichtigung von nichteinheimischen Pflanzenarten, von "Kulturflüchtlingen" sowie von Pflanzenvorkommen auf Sekundärstandorten bei der Aufstellung Roter Listen. Schriftenreihe für Vegetationskunde 23:175–190

Landeshauptstadt Dresden (ed) (2014) Landschaftsplan Dresden. Bürgerinformation zur Öffentlichkeitsbeteiligung. https://www.dresden.de/media/pdf/umwelt/LP_Faltblatt_Buergerflyer.pdf. Accessed 3 Sept 2018

Landy F (ed) (2018) From urban national parks to natured cities in the global south. The quest for naturbanity. Springer, Heidelberg

Li Y, Peng C (2015) Chengdu's ecological belt: new water system gets afloat. The Telegraph/China Daily. www.chinadaily.com.cn, https://www.telegraph.co.uk/.../chengdu-ecological-belt.html. Accessed 5 Sept 2018

MacKinnon K, Hickey V (2009) Nature-based solutions to climate change. Oryx 43(1):13–16

MacKinnon K, Dudley N, Sandwith T (2011) Natural solutions: protected areas helping people to cope with climate change. Oryx 45(4):461–462

Maes J, Jacobs S (2015) Nature-based solutions for Europe's sustainable development. Conserv Lett. 10: S 121–124., https://onlinelibrary.wiley.com/toc/1755263x/10/1. Accessed 26 Oct 2018

Magistrat Linz (ed) (2018) ÖKO.L – Zeitschrift für Ökologie, Natur- und Umweltschutz. https://www.linz.at/umwelt/4043.asp. Accessed 26 Oct 2018

Manea G, Matei E, Vijulie J, Tîrlă L, Cuculici R, Cocoş O, Tişcovschi A (2016) Arguments for integrative management of protected areas in the cities – case study in Bucharest city. Procedia Environ Sci 32:80–96

Marshall A (2006) Best of Asia: best urban oasis. Time Journal, 15th May. https://www.content.time.com/time/magazine/article/0,9171,1194117,00. Accessed 28 Oct 2018

Mathey J, Rößler S, Lehmann I, Bräuer A, Goldberg V, Kurbjuhn C, Westbeld A, Hennersdorf J (2011a) Noch wärmer, noch trockener? Stadtnatur und Freiraumstrukturen im Klimawandel, Naturschutz und Biologische Vielfalt 111. Landwirtschaftsverlag, Münster

Mathey J, Rößler S, Lehmann I, Bräuer A, Goldberg V (2011b) Anpassung an den Klimawandel durch Stadtgrün – klimatische Ausgleichspotenziale städtischer Vegetationsstrukturen und planerische Aspekte. In: Böcker R (ed) Die Natur im Wandel des Klimas. Eine Herausforderung für Ökologie und Planung. Darmstadt: Kompetenznetzwerk Stadtökologie, Conturec 4. Darmstadt Kompetenznetzwerk Stadtökologie, pp 79–88

Mc Neely JA (2001) Cities and protected areas. Parks 11(3):1–3. IUCN, Gland, Switzerland. https://www.iucn.org/protected...protected.../urban-conservation-strategies. Accessed 12 Oct 2016

McPhearson T, Pickett STA, Grimm NB, Niemelä J, Alberti M, Elmqvist T, Weber C, Haase D, Breuste J, Quresh S (2016) Advancing urban ecology toward a science of cities. Bioscience 66(3):198–212

Naturkapital Deutschland – TEEB DE (2016) Ökosystemleistungen in der Stadt – Gesundheit schützen und Lebensqualität erhöhen. Kowarik I, Bartz R, Brenck M, Hansjürgens H (eds). Technische Universität Berlin, Helmholtz-Zentrum für Umweltforschung – UFZ, Berlin

Naturschutzbund Deutschland, NABU (2017) 30-Hektar-Tag: Kein Grund zum Feiern. Unser Flächenverbrauch ist noch immer viel zu hoch. https://www.nabu.de. Accessed 24 Aug 2018

Neom (2018) Fact sheet – Neom. www.neom.com/.../NEOM_FACT_SHEET_RGB_100073132_L. Accessed 18 Oct 2018

Nordelta (2015). http://www.nordelta.com/ingles/inicio.htm. Accessed 26 Oct 2015

Norton BA, Coutts AM, Livesley SJ, Harris RJ, Huntera AM, Williams NSG (2015) Planning for cooler cities: a framework to prioritise green infrastructure to mitigate high temperatures in urban landscapes. Landsc Urban Plan 134:127–138

Pauleit S (2011) Stadtplanung im Zeichen des Klimawandels: nachhaltig, grün und anpassungsfähig. Conturec, Darmstadt 4:5–26

Penn-Bressel G (2018) "Urban, kompakt, durchgrünt" – Strategien für eine nachhaltige Stadtentwicklung. Umweltbundesamt (ed). https://www.umweltbundesamt.de/sites/default/files/medien. Accessed 3 Sept 2018

Plachter H (1991) Naturschutz (=UTB für Wissenschaft: UNI-Taschenbücher; 1563). Fischer, Stuttgart

Register R (1987) Ecocity Berkeley: building cities for a healthy future. North Atlantic Books, Berkely

Register R (2006) Ecocities: rebuilding cities in balance with nature. New Society Publishers, Gabriola Island

Reidl K, Schemel H-J, Blinkert B (2005) Naturerfahrungsräume im besiedelten Bereich. Ergebnisse eines interdisziplinären Forschungsprojektes. Nürtinger Hochschulschriften 24, 283pp

Ribbeck E (2007) Rasches Wachstum, schwache Planung, städtische Armut. Bundeszentrale für politische Bildung. Dossier Megastädte. 8.5.2007. www.bpb.de/internationales/weltweit/megastaedte/.../stadtplanung-in-megastaedten?p. Accessed 30 Aug 2018

Rittel K, Wilke C, Heiland S (2011) Anpassung an den Klimawandel in städtischen Siedlungsräumen – Wirksamkeit und Potenziale kleinräumiger Maßnahmen in verschiedenen Stadtstrukturtypen. Dargestellt am Beispiel des Stadtentwicklungsplans Klima in Berlin. In: Böcker R (ed) Die Natur der Stadt im Wandel des Klimas – eine Herausforderung für Ökologie und Planung, Conturec, Bd 4. Kompetenznetzwerk Stadtökologie, Darmstadt, pp 67–78

Roloff A, Thiel D, Weiss H (eds) (2010) Urbane Gehölzverwendung im Klimawandel und aktuelle Fragen der Baumpflege. Forstwissenschaftliche Beiträge Tharandt/Contribution to Forest Sciences Beiheft 9:63–81

Rusche K, Fox-Kämper R, Reimer M, Rimsea-Fitschen C, Wilker J (2015) Grüne Infrastruktur – eine wichtige Aufgabe der Stadtplanung. ILS – Institut für Landes- und Stadtentwicklungsforschung (Hrsg), ILS-Trends 3/2015, S 1–8. https://www.ils-forschung.de/files_publikationen/pdfs/ILS-TRENDS_3_15.pdf. Accessed 3 Sept 2018

Schemel HJ (1998) Naturerfahrungsräume. Ein humanökologischer Ansatz für naturnahe Erholung in Stadt und Land. = Angewandte Landschaftsökologie, No 19. BfN-Schriften-Vertrieb, Münster

Schemel H-J, Wilke T (2008) Kinder und Natur in der Stadt. Spielraum Natur: Ein Handbuch für Kommunalpolitik und Planung sowie Eltern und Agenda-21-Initiativen. BfN-Skripten 230. Bundesamt für Naturschutz (ed), Bonn–Bad Godesberg. http://www.bfn.de/fileadmin/MDB/documents/service/skript230.pdf. Accessed 28 Dec 2015

Schwarz C, Sturm U (2008) Garden Cities of To-Morrow – oder die schrumpfende Stadt und der Garten. Forum der Forschung, BTU Cottbus. Eigenverlag, Cottbus, pp 99–104

Sociedade dos Amigos Parque Nacional da Tijuca (ed) (2004) Trilhas. Parque Nacional da Tijuca, Rio de Janeiro

Statistisches Bundesamt (2017) Bodenfläche nach Art der tatsächlichen Nutzung vom 15.11.2017, Berechnungen des Umweltbundesamtes, Fachserie 3, Reihe 5.1, 2016, Wiesbaden

Stopka I, Rank S (2013) Naturerfahrungsräume in Großstädten. Wege zur Etablierung im öffentlichen Freiraum. BfN-Skripten 345. Bundesamt für Naturschutz (ed), Bonn-Bad Godesberg

Sukopp H, Sukopp U (1987) Leitlinien für den Naturschutz in Städten Zentraleuropas. In: Miyawaki A, Bogenrieder A, Okuda S, White J (eds) Vegetation ecology and creation of new environments. Tokai University Press, Tokyo, pp 347–355

Sukopp H, Weiler S (1986) Biotopkartierung im besiedelten Bereich der Bundesrepublik Deutschland. Landschaft und Stadt 18(1):25–38

Taubenböck H, Esch T, Felbier A, Wiesner M, Roth A, Dech S (2012) Monitoring of mega cities from space. Remote Sens Environ 117:162–176

Taubenböck H, Wurm M, Esch T, Desch S (eds) (2015) Globale Urbanisierung. Perspektive aus dem All. Springer, Berlin

TEEB (2011) TEEB manual for cities: ecosystem services in urban management. http://www.naturkapitalteeb.de/aktuelles.html. Accessed 26 Aug 2014

The Green City (2015) France adopted a new green rooftop law. http://thegreencity.com/france-adopted-a-new-green-rooftop-law/. Accessed 17 Feb 2016

Trepl L (1991) Forschungsdefizit: Naturschutz, insbesondere Arten- und Biotopschutz, in der Stadt. In: Henle K, Kaule G (eds) Arten- und Biotopschutzforschung für Deutschland, vol 4. Forschungszentrum Jülich, Jülich, pp 304–311. (=Berichte aus der ökologischen Forschung)

Trzyna T (2014a) Urban protected areas: important for urban people, important for nature conservation globally. Claremont. www.thenatureofcities.com/…/urban-protected-areas-important-for-urban-people. Accessed 12 Oct 2016

Trzyna T (2014b) Urban protected areas: profiles and best practice guidelines. Best practice protected area guidelines series, no. 22, IUCN, Gland, Switzerland. https://www.iucn.org/protected…protected…/urban-conservation-strategies. Accessed 18 Oct 2018

TU Wien (2018) Europeansmartcities. www.smart-cities.eu. Accessed 4 Sept 2018

Umweltbundesamt (2018) Siedlungs- und Verkehrsfläche. https://www.umweltbundesamt.de/daten/flaeche-boden…/siedlungs-verkehrsflaeche. Accessed 24 Aug 2018

Unger F (1995) Wie Detroit so das ganze Land. Stadtbauwelt 127(36):1986–2003

University of Manchester, Centre for Urban and Regional Ecology (2006) Adaptation Strategies for Climate Change in the Urban Environment (ASCCUE) (2006). Draft final report of the National Steering Group, Manchester

Vanolo A (2016) Is there anybody out there? The place and role of citizens in tomorrow's smart cities. Futures 82:26–36

Wittig R, Breuste J, Finke L, Kleyer M, Rebele F, Reidl K, Schulte W, Werner P (1995) Wie soll die aus ökologischer Sicht ideale Stadt aussehen? – Forderungen der Ökologie an die Stadt der Zukunft. Zeitschrift f Ökol und Naturschutz 4:157–161

Supplementary Information

Index – 387

© Springer-Verlag GmbH Germany, part of Springer Nature 2022
J. Breuste, *The Green City*,
https://doi.org/10.1007/978-3-662-63976-4

Index

A

Aalborg Commitments 302
Acceptance 20, 26, 64, 86, 104, 118, 167, 186,
    188, 201, 203, 268, 283, 285, 286, 288,
    292–295, 298, 319, 334, 347, 366
Agricultural area 20, 176, 178, 199, 336
Agricultural landscape 8, 10, 12, 297
Agriculture 53, 78, 88, 93, 177, 179, 181, 238,
    245, 259, 272, 281, 377
Allotment
– garden 6, 9, 15, 16, 21, 52–57, 175, 177–182,
    185–186, 188–190, 194, 195, 198, 272, 294,
    335
– gardening 53, 54, 262
Apophytes 11
Aravalli Biodiversity Park 353, 354

B

Balance with nature 76, 173, 256, 307, 308, 368
Biocenoses 234
Biodiversity
– conservation 231, 242, 260, 263, 264, 266,
    267, 269, 270, 272, 275, 281, 282, 286, 288,
    338
– hotspot 147, 232, 353
– indicator 274
– native 181, 212, 230, 231, 241, 247, 273–277
– strategy 266–268, 270, 271, 279, 281, 282,
    286
– subjective 244
– urban 114–115, 147, 230–247, 249–251, 256,
    266, 286, 296
Biodiversity region 277
Biodiversity strategy 231
BioFrankfurt 232, 233, 249, 271, 277
BioFrankfurt biodiversity region 247
Biotope mapping 233, 235, 236, 246, 265, 266
Birkenhead Park 36
Bonn Biodiversity Report 280
Boulevard 30–31
Breeding bird 207, 238–243
British Royal Horticulture Society 190
Bucharest 354–356

C

Cape Town 43–44, 91–93, 202, 334, 335, 348,
    349, 356–357

Carbon sequestration 107, 109, 110, 159,
    160, 166
Catchment area 47, 133–135, 138, 213, 295
Central park 36, 37, 40, 41, 43, 45, 46, 94, 133,
    274, 374, 375
Cheonggyecheon 216–217
Chinese gardens 42, 97–98, 211
Cholera epidemic 64
City
– allochthonous 72
– autochthonous 72
– European 22, 29, 34, 38, 58, 59, 73, 76, 93,
    134, 155, 158, 160, 199, 302, 330
– healthy 76, 79, 110, 368
City Biodiversity Index (CBI) 230, 274–277
City park 35, 38–40, 43, 94, 105, 133, 139, 202,
    238, 267, 334, 343
Cityscape 60, 132, 135, 146, 156, 166, 204, 263,
    291, 334
Climate 2, 4, 5, 7, 12, 71, 79, 80, 83, 102, 107,
    109, 110, 114, 115, 120, 122, 140, 141,
    146–148, 154, 157, 158, 164, 166, 168, 175,
    187–189, 194, 200, 256, 258, 260–262, 266,
    269, 270, 273, 275, 279, 280, 282, 303, 304,
    306, 317, 321, 359, 360, 363, 364
Climate moderation 156, 176, 204, 359, 361
CO_2 sequestration 141, 158, 275
Community gardening 108, 178, 182
Compact city 306, 325
Compact city in an ecological network 306, 325
Conservation 6, 20, 31, 53, 165, 202, 207, 230,
    231, 235, 250, 260, 266, 267, 270–273, 281,
    283, 285, 288, 289, 294, 296, 317, 333, 346,
    348, 349, 353, 356, 364–366
Conservatism 283
Convention on Biological Diversity (CBD) 230,
    267–269, 273–276, 280
Cultural services 105, 107–109, 118, 121, 187,
    188

D

Delhi 79, 351, 353, 354
Desert city 71, 78, 79
Disservices 107, 109, 122
District forest 46, 49, 154, 169–172
District park 135, 139, 282
Diversity 6, 41, 110, 133, 135, 204, 230, 256,
    260, 291, 339, 348, 353

Dresden 48, 146, 147, 180, 291, 306, 319, 325, 360
Drought stress 164

E

Eco-city 256, 307, 367–369, 374
Ecosystem
– benefits 103
– function 8, 103, 264
– properties 103
– services 49, 102, 124, 125, 139, 140, 143, 147, 148, 152, 156, 157, 164–166, 171, 173, 175, 178, 187–189, 194, 199, 204, 205, 208, 230, 231, 238, 244, 249–251, 262–264, 269–271, 273, 275, 283, 288, 304, 305, 322, 328, 333, 334, 336, 348, 357, 359, 362, 363, 365, 369
Edible green infrastructure 175
English garden 35, 39, 94, 95, 133, 202
Environmental education 6, 108, 114, 176, 184, 189, 273, 275, 281, 285, 288, 294, 302, 339, 343, 365
Environmental justice 113, 261, 281, 328
European
– Landscape Contractors Association 256, 303
– Sustainable Cities & Towns Campaign 302
Exposure 244, 360, 366

F

"Federal Capital of Biodiversity" 281
Flood protection 60, 104, 200, 201, 203, 208, 210, 211, 366
Flood risk 203, 364, 365
Food 2, 11, 20, 21, 23, 43, 48, 54, 75, 76, 83, 88, 97, 107–109, 114, 118, 120, 121, 174–176, 178, 181–183, 187–190, 196, 198–200, 250, 336, 364, 373
Forest
– park 45, 47, 49, 50, 111, 133, 135, 140, 162, 301
– urban 6, 8, 47–50, 79, 81, 107, 111, 117, 118, 132–134, 147–174, 301, 335, 341, 368
– vertical 171, 172, 256
Forest city 49, 77, 78, 169
Front yard 177

G

Garden, English 35, 39, 94, 95, 133, 202
Garden city 36, 164, 256, 259, 368
"Global picturesque" 95

Golf course 20, 43, 95, 276
Grassland 250
Green
– infrastructure 3, 38, 110, 111, 119, 250, 256, 258, 263, 271, 286, 288–290, 305, 309, 336, 364, 370, 371
– lungs 26–29, 377
– planning 21, 23–26, 42, 50, 286
– rivers 365
Green building 79, 259, 262, 365
Green city 2, 3, 5–7, 15, 25, 38, 71, 75, 79, 86, 120, 124, 173, 317–319, 321, 322, 328, 330, 334–336, 338, 341, 343, 346, 347, 349, 350, 352, 354, 358, 360, 362, 366–370, 372–374, 378
– concept 6–7, 256, 257, 259–263, 265, 267, 269–273, 275, 277, 279, 282, 284, 286, 288, 292, 294, 296, 298, 301, 302, 304, 306, 308
Green Guerillas 183
Green space(s) 6, 13, 16, 27–29, 38, 42–44, 52, 105, 107, 110–113, 125, 126, 132–135, 146, 149, 157, 158, 165, 169–171, 176, 177, 182, 188, 194, 199, 202, 237, 239, 245, 248, 257–265, 270–272, 282, 283, 291, 294, 303, 304, 306, 308, 317, 318, 322, 328, 330, 331, 335, 359, 369, 373
– supply 330
Greenbelt 168, 294, 350, 375
Green-blue network 325, 336

H

Habitat III 304
Habitat structures 237
Habitat supply 103, 118, 143, 160
Health benefit 118
Heat islands 325, 360
Home garden(s) 14, 52, 175–179
Home gardeners 198
"Heat islands" 264
Hour of Garden Birds 242, 243
Hyde Park 27–29, 34, 40–41, 46, 133

I

Indicator 235–241, 244, 274, 275, 370
Indicator for biodiversity 236
Industrial city 26, 28, 76, 173, 283
Infrastructure
– green 3, 263–265, 371
– green and blue 202, 264, 318, 328
"Initiation wildernesses" 92
International Union for Conservation of Nature (IUCN) 348, 363

International Union for Conservation of Nature
and Natural Resources (IUCN) 283

K

Kitchen garden(s) 10, 14, 21, 23, 51, 52, 71,
176, 177, 189, 200

L

Land
– consumption 87, 245, 258, 321, 322
– fragmentation 258
– use 114, 115, 119, 120, 123, 125, 181, 183,
185, 235, 264, 279, 321, 370, 377
Landscape
– cultural 8, 72, 75, 76, 165, 306, 368
– garden 9, 14, 26, 36, 44–46, 94
– park 14, 133, 237, 298, 299
– urban 15, 202, 260, 272, 293, 353
Lawn 36, 40, 47, 78, 94–96, 117, 121, 132, 147,
180, 248, 295, 346, 360
Learning site
– about nature and its processes 342
– green 341
– urban 342
Leipzig Charter 303
Life reform 53
Lifestyle(s) 23, 29, 33, 41, 112, 137, 178, 180,
182, 195, 196, 199, 245, 338
Local Environmental Initiatives (ICLEI) 267
Local park 135, 139
Lung, green 28, 377

M

"Mainstreaming biodiversity" 269
Malaria 77, 203
Mangrove rehabilitation 365
Meiji Shrine 89, 90
Millennium Ecosystem Assessment
Report 103, 304
Mountain city 82, 83, 87, 88

N

Nagoya declaration 268, 269, 274, 277
Nagoya Strategy for Biodiversity 308
National strategy
– on biological diversity 166, 244, 257, 270,
288
– to the Green City 166

Natural forest 10, 11, 47, 148, 168, 340, 378
Nature
– awareness 195, 245, 246, 257, 294, 338
– capital 102, 103, 271
– category 121, 134
– components 20, 103, 138, 204, 294
– conservation 6, 20, 31, 53, 92, 148, 155, 156,
166, 202, 207, 208, 212, 231, 235, 237, 241,
244, 249, 257, 259, 265–267, 273, 277, 279,
288, 293–295, 302, 317, 338, 343, 347, 349,
350, 352, 353, 355, 365, 369, 377
– experience 15, 109, 121, 156, 157, 176, 189,
204, 207, 208, 211, 212, 258, 271, 285, 289,
302, 334, 336, 338–341
– type 11, 14, 110, 118, 121, 132, 304
– urban 2, 3, 8–10, 13, 20–22, 24, 25, 27–29,
120, 318, 320, 328, 339, 346, 369, 371, 372
Nature Awareness 244
Nature Park Schöneberger Südgelände 301
Nature-Based Solutions (NbS) 3, 76, 250, 317,
362–366
Neighborhood forest 154
Neighbourhood Park 135
Neobiota 11
Networking 260, 262, 273, 317, 325, 334, 353
New Urban Agenda 304, 306
New urban forest 169, 170
NGOs 196, 213, 267, 318
Noise abatement 363
Novel urban wilderness 285, 288
Novel urban wildernesses 14

O

Ohlsdorf Cemetery 46–47
Orchard(s) 14, 20–22, 52, 93, 94, 134, 148, 177,
350, 372
Orobiomes 83, 86

P

Paradise garden(s) 55, 71
Paraná Delta 212
Park forest 30, 154, 166, 188
Park cemetery 44–46
Parque Central 42
Parque Nacional da Tijuca 358
People-biodiversity paradox 244
People's park 35, 50
Principal/City/ Metropolitan Park 135
Project "Cities dare wilderness" 277, 293–294
Promenade 29–31, 259
Provisioning services 105, 107, 108, 115, 187

Q

Quality of life concept 114, 269

R

RAMSAR Convention 200, 201
Recreation
– forest 40
– garden 176, 189, 199
Regulating services 105, 107, 109, 115, 187
Relaxation 40, 112, 138, 139, 145, 146, 156,
    157, 159, 166, 186–188, 190, 199, 216
Renaturation 201, 208, 210, 211, 216, 281, 336
Requirements (demands) 112
Restoration 208, 216, 270, 318, 349, 350,
    356–357, 364–366
Rewilding 267, 288
Rio de Janeiro 202, 268, 304, 348, 349, 358
Rio Matanza-Riachuelo 62, 213–215

S

Self-sufficiency 54, 175, 178, 182, 192, 199
Sensitivity 190, 346, 360, 365
Service providing units 103, 117, 119
Settlement and traffic area 272
Sewer, industrial 59
Shanghai 33, 43, 62, 144, 163, 172, 173, 196,
    202, 242, 248, 268
Shintoism 89
Shrine forest 89, 90, 334
Smart
– city 362, 369, 370, 374, 375
– compact green city 370, 371
– green city 370
Species diversity 120, 230–232, 234, 237–241,
    244, 248, 305
Square 22, 27–29, 40, 42, 43, 95, 133, 149, 207,
    259, 335
Street tree(s) 81, 82, 105, 117, 148, 149, 161,
    164, 168, 259, 272
Structural diversity 11, 112, 135, 234, 236, 247,
    248, 292
Succession, forest 154
Sustainability 2–4, 6–7, 114, 179, 249, 278,
    302–305, 339

T

Tabriz 135, 138
TEEB approach 104
Tempelhofer Feld park 295, 296
Temperature moderation 359, 361, 362

Therophyte 12
Tokai forest 356–357
Transpiration 158
Tree
– population 31, 45, 81, 91, 148, 149, 164,
    186, 194
– urban 147–174

U

Urban
– agriculture 174–177, 183, 365
– development 2, 3, 24–26, 33, 36, 37, 51, 60,
    72, 73, 76, 102, 166, 169, 170, 179, 186, 202,
    216, 217, 233, 256, 257, 260–263, 265, 282,
    286, 294, 303, 305–307, 318, 321, 323–325,
    328, 336, 348, 362, 367, 369–371
– dweller 8, 16, 20–23, 26, 28, 36, 52, 64, 105,
    132, 137, 143, 148, 150, 152, 157, 159, 174,
    175, 179, 243, 244, 267, 280, 286, 294, 317,
    318, 325, 328, 330, 338, 342, 346, 347, 350,
    352, 362
– ecosystem services 103, 105, 106, 108, 110,
    114–121, 124, 125, 250, 251
– forest 81, 148, 150, 160, 166, 168–171, 173,
    239
– Forest Management Plan 155, 168
– forestry 149, 152, 153, 155
– garden 13, 14, 24, 42, 95, 107, 121, 132, 149,
    174–200
– gardening 176, 178, 180–183, 188, 190, 193,
    196, 272
– green 6, 42, 43, 111–113, 126, 132–134, 143,
    149, 155, 194, 257–262, 265, 282, 288, 303,
    304, 318, 330
– green infrastructure 38, 263–265, 370
– greening strategy 256, 303, 304
– nature 2, 20, 71, 73, 75–77, 79, 81, 83,
    85–86, 89, 92, 94, 95, 97, 98, 102–105, 107,
    108, 110, 111, 113, 115, 117, 119, 120,
    122–124, 126, 132, 139, 142, 147, 155, 188,
    194, 195, 200, 203, 208, 216, 230, 231, 233,
    244–247, 256, 257, 259–263, 265, 267,
    269–273, 275, 277, 279, 282, 284, 286, 288,
    292, 294, 296, 298, 301, 302, 304, 306, 308,
    317, 318, 321, 322, 324–326, 328, 331–346,
    348, 359–363, 367–372
– nature conservation 166, 302, 306, 343, 347,
    349, 350, 353
– park 13, 14, 44, 95, 118, 132–144, 146, 147,
    159, 165, 196, 234, 236, 294, 341, 355, 360
– and peri urban agriculture 175
– planning 16, 24, 26, 31, 38, 55, 102, 120,
    233, 235, 257, 260, 303, 304, 318, 333, 369

- protected areas 110
- structure 9, 12, 75, 83, 93, 132, 143, 146, 230, 237–239, 286, 303, 348, 370, 377
- structure type 120, 121, 123, 124, 238
- wild spaces 284
- wilderness 9, 14, 71, 91–92, 119, 121, 154, 272, 277, 283–294, 297, 301, 302, 334, 339–341, 354
- wildscape 284
- woods and woodlands 148
Urban nature reserves 348–352
Urban protected areas 346, 348, 352
Urban wildernesses 340
Utility 21
UV light 160

V

Văcărești Nature Park 354–355
Victorian gardenesque 95
Vision and concept 257
Vulnerability 360, 361

W

Wasteland 9, 256, 284
Wastewater treatment 59, 114, 117, 121
Water management, urban 303
Water pollution 58, 59, 107, 203, 213, 348, 349

Watercourse restoration 208
Waters
- open 201
- quality 203, 208, 211, 213, 271
- running 59
- still 201, 208
- urban 20, 58, 59, 118, 132, 200–208, 335, 341, 375
Wetland(s) 8, 11, 48, 77, 117, 118, 121, 200–204, 207, 208, 237, 250, 271, 283, 290, 335, 336, 341, 354, 355, 364–366, 375–377
White Paper on Urban Greening 256, 260
Wild nature 20, 21, 47, 96, 298, 301
Wilderness 71, 79, 89, 154, 241, 245, 256, 283–286, 288, 289, 293, 294, 297–301, 322, 324, 340, 348, 350, 353
Wilderness area, urban 284, 286, 298
Wilderness concept, urban 285, 286, 288, 292
Wildlife 143, 173, 244, 294–295, 302, 351
Wildlife gardening 180
Woodland 10, 36, 120, 133, 150, 154, 155, 157, 158, 160, 165–167, 170, 171, 259, 290, 293, 297, 298, 308, 346
- urban 6, 14, 48, 72, 147–155, 157–160, 165–167, 170, 238, 239, 250, 343
WWT London Wetland Centre 207

Z

Zonobiomes 71, 72, 83, 150

The manufacturer's authorised representative in the EU is Springer
Nature Customer Service Centre GmbH, Europaplatz 3, 69115 Heidelberg,
Germany. If you have any concerns regarding our products, please
contact ProductSafety@springernature.com

Printed and bound by CPI Group (UK) Ltd, Croydon, CR0 4YY
24/04/2026
02096317-0012